ELASTIC FILAMENTS
OF THE CELL

ADVANCES IN EXPERIMENTAL MEDICINE AND BIOLOGY

ELASTIC FILAMENTS OF THE CELL

Edited by

Henk L. Granzier

Washington State University
Pullman, Washington

and

Gerald H. Pollack

University of Washington
Seattle, Washington

SPRINGER SCIENCE+BUSINESS MEDIA, LLC

Library of Congress Cataloging-in-Publication Data

Elastic filaments of the cell / edited by Henk Granzier and Gerald H. Pollack.
    p. cm. -- (Advances in experimental medicine and biology ; v. 481)
    "Proceedings of the conference Elastic Filaments of the Cell, held June 16-19, 1999, in
Seattle, Washington"--T.p. verso.
    Includes bibliographical references.
    ISBN 978-0-306-46410-2   ISBN 978-1-4615-4267-4 (eBook)
    DOI 10.1007/978-1-4615-4267-4

    1. Cytoplasmic filaments--Congresses. 2. Intermediate filament proteins--Congresses.
3. Muscle contraction--Congresses. 4. Myofibroblasts--Congresses. I. Granzier, Henk.
II. Pollack, Gerald H. III. Series.

QH603.C95 E43 2000
571.6'5--dc21
                                                                                00-042439

Proceedings of the conference Elastic Filaments of the Cell, held June 16–19, 1999, in Seattle,
Washington

ISBN 978-0-306-46410-2

©2000 Springer Science+Business Media New York
Originally published by Kluwer Academic / Plenum Publishers, New York in 2000

http://www.wkap.nl/

10 9 8 7 6 5 4 3 2 1

PREFACE

This volume contains the proceedings of the conference *Elastic Filaments of the Cell*, held June 16-19, 1999, on the campus of the University of Washington (Seattle, WA). The conference was organized to focus on the rapid progress made in understanding elastic protein filaments, and to celebrate the career of Dr. Charles Trombitás whose work, spanning almost four decades, has helped provide the basis for this progress.

Dr. Trombitás's early research career in Hungary focused on the study of insect flight muscle. He and his colleagues were among the first to discover elastic connecting filaments in this muscle type. These filaments connect the thick filaments to the Z-lines and provide the high resting stiffness that allows insect flight muscles to oscillate at high frequencies. This early work is filled with beautiful electron micrographs and novel, forward-looking interpretations. It provided an important impetus for subsequent research on vertebrate muscle.

It is now well established that vertebrate striated muscle also contains elastic filaments. These filaments consist of the protein titin (also known as connectin). Titin is a giant filamentous polypeptide of multi-domain construction spanning between the Z- and M-lines of the sarcomere. Titin provides long-range elasticity to muscle, it has been implicated in myofibrillogenesis as a thick filament scaffold, and it may function as a cell-signaling molecule. Furthermore, recent studies indicate that titin-like proteins are also present in nonmuscle cells and that titin may provide structure and elasticity to chromosomes.

Many of the earlier and newer findings in the titin field are presented in the conference proceedings. Presentations focused on, for example, the recent discoveries of titin isoform sequences, molecular properties of single titin molecules using atomic force microscopy and laser tweezers, and the physiological role of titin in the cell. Highlights of the meeting were also the work on titin-like proteins in nonmuscle and smooth muscle cells and in chromosomes. The lively discussions that followed the presentations are also included in the proceedings.

Research on elastic filaments of the cell is currently a mature and rapidly developing area with laboratories from around the world making important contributions. Many factors contributed to the successes of the field, not the least of which is the exceptional work of Dr. Trombitás. This work has stimulated a great number of scientists to enter the field, many of whom were present at the meeting.

We would like to acknowledge the financial support received from the University of Washington and Washington State University. We are especially grateful for the generous support from the Whitaker Foundation, which made a difference. Many thanks also to the "Pollack lab crew" for helping to make the conference so memorable and successful. Finally, a special word of thanks for the superb assistance received from Jeanne Jensen in formatting this volume for publication.

Henk L. Granzier
Gerald H. Pollack

CONTENTS

TITIN-LIKE PROTEINS

FUNCTIONAL ROLE OF ELASTIC FILAMENTS

CONNECTING FILAMENTS: A HISTORICAL PROSPECTIVE

Károly Trombitás

Department of Veterinary and Comparative Anatomy, Pharmacology and Physiology, Washington State University, Pullman, WA

Abstract: This short review covers the development of the extensible filament research from the very beginning until the most recent results. This work emphasizes the milestones of discovery, which led us from initial observations that were solely ultrastructural to the molecular understanding of the extensible process of these filaments.

Introduction

The motivation behind this brief review is that I am one of the few researchers who has actively participated in the field of elastic filament research from its beginning, and I would like to pass on my experiences. Many exciting results have recently been obtained, and this has rapidly accelerated progress and attracted many young researchers to the field. These new members, however, have no personal experience in this field's heroic beginnings when we searched for answers with less well-developed methods and instruments. It might be useful to know some of our history and to learn from these past experiences. As such, an aim of this chapter is to emphasize that one of the most important properties of a scientist is open-mindedness.

This review is not to show the complete history of this research in full detail; its main goal is to highlight the milestones of discovery that opened our minds to a better understanding of the mechanisms of muscle function. This review will cover how the idea of the connecting filament developed from research on insect flight muscle, the emotional debates that followed as the find-

Elastic Filaments of the Cell, edited by Granzier and Pollack
Kluwer Academic/Plenum Publishers, 2000

1

ings were not generally accepted by the mainstream, and how supportive the concept of connecting filaments was in the later analysis of the third filament system in the vertebrate muscle. The early results and debates will be emphasized, as will be a few of the recent advances in this field.

Results and Discussion

In discussing the elastic filaments, it is unavoidable to at least briefly discuss the theories of muscle contraction that were developed during the last fifty years. Before the era of electron microscopy, the general concept was that the myofibrils elongated and shortened by the unfolding and refolding of contractile proteins. The first electron micrographs revealed that the myofibrils consisted of continuous filaments, thick in the A-band and thin in the I-band (e.g. Draper and Hodge, 1949; Rozsa *et al.*, 1950), and that the filaments could shorten and elongate with the above mechanism. It was assumed that the contractile proteins had elastic properties too. Thus, in the beginning there was a one-filament sarcomere model that already took muscle elasticity into account (Fig. 1; Hodge, 1955).

Due partly to the quick development of electron microscopical (EM) methods, and partly to the improvement of optical microscopy, in the mid-1950s the sliding-filament theory of muscle contraction was developed (H.E. Huxley and Hanson, 1954; A.F. Huxley and Niederke, 1954). According to this two-filament sarcomere model, the discontinuous, and inextensible, thick and thin filaments overlap in the middle of the sarcomere. Sarcomere length change was attributed to the varying degree of filament overlap. In this sarcomere model, there was no elastic component except for the limited elasticity in the cross-bridges. The model was very elegant, and it could successfully explain many features of muscle contraction (Figs. 2-3). However, it failed to explain the long-range elasticity of muscle.

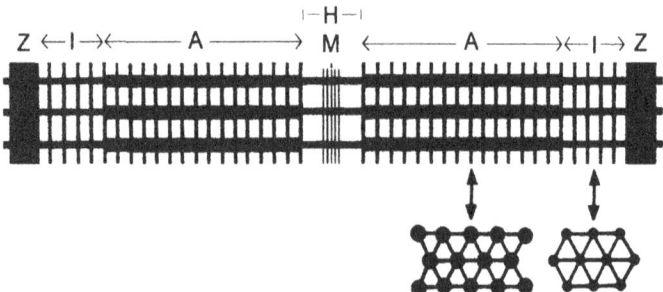

Figure 1. Single filament sarcomere model with longitudinal and transversal views (adapted from Hodge, 1955).

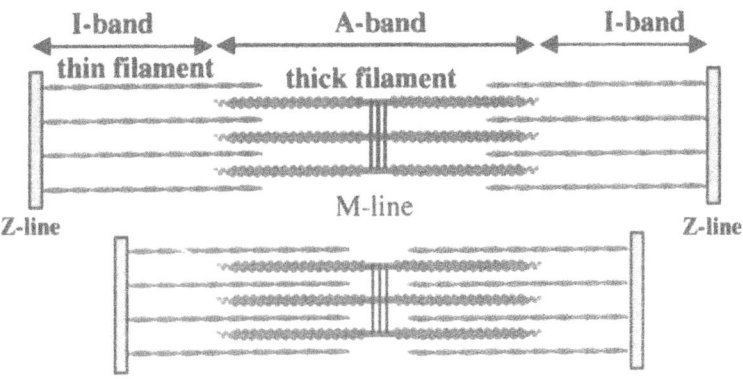

Figure 2. *The sliding (two) filament sarcomere model with sarcomere length change. (Adapted from Alberts* et al., Molecular Biology of the Cell, *Garland Publishing, Inc., New York 3rd Ed., 1994, Fig. 16-89.)*

Figure 3. *Local contraction of human tibialis posterior muscle illustrates the sliding filament theory. All possible variation of sarcomere lengths can be found in this single micrograph. In short sarcomeres, double overlap zones of thin filaments developed in the middle of the sarcomeres. In extended sarcomeres, the H-zones and I-bands vary proportionally with sarcomere lengths. At extremely stretched sarcomeres, gaps developed between the thick and thin filaments (Trombitás, unpublished). Scale bar: 1μm.*

This is a serious shortcoming. It was well known that insect muscle is elastic. When single myofibrils were prepared from insect flight muscle, the stretched myofibrils could shorten either in the relaxed or rigor state (e.g., Trombitás and Tigyi-Sebes, 1977). The two filament model also could not explain what kept the myofibrils together, either when myosin was extracted (Hanson and Huxley, 1956), or when the sarcomere was stretched beyond overlap (Huxley and Peachey, 1961; Carlson *et al.*, 1961).

Although not without confusion and conflicting ideas, to address these issues, a third kind of filament was proposed to exist in the sarcomere of skeletal muscle (Fig. 4). These filaments are:

a) The S-filaments, which connect the tip of the thin filaments in the middle of the sarcomere. These filaments were introduced by Hanson and Huxley (1956) to explain sarcomere integrity and elasticity. When the double overlap zone of the thin filaments in very short sarcomere were discovered (see Fig. 3), the S-filament concept was abandoned;

b) The T-filaments (Hoyle, 1968; McNeill and Hoyle, 1967), which run through the whole sarcomere connecting the Z-lines. (Note: there is no convincing ultrastructural evidence for T-filaments in vertebrate skeletal muscle.)

c) The gap filaments (Fig. 5), which i) either make connections between thick and thin filaments (Sjöstrand, 1962), or which ii) make asymmetrical connections between two thick filament through the Z-line (Locker and Leet, 1976a). (Note: the asymmetrically arranged elastic filaments could not keep thick filaments in highly stretched sarcomeres in register.)

Auber and Couteaux (1963), investigating the Z-line structure in cross sections in the asynchronous insect flight muscle, demonstrated connections between the thick filaments and the Z-line. Garamvölgyi (1963) confirmed these connections using longitudinal sections from bee flight muscle. The bee's

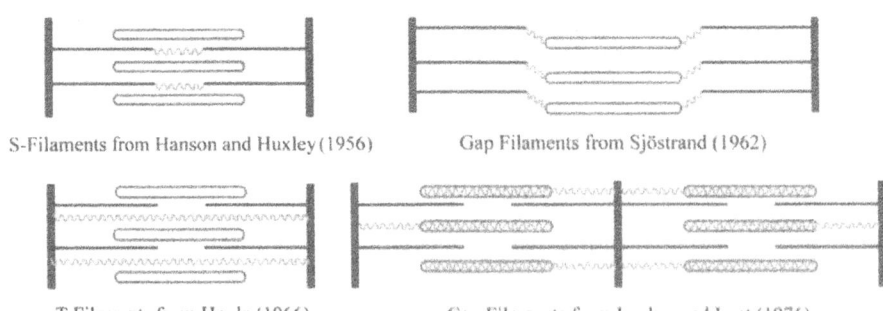

S-Filaments from Hanson and Huxley (1956)   Gap Filaments from Sjöstrand (1962)

T-Filaments from Hoyle (1966)   Gap Filaments from Locker and Leet (1976)

Figure 4. *Three filament sarcomere models for vertebrate skeletal muscles. The elastic or extensible filaments are superimposed on the sliding filament sarcomere model (adapted from Wang, 1985).*

flight muscle, like the skeletal muscle, is highly extensible; in living state the muscle is extensible to beyond overlap (Figs. 5, 6). Furthermore, at extreme long lengths, the thick filaments became elongated. This finding was attributed to the pulling force of a third kind of elastic filament, which made connections between the tip of the thick filaments and the Z-line. Based on this idea, a three-filament sarcomere model was created in which the I-band contained two kinds of thin filaments: the actin-containing filaments, which overlap the thick filaments; and elastic filaments, which connect the thick filaments to the Z-line (Garamvölgyi, 1965, 1971). These filaments were named connecting filaments by Pringle (1967). This filament model seemed to most suitably describe the elastic properties of the sarcomere (Fig. 7).

The three-filament sarcomere model was presented by Hoyle and Garamvölgyi using ultrastructural data, and by Guba extracting the fibrillin, an elastic filament protein from myofibril ghost, and extensively discussed at the 1966 international muscle symposium in Budapest (*Symposium on Muscle*, Ernst and Sraub eds., 1968). Although there was a general consensus about the appearance of the gap filaments in overstretched vertebrate skeletal muscle, the evidence was not convincing enough to firmly establish a third type of filament (see the comments of H.E. Huxley and J. Hanson in *Symposium on Muscle*, Ernst and Straub eds., 1968). Because of the identity problem of the gap filaments, A.F. Huxley (1974) later came to the same conclusion as H.E. Huxley

Figure 5. *Gap filaments in troponin antibody labeled frog semitendinosus muscle. The characteristic troponin pattern is visible in the I-band on the thin filaments but not on the gap filaments (Trombitás et al., 1990). Scale bar: 1μm.*

Figure 6. *Highly stretched bee flight muscle structure in single filament layer. The connecting filaments reached their extensible limit and elongated the thick filaments in the gap. The dense zones at the tip of the thin filaments represent the remaining overlap zones (Trombitás, unpublished). Scale bar: 1μm.*

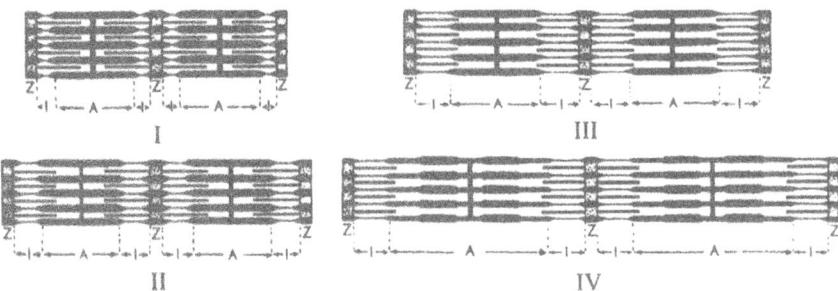

Figure 7. *Three-filament sarcomere model for insect flight muscle. Figure 12 is represented by sarcomere version I; Figure 6 is represented by sarcomere version IV (adapted from Garamvölgyi, 1965).*

and Hanson. Furthermore, Hanson (see the comment in *Symposium on Muscle*, Ernst and Straub eds., 1968) misunderstood Garamvölgyi's results when she compared her rigor-stretched myofibrils to the overstretched bee muscle, stretched in the relaxed state. Although she interpreted these results as stretching artifacts (Hanson, ibid.), the thick filament elongation in long sarcomeres was confirmed in bee muscle (Trombitás and Tigyi-Sebes, 1977). The bee muscle contains a considerable amount of paramyosin (Trombitás *et al.*, 1977; Bullard, 1983), and just recently direct mechanical measurements using a nanofabricated force transducer on single, isolated, invertebrate thick filaments with high paramyosin content showed that the thick filaments are stretchable (Neumann *et al.*, 1998). Thus, since the thick filaments with high paramyosin content are stretchable, Garamvölgyi's result still stands. Furthermore, applying stretch to vertebrate skeletal muscle in rigor, thick filament elongation was reported (Suzuki and Sugi, 1983).

The rejection of a third kind of filament was not surprising. In a brilliant EM study, H.E. Huxley (e.g., 1957) produced extra thin longitudinal sections that contained one filament layer. The obtained micrographs presented clear evidence for just two discontinuous filament sets. Furthermore, in serial cross sections, he counted the number of filaments in the H-zone, the overlap zone and the I-band in the same myofibril. Huxley could not find, in these technically perfect experiments, extra filaments in the I-band; the number of the filaments in the overlap zone was about the same as the sum of the number of thick filaments in the H-zone and the thin filaments in the I-band.

Since the sliding filament theory explained muscle contraction so well, the muscle field became conservative and, in my experience, this hindered progress in elastic filament research. It became very difficult to publish new results from this field in high-ranking journals. Referees were not sufficiently open-minded (Trombitás, personal experience).

Although the three-filament sarcomere model concept was turned down in the middle of the sixties, additional proof for myofibrillar elasticity soon came from the area of muscle mechanics. When glycerinated muscle fiber was used for mechanical measurement, the same high resting tension was observed as in the intact fiber (White, 1967). Since the glycerinated insect flight muscle fiber contained only myofibrils, the elasticity should have only resided in the myofibrils themselves. This observation provided strong indirect evidence for the presence of connecting type filaments. Furthermore, when these fibers were stretched in rigor, the thin filaments were broken along the Z-line, and pulled away with the A-band, leaving a clear zone between the A-band and the Z-line. In this gap, very thin filaments connected the tip of the thick filaments to the Z-line (Fig. 8; White, 1967; Reedy, 1971; Trombitás and Tigyi-Sebes, 1977; White, 1983). These filaments were identified as connecting filaments (Pringle, 1967). The connecting filament idea was supported using ultrastructural investigations by Zebe *et al.* (1968).

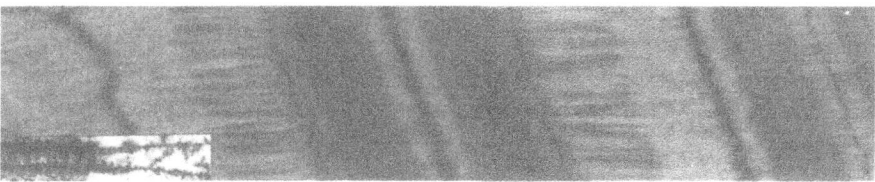

Figure 8. *Bee flight muscle stretched in relaxed state, transferred into rigor, stretched again and activated. Since the rigor bridges prevented the sliding motion during the rigor stretch, the thin filaments were broken along the Z-line, occasionally only one side of the sarcomeres. The sarcomere integrity was held by the extended connecting filaments in the broken half sarcomeres. On the effect of the activation, the broken thin filaments were propelled into the intact half-sarcomeres. The thin filament sliding stopped when the broken tip of the thin filaments reached the last correctly polarized bridges. Therefore, they have never entered into the intact I-band. [The reader should remember: in the* in vitro *motility assay, the thin filament movement was characterized by bidirectional movement, namely the same bridges could move the thin filaments in both directions, forth and back, although with a different velocity. In the intact filament lattice, the broken thin filaments have never been influenced by the oppositely polarized bridges; therefore, the bidirectional movement has no physiological significance. This experiment was the first* in vitro *motility assay that was perfected in the intact filament lattice (Trombitás and Tigyi-Sebes, 1977, 1984, 1985; Trombitás and Pollack, 1995)]. Since the thin filaments were removed from the broken half sarcomeres, the connections with the extended connecting filaments and the thick filaments could be clearly visualized. Every thick filament joined to a connecting filament, which anchored it to the Z-line (see the inset). Scale bar: 1µm.*

Then, Ashhurst (1967, 1971) investigated the Z-line structure of the Lethocerus flight muscle in cross section. In sarcomeres at resting length, she was unable to detect filaments near the Z-line in the thick filament lattice positions, and she concluded that connecting filaments did not exist in Lethocerus flight muscle (Fig. 9). As she stated later (Ashhurst, 1977): "Why these filaments are visible only after stretching muscle fibers in rigor remains, as of yet, a mystery." So, the connecting filaments were again considered as artifacts of rigor stretching. Ashhurst's results were so good and convincing that even Garamvölgyi, who was fanatically convinced about the presence of the connecting filaments in the insect flight muscle, tried to find other explanations for his earlier observations (Garamvölgyi *et al.*, 1973, 1974). Ashhurst's results also convinced Pringle. He developed a new theory: that the connecting filament's material forms a gel-like structure near the Z-line in resting muscle, and when the muscle is stretched in rigor, thin connecting filaments developed from this gel-like material (named connecting material; Pringle, 1977).

Thus, the question was raised as to whether, in myofibrils stretched in the relaxed state, the I-band contains connecting filaments that function in muscle contraction, or do the connecting filaments relate solely to rigor stretching and are artifacts. The job was to find the connecting filaments in a myofibril stretched in the relaxed state. However to distinguish in an intact I-band, the connecting filaments from the thin filaments is in no way a straightforward task. To overcome this problem, a successful attempt was made to selectively remove the thin filaments from bee flight muscle stretched in relaxed state by a fixation process and subsequent washing (Trombitás and Tigyi-Sebes 1972, 1974). In the absence of thin filaments, the thick filaments remained anchored to the Z-line on both sides of the Z-line by the connecting filaments (Fig. 10). This was the first direct evidence for the presence of connecting filaments in the insect flight muscle. [The reader should be reminded that the same idea was used later, when the intact titin filaments were shown in the heart muscle I-band; the thin filaments were selectively removed using gelsolin (Funatsu *et al.*, 1993; Trombitás and Granzier, 1997; Granzier *et al.*, 1997)].

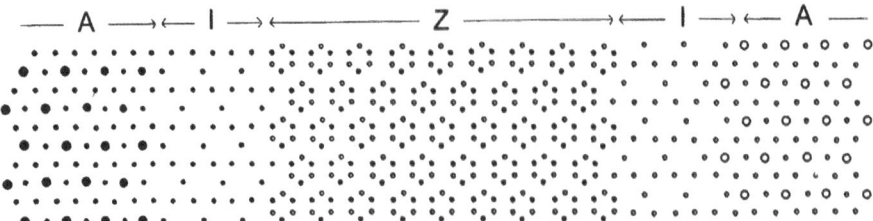

Figure 9. *The filament lattice at the Z-line region in neighboring sarcomeres in Lethocerus muscle (adapted from Ashhurst, 1967).*

Since the above results conflicted with Ashhurst's findings, Lethocerus muscle near resting length was investigated in thin longitudinal sections containing one filament layer. Figure 11 clearly shows that the connecting filaments join the thin filaments just before the thin filaments run into the Z-line. Thus, if the section plane crosses this near Z-line region, then only the thin filaments can be seen in the oblique cross section. This micrograph explains the mystery: while Ashhurst's findings were correct, in fact the connecting filaments could still exist in the Lethocerus muscle. [The reader should be reminded that the titin filaments are bound tightly to the thin filaments in skeletal

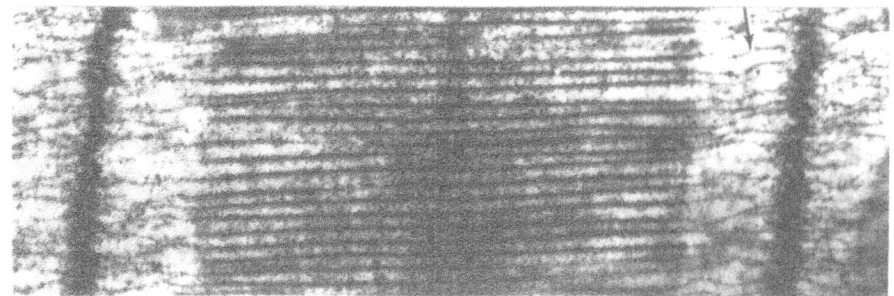

Figure 10. *Bee flight muscle was stretched in relaxed state, then the actin-containing thin filaments were selectively removed. The overlap zones disappeared, and the I-band contained only connecting filaments, which anchored the thick filaments to the Z-line (arrow). Scale bar: 1µm.*

Figure 11. *Lethocerus flight muscle near rest length in thin section containing only single filament layer. The connecting filaments in the I-band do not go directly to the Z-line; instead they attach to the thin filaments, where they have a small bulb (arrowheads, Trombitás, unpublished). Scale bar: 0.5 µm.*

and heart muscle in an ~100 nm-wide zone near the Z-line (Trombitás *et al.* 1993, 1995; Trombitás and Pollack, 1993).] It was shown that the bee connecting filaments run directly to the Z-line (Trombitás and Tigyi-Sebes, 1975), thus the bee flight muscle Z-line appears to be different from the Lethocerus muscle Z-line (Fig. 12).

Although not without confusion, the concept of connecting filaments was accepted in the 1977 *Insect Flight Muscle Symposium* at Oxford. This was the first milestone on the long road toward accepting the three-filament sarcomere model, at least for the insect flight muscle. The existence of the connecting filaments and their elastic properties were convincingly demonstrated in ultrastructural studies (Trombitás and Tigyi-Sebes, 1977).

Figure 10 is a micrograph of a sarcomere, stretched in a relaxed state, then the thin filaments selectively removed. This figure was later erroneously referred to as being an example of rigor stretched muscle (Ashhurst, 1977). Thus, to confirm that the I-band contained connecting filaments, bee muscle was stretched first in relaxed state (from freshly prepared living tissue) when wide I-bands and H-zones develop, then the muscle was transferred into rigor solution and stretched again. Upon this second stretch, the thin filaments broke regularly along the Z-line, independently of the existing wide I-bands. Between the tip of the thin filaments, projecting out of the A-band, and the Z-line, a gap developed, filled with connecting filaments, seemingly connecting the broken tip of the thin filaments and the Z-line (Fig. 13). The regularity of the break indicated that the integrity of the sarcomere was not destroyed and the appearance of the filaments within the gap proved the point (Trombitás and Tigyi-Sebes, 1977). Thus, the connecting filament is present in the living functioning muscle.

To investigate whether the bee flight muscle myofibrils have internal elasticity, the fibers were stretched in the relaxed state to about 150% of their resting length, and then they were transferred into rigor solution. After rigor development, the stretch was terminated. In spite of the rigor connection between the thick and thin filaments, the muscle contracted and shortened almost to its resting length. Ultrastructural studies revealed that during the stretch, the

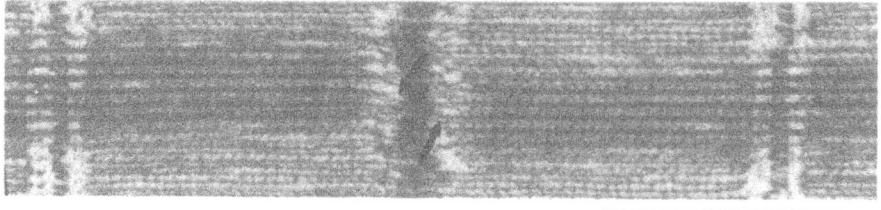

Figure 12. *Single filament layer of bee flight muscle. The thick filaments run directly to the Z-line (arrows, Trombitás, unpublished). Scale bar: 1 μm.*

connecting filaments gained so much elastic energy as to shorten the I-band, crumpling the thin filaments into a narrow band along the Z-line, all while the overlap zones and H-zone remained unchanged (Fig. 14). This was the first direct evidence for the elasticity that resides in myofibrils (Trombitás and Tigyi-Sebes, 1977). [The reader should remember: in the two-filament sarcomere model, sarcomere contraction was forbidden in the absence of Mg-ATP and free Ca^{2+} ions. Therefore, the passive sarcomere shortening strongly suggested the presence of a third kind of elastic filament inside the sarcomere.]

The conference presented other important new findings. A 180 kDa protein component of the connecting filaments had been identified by immunoelectron microscopy (Bullard *et al.*, 1977). The antibodies labeled a

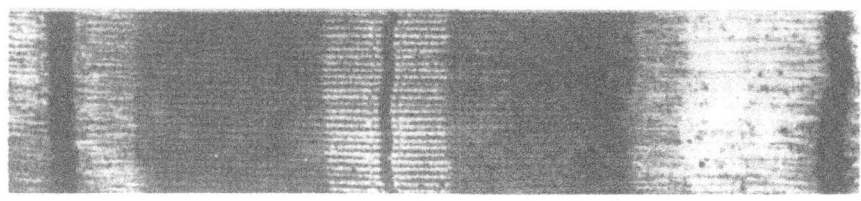

Figure 13. Bee flight muscle stretched in relaxed state, then again in rigor. The connecting filaments appeared in the gap, so they were present in the I-band after the muscle was stretched in relaxed state. Scale bar: 1 μm.

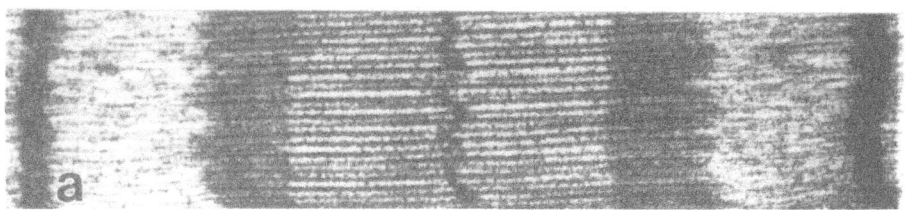

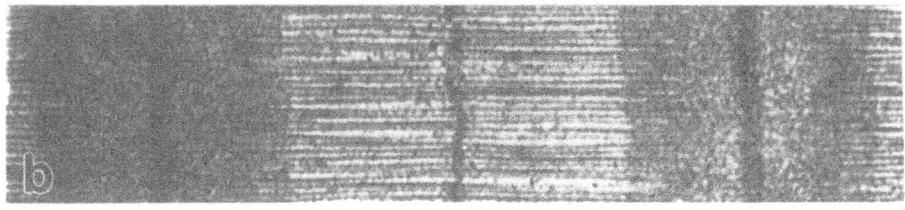

Figure 14. Bee flight muscle stretched in relaxed state, then (a) fixed in stretched state, (b) the stretched muscle was transferred into rigor and the stretch was terminated. The muscle shortened, keeping the H-zones intact, but crumpling the thin filaments in both sides of the Z-line, due to the elastic energy, gained the connecting filaments during the stretching. Scale bar: 1μm.

narrow band along both sides of the Z-line, just where the connecting filaments were expected. Later the 360 kDa doublet of this protein was extracted and named projectin (Saide, 1981). The antibodies against projectin labeled the sarcomere in exactly the same region as above. An example of the projectin-labeled Lethocerus flight muscle is shown in Figure 15. The antibodies labeled two distinct epitopes on both sides of the Z-line in slightly stretched sarcomeres.

Further evidence for the physiological role of the connecting filaments was presented in the conference by mechanical measurements. Experiments showing the high resting tension and the high stretch activation of the insect flight muscle also supported the concept of the elastic filament (Pringle, 1977; White, 1977; Steiger, 1977).

In summary, a big step was made at the Oxford *Insect Flight Muscle Symposium* toward understanding the elasticity of the insect flight muscle, with revealing ultrastructural, immunoelectron microscopic, and mechanical results.

As Ashhurst pointed out (Ashhurst, 1971), the Z-line structure is a determining factor in the presence, or absence, of connecting filaments. However, usually the Z-line structure was investigated using slightly oblique cross sections (Auber and Couteaux, 1963; Ashhurst, 1967, 1971, 1977; Cullen 1971; Saide and Ulrrick, 1973), while only a few studies were done using both longitudinal and oblique cross sections (Garamvölgyi, 1965; Trombitás and Tigyi-Sebes, 1975). As already discussed above, this led to Ashhurst explicitly denying any presence of connecting filaments in the Z-line (Ashhurst 1967,1971,1977). Other misunderstandings also slowed the construction of an accurate elastic filaments model.

Saide, who strongly supported the presence of the connecting filaments in the bee flight muscle, denied the presence of connecting filaments inside the Z-line, based on immunoelectron microscopy results (Saide, 1981). But the fact alone that the projectin antibody was not found in the Z-line, does not

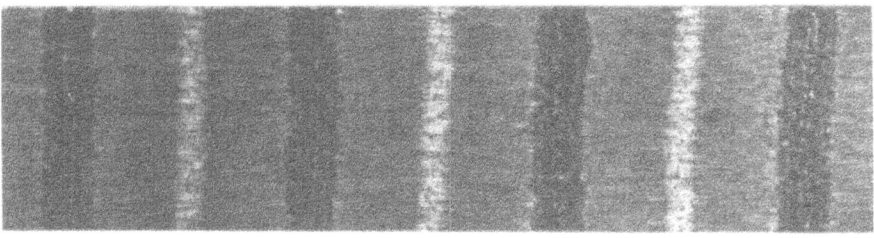

Figure 15. Lethocerus flight muscle labeled with projectin antibody. Two distinct epitopes were labeled, one in the I-band near the Z-line, the other in the A-band near the tip of the thick filaments (Trombitás, unpublished). Scale bar: 1 μm.

necessary mean that the connectin filaments are not present in the Z-line. The projectin is a long protein — the bee projectin contour length is about 300 nm (Nave and Weber, 1990; Daley *et al.*, 1998) — and probably every antibody against projectin has its own particular antigenic labeling site. At least this is a general rule regarding the titin antibodies. There is only one exception: the 9D10 antibody (S.M. Wang and Greaser, 1985), which labels the whole PEVK segment in the skeletal and heart muscle (Trombitás *et al.*, 1998b).

For the three-dimensional reconstruction of the bee flight muscle Z-line, an elegant work was done by Deatherage and his colleagues, based on cross sections of the Z-line (Cheng and Deathrage, 1989; Deatherage *et al.*, 1989). The model left open the connecting filament origin and localization inside the Z-line. This model was later modified by Reedy's group (Edwards *et al.*, 1994) and the connecting filament origin was determined to be located in the middle of the Z-line.

Nevertheless, the longitudinal view of the Z-line can not be ignored in constructing a Z-line model. Let us explore some examples: when the bee or wasp muscle was stretched with relatively high speed in rigor, the whole Z-line structure extended suggesting that the connecting filaments from the neighboring sarcomeres overlap the Z-line (Fig. 16). This impression was supported by studies of isolated thick filament assemblies from opposing sarcomeres still linked together at their tips by connecting filaments (Fig. 17; Trombitás and Tigyi-Sebes, 1979,1986) and longitudinal views of the Z-line structure prepared by a freeze-fracture method (Fig. 18). These linkers are clearly seen, as "Z-filaments" in a longitudinal section of an obliquely stretched Z-line, in a fiber that was first stretched in relaxed state, then again in rigor (Fig. 19). Seemingly, the thick filaments are continuous from one sarcomere to the next, but actually thc overlapping projectin may bind together two thick filaments and also involve other Z-line proteins.

The connecting filaments may play the same role in the insect Z-line, as titin plays in the skeletal muscle. It was recently demonstrated that the titin filaments from opposite sarcomeres completely overlapped the Z-lines in human soleus muscle (Gregorio *et al.*, 1998). Titin may serve as a molecular template for assembling the sarcomere including the M- and Z-lines (Review by Gregorio *et al.*, 1999). The connecting filaments may play a role similar to that of titin, namely they may serve as a molecular template of the insect flight muscle Z-line assembly.

In the middle 1970s, the presence of an elastic filament was attributed as the peculiarity of the insect flight muscle, but such filaments were denied in the skeletal muscle. However, the elastic filament idea was kept alive with reinvestigation of the gap filaments in overstretched skeletal muscle (Locker and Leet, 1975, 1976a,b; Locker *et al.*, 1976; Locker *et al.*, 1977). Nevertheless, a set-

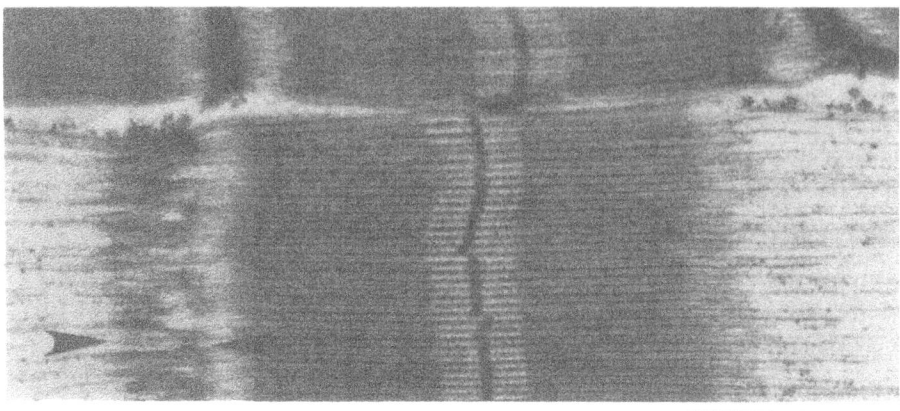

Figure 16. *Bee flight muscle was stretched in relaxed state, then with relatively high speed in rigor again. Some of the Z-line elongated, where the thin filaments were broken (arrowheads, Trombitás, unpublished). Scale bar: 1 μm.*

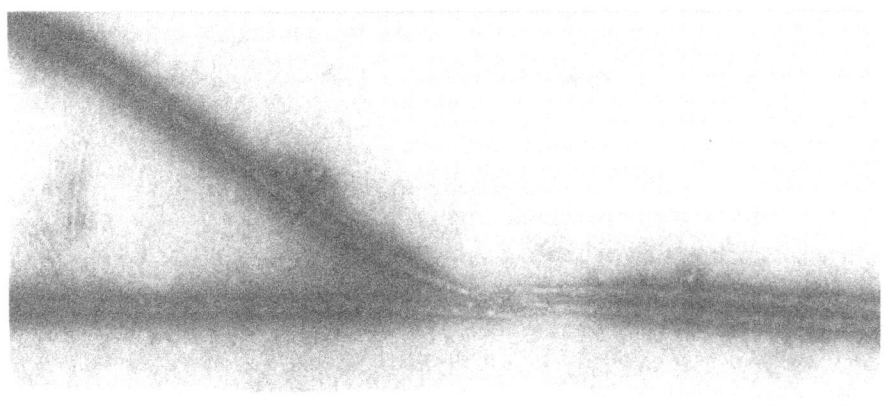

Figure 17. *Isolated thick filaments from neighboring sarcomeres (bee flight muscle). The tip of the thick filaments are connected by a zig-zag-like structure in the Z-line region. Scale bar: 0.5 μm.*

back was that the gap filaments were not found in the I-band in cross section again; the theory could not gain general acceptance (Ullrick *et al.*, 1977).

The second milestone of acceptance of the three filament sarcomere model related partly to the discovery of a giant myofibrillar protein titin, also named connectin (Wang *et al.*, 1979; Maruyama *et al.*, 1981), and partly to the discovery of the fact that the elasticity of skeletal muscle resided mainly in the myofibrils (Magid *et al.*, 1984; presented first in *Contractile Mechanism of Muscle*, Pollack and Sugi eds., 1982). However, the real turning point—when the scientific community seriously re-examined the elastic filament problem—occurred

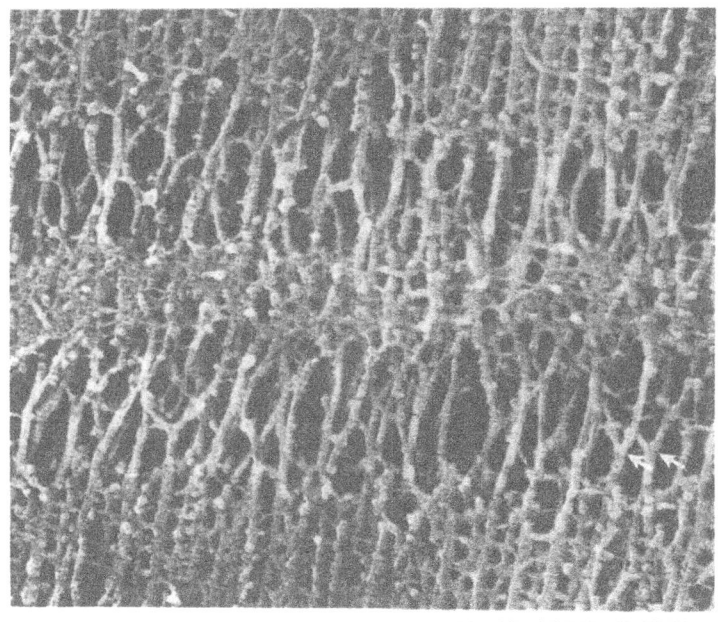

Figure 18. *The image of Z-line region from bee flight muscle stretched in rigor using freeze-fracture method. See the bifurcated connecting filaments (arrows), which were occasionally pulled out from the Z-line area, that resemble the links between isolated thick filaments from opposite sarcomeres (see Fig. 17). Between the "Z-filaments" there are radial interconnections and amorphous material (Trombitás and Baatsen, unpublished). Scale bar: 0.5 μm.*

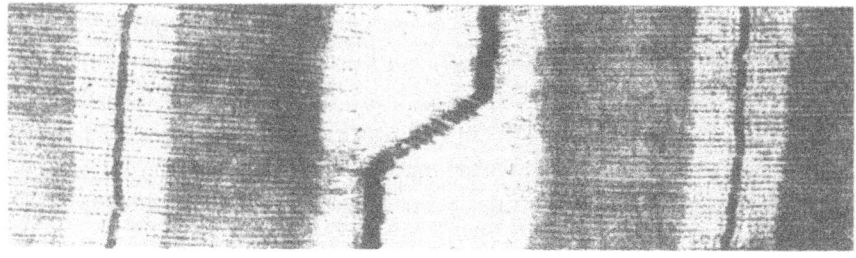

Figure 19. *Bee flight muscle stretched in relaxed state, then again in rigor. Asymmetrically broken thin filaments created an obliquely stretched Z-line, where these "Z-filaments" became observable. Scale bar: 1 μm.*

in the *Contractile Mechanism of Muscle Symposium* held in Seattle in 1982. Wang presented here one of his first immunofluorescent studies using titin antibodies. From the results, he composed the first hypothetical three-filament sarcomere model for vertebrate skeletal muscle where the titin filaments were incorporated as elastic components (Wang, 1984).

Furthermore, in this conference Magid presented mechanical studies on vertebrate skeletal muscle, which were followed with great attention. He used the experimental procedure that was developed for the insect flight muscle (Trombitás and Tigyi-Sebes, 1977) using skeletal muscle fibers. The fibers were stretched beyond overlap in relaxed state, and then they were transferred into rigor. After the rigor developed, the stretch was terminated. In spite of the high sarcomere length, the muscle fibers contracted almost to their resting length, as was shown in the insect flight muscle, which even had an overlap zone (Trombitás and Tigyi-Sebes, 1977). This study demonstrated again that muscle elasticity mainly resided in the myofibrils, and that the elasticity was represented by the gap filaments. He also showed the different natures of the gap filaments and the thin filaments, since the myosin heads could not decorate the gap filaments (Magid *et al.*, 1984). There was now increasing evidence that the gap filaments were composed of titin (e.g., Wang, 1985; Trombitás *et al.*, 1991).

Since titin has good antigenic properties, by the second half of the 1980s numerous antibodies were developed (reviewed by Maruyama, 1994) that labeled titin epitopes in the sarcomere from M-line to Z-line (Fürst *et al.*, 1988). Using these antibodies in immunoelectron microscopic studies, titin's localization, elastic properties and physiological role were revealed (see the recent reviews of Maruyama, 1997; Trinick, 1996; Wang, 1996, Labeit *et al.*, 1997; Horowits, 1999).

The third milestone is connected to the elucidation of the complete sequence of the human soleus and heart muscle titin (Labeit and Kolmerer, 1995). The primary structure indicates that in the elastic I-band, titin is composed mainly of two types of segments: tandem Ig segments, consisting of serially linked Ig-like domains, and the PEVK segment that has a unique sequence and composition (70% of its residues are P-proline, E-glutamine, V-valine, and K-lysine). There is an essential difference between the proximal tandem segment and the PEVK segment in skeletal and heart muscle, namely these segments are considerably shorter in the heart muscle than in the skeletal muscle.

The knowledge of the primary structure of titin led us toward a molecular understanding of the elasticity of titin filaments. The possibility to develop sequence-specific titin antibodies, which could mark the ends of different segments of titin, now became a reality. To follow the elastic properties of I-band titin, four sequence-specific antibodies were used that mark the ends of the tandem Ig and PEVK segments. By following the extension of the segments as

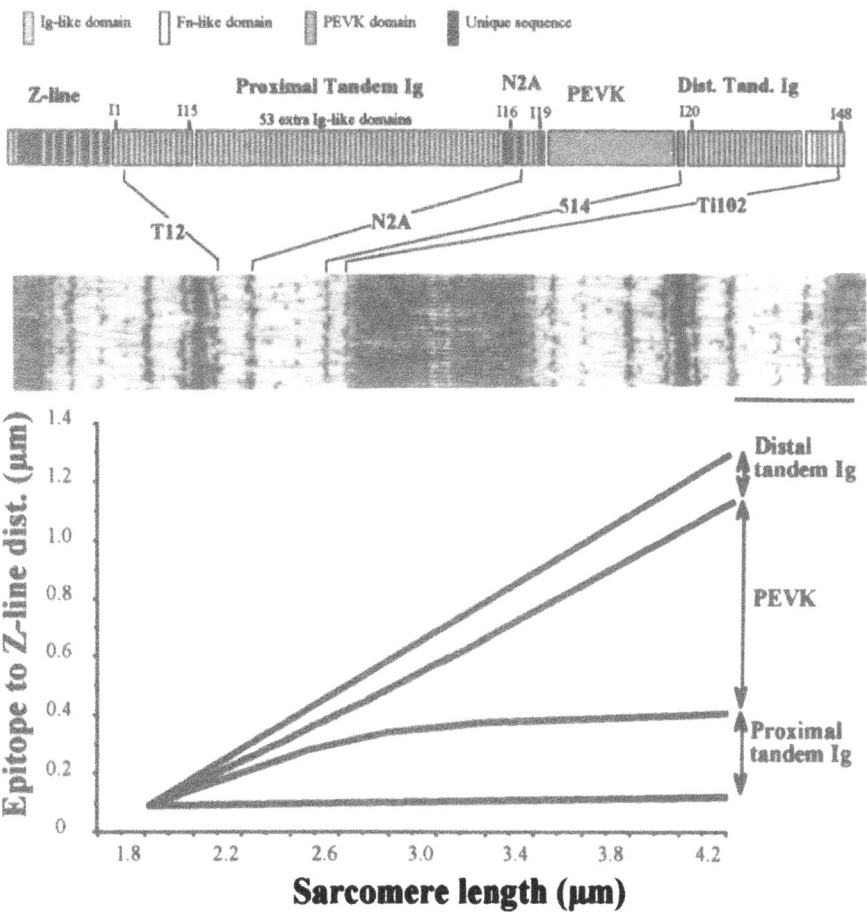

Figure 20. *The tandem Ig and PEVK segments extension as a function of the sarcomere length. Scale bar: 1 μm.*

a function of sarcomere length in human soleus muscle, their respective contribution to titin's elastic behavior was established. It was found that in short sarcomeres, the tandem Ig and PEVK segments were contracted. Upon stretching sarcomeres from about 2.0 μm to 2.7 μm, the "contracted" tandem Ig segments straightened, while their individual Ig domains remained folded. When sarcomeres were stretched beyond about 2.7 μm, the tandem Ig segments did not further extend; instead PEVK extension was more dominant (Figs. 20-21; Trombitás *et al.*, 1998a). Modeling tandem Ig and PEVK segments as entropic springs with different bending rigidities (Kellermayer *et al.*, 1997), indicated that in the physiological sarcomere range the Ig-like domains of tandem Ig

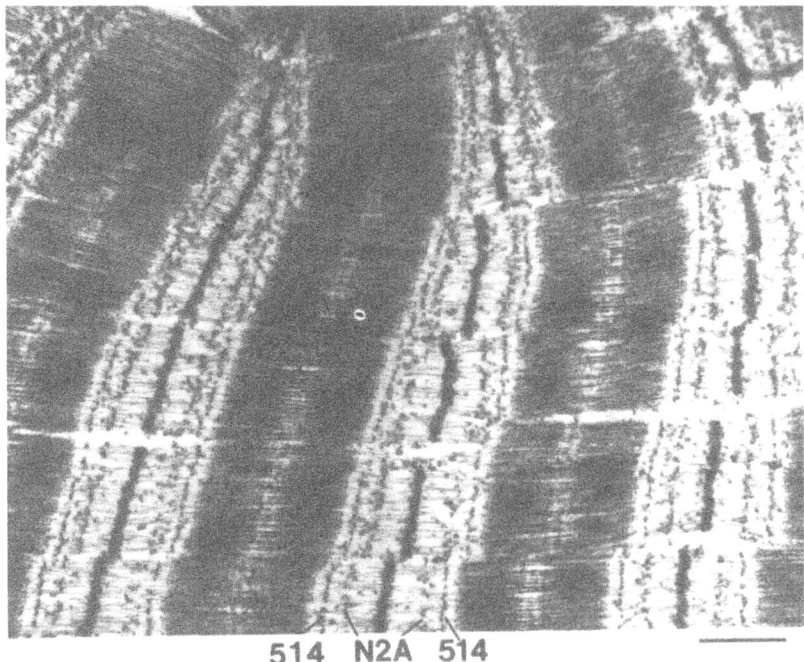

Figure 21. *Direct visualization of the tandem Ig and the PEVK segment length extension as a function of sarcomere length in locally contracted human soleus muscle using N2A and 514 titin antibodies (Trombitás and Granzier, unpublished). Scale bar: 1 μm.*

segment remain folded and that the PEVK segment behaves as a permanently unfolded polypeptide. This arrangement of the titin protein was represented as a serially linked two-spring system (Trombitás *et al.*, 1998a).

The entropic spring nature of the tandem Ig segments was confirmed by Linke *et al.* (1998a), but for the PEVK extension a combined entropic and enthalpic spring nature was suggested (Linke *et al.*, 1998b).

In the heart muscle, the PEVK segment is very short, but cardiac titin has an N2B segment with a large unique sequence. The short PEVK segment allows very limited elongation without Ig domain unfolding. Very recently, this N2B segment was shown to be extensible. After the Ig segments straighten and PEVK segment elongates, the N2B segment begins to greatly extend, allowing the sarcomere more freedom to accommodate a wide range of sarcomere lengths. Thus, the heart titin can be characterized as a serially linked three-spring system (Helmes *et al.*, 1999; Linke *et al.*, 1999; Trombitás *et al.*, 1999).

The fourth milestone is just before us, and we can see it shining on the horizon. There is a well-founded hope that the use of transgenic technology will allow elimination of part of titin's primary structure. This may make it possible to directly test the role of titin's subdomains in, for example, thick filament length regulation or in cell-signaling processes. It is likely that this and other ongoing work will keep our field exciting for many years to come.

During the last ten years the elastic filament research community has made enormous progress, as this (Seattle 1999) conference will most certainly show. As the *Insect Flight Muscle* book did in the past (*Insect Flight Muscle*, Tregear J. ed., Elsevier, Amsterdam, 1977), I hope the proceedings of this conference will serve as an important reference in the future of elastic filament research.

Acknowledgment

The author thanks Henk Granzier for helpful discussions, and Danielle Higgins and Mark MacNab for technical support. This work was supported by the Hungarian OTKA (T6280 to K.T.) and by the National Institutes of Health National Heart, Lung, and Blood Institute (HL1497 and HL62881 to H.G.)

References

Ashhurst DE. Z-line of the flight muscle of belostomatid water bugs. *J Mol Biol* 1967;27:385-389.

Ashhurst DE. The Z-line in Insect flight muscle. *J Mol Biol* 1971;27:385-389.

Ashhurst DE. "The Z-line: Its structure and evidence of connecting filaments." In *Insect Flight Muscle*, RT Tregear, ed. Amsterdam: Elsevier/North Holland Biomedical Press, 1977.

Auber J, Couteaux R. Ultrastructure de la striae Z dans des muscules de diptéres. *J Microsc* 1963;2:309-316.

Bullard B. Contractile proteins of insect fight muscle. *Trends Biochem Sci* 1983;8:68-70.

Bullard B, Hammond KS, Luke BM. "Immunological investigation of proteins associated with thick filaments of insect flight muscle." In *Insect Flight Muscle*, RT Tregear, ed. Amsterdam: Elsevier/North Holland Biomedical Press, 1977.

Carlson F, Knappeis GG, Buchtal F. Ultrastructure of the resting and contracted striated muscle fiber at different degrees of stretch. *J Biophys Biochem Cytol* 1961;11:91-117.

Cheng N, Deatherage JF. Three dimensional reconstruction of the Z disk of sectioned bee flight muscle. *J Cell Biol* 1989:108:1761-1764.

Cullen MJ. *Doctoral Thesis*, Oxford University, Oxford, England, 1971.

Daley J, Southgate R, Ayme-Southgate A. Structure of the *Drosophila* projectin protein: isoforms and implication for projectin filament assembly. *J Mol Biol* 1998;279:201-210.

Deatherage JF, Cheng N, Bullard B. Arrangement of filaments and cross-links in the bee flight muscle Z disk by image analysis of oblique section. *J Cell Biol* 1989;108:1775-1782.

Draper MH, Hodge AJ. Studies on muscle with the electron microscope. I. The ultra structure of toad striated muscle. *Aust J Exp Biol Med Sci* 1949;27;465-503.

Edwards RF, Lucaveche C, Reedy MK. The A-bee-Z problem of actin filament rotation in insect flight muscle, *Biophys J* 1994;A190 (Abstract).

Ernst E, Straub FB. *Symposium on Muscle*, Budapest, Adademiai, Kiado, 1968.

Funatsu T, Kono E, Higuchi S, Kimura S, Ishiwata S, Yoshioka T, Maruyama K, Tsukita S. Elastic filaments *in situ* in cardiac muscle: deep-etch replica analysis in combination with selective removal of actin and myosin filaments. *J Cell Biol* 1993;120:711-724.

Fürst DO, Osborn M, Nave R, Weber K. The organization of titin filaments in the half-sarcomere revealed by monoclonal antibodies in immunoelectron microscopy: a map of ten nonrepetitive epitopes starting at the Z line extends close to the M line. *J Cell Biol* 1988;106:1563-1572.

Garamvölgyi N. Observations preliminaries sur la structure de las striae Z dans le muscle alaire de l'Abeille. *J Microsc* 1963;2:107-112.

Garamvölgyi N. The arrangement of the myofilaments in the insect flight muscle. I. *J Ultrastruc Res* 1965;13:409-424.

Garamvölgyi N. "The functional morphology of muscle." In *Contractile Proteins and Muscle*, K Laki, ed. New York, NY: Dekker, 1971.

Garamvölgyi N, Biczo G, Ladik J, Eöry A. Forces acting between muscle filaments. II. A theoretical computation of the resting elasticity curve. *Acta Biochim Biophys Acad Sci Hung* 1973;8:57-67.

Garamvölgyi N, Biczo G, Eöry A, Suhai S. Forces acting between muscle filaments. III. A mathematical computation of the resting elasticity of bee wing muscle. *Acta Biochim Biophys Acad Sci Hung* 1974;9:233-238.

Granzier H, Kellermayer M, Helmes M, Trombitás K. Titin elasticity and mechanism of passive force development in rat cardiac myocytes probed by thin-filament extraction. *Biophys J* 1997;73:2043-2053.

Gregorio CC, Trombitás K, Centner T, Kolmerer B, Stier G, Kunke K, Suzuki K, Obermayr F, Herrmann B. The NH2 terminus of titin spans the Z-disk: its interaction with a novel 19-kD ligand (t-cap) is required for sarcomeric integrity. *J Cell Biol* 1998;143:1013-1027.

Gregorio CC, Granzier H, Sorimachi H, Labeit S. Muscle assembly: a titanic achievement? *Curr Opin Cell Bio* 1999;11:18-25.

Hanson J. "General Discussion." In *Symposium on Muscle*, E Ernst, FB Straub, eds. Budapest, Adademiai, Kiado, 1968.

Hanson J, Huxley HE. The structural basis of contraction in striated muscle. *Symp Soc Exp Biol* 1956;9:228-264.

Helmes M, Trombitás K, Centner T, Kellermayer M, Labeit S, Linke WA, Granzier H. Mechanically driven contour-length adjustment in rat Cardiac titin's unique N2B sequence titin is an adjustable spring. *Circ Res* 1999;84:1339-1352.

Hodge AJ. Studies on the structure of muscle. III. Phase contrast and electron microscopy of dipteran flight muscle. *J Bioysic and Biochem Cytol* 1955;1:361-380.

Horowits R. The physiological role of titin in striated muscle. *Rev Physiol Biochem Pharm* 1999; 138:57-96.

Hoyle G. "General Discussion." In *Symposium on Muscle*, E Ernst, FB Straub, eds. Budapest, Adademiai, Kiado, 1968.

Huxley HE. "General Discussion." In *Symposium on Muscle*, E Ernst, FB Straub, eds. Budapest, Adademiai, Kiado, 1968.

Huxley AF. Muscular contraction. *J Physiol (Lond)* 1974;243:1-43.

Huxley AF, Niedergerke R. Structure changes in muscle during contraction. *Nature* 1954;173:971-972.

Huxley AF, Peachey LD. The maximum length for contraction in vertebrate striated muscle. *J Physiol (Lond)* 1961;156:150-165.

Huxley HE, Hanson J. Changes in the cross striations of muscle during contraction and stretch and their structural interpretation. *Nature* 1954;173:973-976.

Huxley HE. The double array of filaments in cross-striated muscle. *J Biophysic Biochem Cytol* 1957;3:631-647.

Kellermayer MS, Smith SB, Granzier HL, Bustamante C. Folding-unfolding transitions in single titin molecules characterized with laser tweezers. *Science* 1997;276:1109-1112.

Labeit S, Kolmerer B. Titins: giant proteins in charge of muscle ultrastructure and elasticity. *Science* 1995;270:293-296.

Labeit S, Kolmerer B, Linke WA. The giant protein titin. Emerging roles in physiology and pathophysiology. *Circ Res* 1997;80:290-294.

Linke WA, Ivemeyer M, Mundel P, Sockmeier MR, Kolmerer B. Nature of PEVK-titin elasticity in skeletal muscle. *Proc Natl Acad Sci USA* 1998a;95:8052-8057.

Linke WA, Stockmeier MR, Ivemeyer M, Hosser H, Mundel P. Characterizing titin's I-band Ig domain region as an entropic spring. *J Cell Sci* 1998b;111:1567-1574.

Linke WA, Rudy DE, Centner T, Gantel M, Witt C, Labeit S, Gregorio C. I-band titin in cardiac muscle is a three element molecular spring and is crtical for maintaining thin filament structure. *J Cell Biol* 1999;146:631-644.

Locker RH, Leet NG. Histology of highly-stretched beef muscle. I. The fine structure of grossly stretched single fibers. *J Ultrastruc Res* 1975;52:64-75.

Locker RH, Leet NG. Histology of highly-stretched beef muscle. II. Further evidence on the location and nature of gap filaments. *J Ultrastruc Res* 1976a;55:157-172.

Locker RH, Leet NG. Histology of highly stretched beef muscle. IV. Evidence for movement of gap filaments through the Z-line, using the N_2-line and M-line as markers. *J Ultrastruc Res* 1976b;56:31-38.

Locker RH, Daines GJ, Leet NG. Histology of highly-stretched beef muscle. III. Abnormal contraction patterns in ox muscle, produced by overstretching during pre-rigor blending. *J Ultrastruct Res* 1976;55:173-181.

Locker RH, Daines GJ, Carse WA, Leet NG. Meat tenderness and the gap filaments. *Meat Sci* 1977;1:87-104.

Magid AD, Tings-Beal HP, Casvel M, Koutis T, Urcareche C. "Connecting filaments, core structures to the sliding filament model." In *Contractile Mechanism of Muscle*, GH Pollack, H Sugi, eds. New York, NY: Plenum Press, 1984.

Maruyama K. Connectin, an elastic protein of striated muscle. *Biophys Chem* 1994;50:73-85.

Maruyama K. Connectin/titin, giant elastic protein of muscle. *FASEB J* 1997;11:341-345.

Maruyama K, Kimura S, Ohashi K, Kuwano Y. Connectin, an elastic protein of muscle. Identification of "titin" with connectin. *J Biochem* 1981;89:701-709.

McNeill PA, Hoyle G. Evidence for superthin filaments. *Am Zool* 1967;7:483-498.

Nave R, Weber K. A myofibrillar protein of insect muscle related to vertebrate titin connects Z band and A band: purification and molecular characterization of invertebrate mini-titin. *J Cell Sci* 1990;95:535-544.

Neumann T, Fauver M, Pollack GH. Elastic properties of isolated thick filaments measured by nanofabricated cantilevers. *Biophys J* 1998;75:938-947.

Pringle JW. The Contractile mechanism of insect fibrillar muscle. *Prog Biophys Mol Biol* 1967;17:1-60.

Pringle JWS. "The mechanical characteristic of insect fibrillar muscle." In *Insect Flight Muscle*, RT Tregear, ed. Amsterdam: Elsevier/North Holland Biomedical Press, 1977.

Reedy MK. "Electron microscope observations concerning the behavior of the cross-bridge in striated muscle." In *Contractility of Muscle Cells and Related Processes*, RJ Podolsky, ed. Englewood Cliffs, NJ: Prentice-Hall, 1971.

Rozsa G, Szent-Györgyi A, Wyckoff RWG. The fine structure of myofibrils. *Exp Cell Res* 1950;1:194-205.

Saide JD. Identification of a connecting filament protein in insect fibrillar flight muscle. *J Molec Biol* 1981;153:661-679.

Saide JD, Ulrrick WC. Fine structure of the honeybee Z disk. *J Mol Biol* 1973;79:329-337.

Sjöstrand F. The connections between A- and I-band filaments in striated frog muscle. *J Ultrastruc Res* 1962;7:225-246.

Steiger GJ. "Stretch activation and tension transients in cardiac, skeletal and insect flight muscle." In *Insect Flight Muscle*, RT Tregear, ed. Amsterdam: Elsevier/North Holland Biomedical Press, 1977.

Suzuki S, Sugi H. Extensibility of the myofilaments in vertebrate skeletal muscle as revealed by stretching rigor muscle fibers. *J Gen Physiol* 1983;81:531-546.

Tregear RT. *Insect Flight Muscle*. Amsterdam: Elsevier/North Holland Biomedical Press, 1977.

Trinick F. Cytoskeleton: titin as scaffold and springs. *Current Biol* 1996;6:258-260.

Trombitás K, Granzier H. Actin removal from cardiac myocytes shows that near Z line titin attaches to actin while under tension. *Am J Physiol* 1997; 273:662-670.

Trombitás K, Pollack GH. Elastic properties of the titin filaments in the Z-line region of vertebrate muscle. *J Musc Res Cell Motil* 1993;14:416-422.

Trombitás K, Pollack GH. Actin filaments in honeybee flight muscle move collectively. *Cell Motil Cytoskeleton* 1995;32:145-150.

Trombitás K, Tigyi-Sebes A. Continuity of thick and thin filaments. *Acta Biochim Biophys Acad Sci Hung* 1972;7:193-194.

Trombitás K, Tigyi-Sebes A. Direct evidence for connecting C filaments in flight muscle of honey bee. *Acta Biochim Biophys Acad Sci Hung* 1974;9:243-253.

Trombitás K, Tigyi-Sebes A. The Z-line of the flight muscle of honey-bee. *Acta Biochim Biophys Acad Sci Hung* 1975;10:83-93.

Trombitás K, Tigyi-Sebes A. Fine structure and mechanical properties of insect muscle. In *Insect Flight Muscle*, RT Tregear, ed. Amsterdam: Elsevier/North Holland Biomedical Press, 1977.

Trombitás K, Tigyi-Sebes A. The continuity of thick filaments between sarcomeres in honey bee flight muscle. *Nature* 1979;281:319-320.

Trombitás K, Tigyi-Sebes A. Cross-bridge interaction with oppositely polarized actin filaments in double-overlap zones of insect flight muscle. *Nature* 1984;309:168-170.

Trombitás K, Tigyi-Sebes A. How actin filament polarity affects crossbridge force in doubly-overlapped insect muscle. *J Muscle Res Cell Motil* 1985;126:2285-2288.

Trombitás K, Tigyi-Sebes A. Structure of thick filament from insect flight muscle. *Acta Biochim Biophys Acad Sci Hung* 1986;21:115-128.

Trombitás K, Tigyi-Sebes A, Pallai G. "The paramyosin content and localization in the honey bee." In *Insect Flight Muscle*, RT Tregear, ed. Amsterdam: Elsevier/North Holland Biomedical Press, 1977.

Trombitás K, Baatsen PH, Lin JJ, Lemanski LF, Pollack GH. Immunoelectron microscopic observations on tropomyosin localization in striated muscle. *J Muscle Res Cell Motil* 1990;11:445-52.

Trombitás K, Baatsen P, Kellermayer M, Pollack GH. Nature and origin of gap filaments in striated muscle. *J Cell Sci* 1991;100:809-814.

Trombitás K, Pollack GH, Wright F, Wang K. Elastic properties of titin filaments demonstrated using a freeze-fracture technique. *Cell Motil Cytoskel* 1993;24:274-283

Trombitás K, Jin JP, Granzier H. The mechanically active domain of titin in cardiac muscle. *Circ Res* 1995;77:856-861.

Trombitás K, Greaser M, Labeit S, Jin JP, Kellermayer M, Helmes M, Granzier H. Titin extensibility *in situ*: entropic elasticity of permanently folded and permanently unfolded molecular segments. *J Cell Biol* 1998a;140:853-859.

Trombitás K, Greaser M, French G, Granzier H. PEVK extension of human soleus muscle titin revealed by immunolabeling with the anti-titin antibody 9D10. *J Struct Biol* 1998b;122:188-196.

Trombitás K, Freiburg A, Centner T, Labeit S, Granzier H. Molecular dissecton of N2B cardiac titin's extensivility. *Biophys J*, 1999;77:3189-3196.

Ullrick WC, Toselli PA, Chase D, Dasse K. Are there extensions of thick filaments to the Z line in vertebrate and invertebrate striated muscle? *J Ultrastruc Res* 1977;60:263-271.

Wang K. "Cytoskeletal matrix in striated muscle: The role of titin, nebulin and intermediate filaments." In *Contractile Mechanisms in Muscle*, GH Pollack, H Sugi, eds. New York, NY: Plenum Press, 1984.

Wang K. Sarcomere-associated cytoskeletal lattices in striated muscle review and hypothesis. *Cell Muscle Motil* 1985;6:315-369.

Wang K. Titin/connectin and nebulin: giant protein rulers of muscle structure and function. *Adv in Biophys* 1996;33:125-132.

Wang K, McClure J, Tu A. Titin: Major myofibrillar components of striated muscle. *Proc Natl Acad Sci USA* 1979;76:3698-3702.

Wang SM, Greaser M. Immunocytoskeletal studies using a monoclonal antibody to bovine cardiac titin on intact and extracted myofibrils. *J Musc Res Cell Motil* 1985;6:293-312.

White DCS. *Doctoral Thesis*, Oxford University, Oxford, England, 1967.

White DCS. "The resting elasticity of insect flight muscle and properties of the cross-bridge cycle." In *Insect Flight Muscle*, RT Tregear, ed. Amsterdam: Elsevier/North Holland Biomedical Press, 1977.

White DCS. The elasticity of relaxed insect fibrillar flight muscle. *J Physiol* 1983;343:31-57.

Zebe E, Meinrenken W, Ruegg JC. Supercontaction of glycerol extracted asychronic insect muscles in the presence of ITP. *Z Zellforsch Mikrosk Anat* 1968;87:603-621.

CONNECTIN: FROM REGULAR TO GIANT SIZES OF SARCOMERES

Koscak Maruyama and Sumiko Kimura

*National Center for University Entrance Examinations, Tokyo and
Department of Biology, Chiba University, Chiba, Japan*

In 1968, the late Professor John W. S. Pringle invited Koscak Maruyama (KM) to work on actomyosin ATPase activity of indirect flight muscle of the waterbug in his ARC Unit of Muscular Mechanics at Oxford. KM spent a few months there in 1968 as well as in 1972. During his stay in Oxford, he came across Charles Trombitás' work on the fine structure of insect flight muscle with special reference to C-filament, now known as projectin filament (Trombitás and Tigyi-Sebes, 1977).

It is a great pleasure for us to pay tribute to Dr. Trombitás' notable contribution to connectin/titin research.

Beginning of connectin

It was in the Department of Biophysics of Kyoto University in 1975 that KM observed under a microscope that a 30 μm strand of a single glycerinated psoas myofibril slowly contracted into a small dot on addition of ATP: he instantly speculated that a sarcomere was longitudinally linked by an elastic filament. Supporting this idea, when a myofibril was mildly treated with trypsin, it contracted into an interrupted line of smaller dots in the presence of ATP. Intact myofibrils were then extracted with Hasselbach-Schneider solution to remove myosin and further treated with 0.6 M KI to extract actin. Ghost myofibrils consisting of deteriorated Z-discs retained their continuity, as Huxley and Hanson (1954) had described earlier. A strand of ghost myofibril attached to a cover glass showed elastic behavior when the free end was mechanically pulled and then released, suggesting that "invisible elastic filaments" linked

Elastic Filaments of the Cell, edited by Granzier and Pollack
Kluwer Academic/Plenum Publishers, 2000

the neighboring Z-discs in the ghost myofibril.   In 1976 we gave this hypothetical filamentous protein the name "connectin."

Now we are able to explain the apparent longitudinal continuity of myosin- and actin-removed sarcomeres. When myosin is solubilized, most of the connectin/titin filaments are set free from the thick filaments and retract toward the Z-discs. However, since a set of connectin filaments extending from the counter Z-disc in a sarcomere overlap around the M line (Obermann *et al.*, 1996), some freed connectin filaments accidentally bind to each other from both sides of the sarcomere (Fig. 1). The number of such linked filaments may be very small, yet it could be enough to prevent the adjacent Z-discs from separating, even after removal of actin filaments. Thus, the continuity of ghost myofibrils is evidently due to "artifact" connection of the freed connectin filaments. This is a revised version of the previous description (Maruyama *et al.*, 1994).

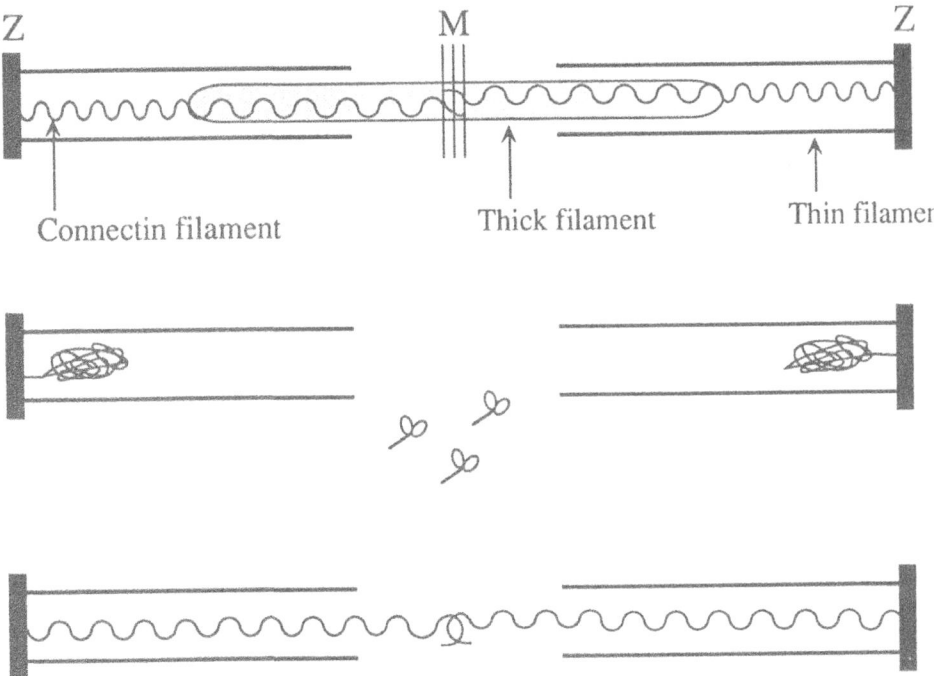

Figure 1. *Behavior of connectin filaments in a rabbit psoas sarcomere when myosin is solubilized. Intact sarcomere (top); when extracted with Hasselbach-Schneider solution, most of freed connectin filaments retract toward Z-disc (middle); a few connectin filaments from both sides of the Z-discs in a sarcomere bind to each other (bottom). This situation does not change on further treatment with 0.6 M KI to remove actin filaments. M, M line; Z, Z-disc. Modified from Maruyama* et al. *(1994).*

Tedius days of isolation

After writing a paper on connectin, the "insoluble" giant protein (Maruyama *et al.*, 1977), we wasted a year and half, simply because we could not believe the presence of a single giant peptide of a few million daltons. The largest peptide at that time was myosin heavy chain of some 200,000 daltons. Therefore, it was expected that there were some crosslinks such as lysinonorleucine in collagen or elastin. We prepared "collagen-free" muscle residues from which salt-, dilute alkali-soluble proteins had been extracted. Grams of yellow rubber-like muscle residues were hydrolyzed in 6 N HCl at 110°C and the hydrolyzate was chromatographed using a big Dowex-50 column. Each time, absence of hydroxyproline was checked. An unknown small peak near the lysinonorleucine position was isolated and finally identified as lysinoalanine by thin layer chromatography. This crosslink was an artifact resulting from the alkaline treatment of muscle!

In 1977 we moved to Chiba from Kyoto. Sumiko Kimura (SK) tested a number of salt solutions to determine whether or not any connectin was solubilized from chicken breast muscle myofibrils, using SDS gel electrophoresis on 3% polyacrylamide gels. Unexpectedly, connectin was observed to be soluble in Guba-Straub solution (0.15 M Na-PO$_4$, pH 6.5, and 0.3 M NaCl). The effects of concentrations of Na-PO$_4$ and NaCl and of pH on connectin solubility were examined: a solution of 0.10 M Na-PO$_4$, pH 6.6 was most suitable, because little myosin was extracted.

It was 1983 when soluble connectin was isolated from chicken skeletal muscle (Kimura and Maruyama, 1983). However, it turned out that "soluble" connectin was hardly solubilized from fresh muscle but was easily obtained from "aged" muscle, i.e. stored overnight at 4°C. An improved SDS gel electrophoresis using 1.8% polyacrylamide gels revealed that the "soluble" connectin band was below the top band, which was abundant in fresh muscle. This observation clearly showed that "soluble" connectin was a proteolytic product of native connectin. We called the former β-connectin and the latter α-connectin. Wang's titin 1 (T1) and titin 2 (T2) (Wang *et al.*, 1979) correspond to α- and β-connectin, respectively. We studied the physicochemical properties of β-connectin in detail (Maruyama *et al.*, 1984).

It took almost five years of trial and error for SK to find a way to isolate α-connectin. It was observed that α-connectin was partially solubilized by 0.20 M Na-PO$_4$, pH 7.0. A number of contaminant proteins were separated by DEAE Toyopearl column chromatography. The last hurdle was how to separate α- and β-connectin. Gel chromatography was not effective to separate 3000 kDa and 2000 kDa proteins. In the presence of 4 M urea, α-connectin was finally separated from β-connectin by DEAE-Toyopearl column chromatography. It was 1989. Apart from molecular mass, its physicochemical properties were very similar to β-connectin (Kimura *et al.*, 1992).

The most serious criticism of our paper (Kimura and Maruyama, 1989) was the following: Was there any possibility that a small portion of the N-terminal and/or C-terminal region was hydrolyzed away during preparation? SDS gel electrophoresis does not distinguish 3000 kDa ± 100 kDa peptides! Fortunately, later it was revealed that antibodies to the N-terminal 65 kDa portion and to the C-terminal 50 kDa portion did react with isolated α-connectin of chicken skeletal muscle (Fig. 2). This was mentioned in a recent review (Maruyama, 1999).

Ten years have elapsed since α-connectin was first isolated. However, even now, use of 4 M urea is essential for its separation from β-connectin (subsequent removal of urea results in complete renaturation).

Wang first estimated MW of connectin/titin to be approximately one million from its mobility on SDS gel electrophoresis, using myosin heavy chain

Figure 2. Immunoblot detection of N-terminal and C-terminal regions of α-connectin isolated from chicken breast muscle. The sample used was before separation of α-connectin from β-connectin. a, Amido Black stain; b, treated with Pc COM1 (antibodies to N-terminal 65 kDa portion; Yajima et al., 1996a); c, treated with Pc 72C (antibodies to C-terminal 50 kDa portion; Yajima et al., 1996b); d, treated with 3B9 (monoclonal antibody to three portions of the A-band binding region; Itoh et al. 1988). α, α-connectin; β, β-connectin.

oligomers as MW markers (Wang *et al.*, 1979). We revised it to 2.8 million (Maruyama *et al.*, 1984), close to the value estimated from the sequence data (Labeit and Kolmerer, 1995). We have shown that myosin heavy chains form dimer, tetramer and so on (not trimer, pentamer etc.) in the absence of SDS (Hu *et al.*, 1988).

Sequence work

Around 1986, when KM was invited to give a special lecture to the medical students of Kyoto University, the late Professor S. Numa, a distinguished molecular biologist, asked him to send a graduate student to start sequencing of connectin using antibodies for screening. Unfortunately, there was no graduate student who accepted this difficult project.

In the fall of 1992, KM's son Kei returned from Canada after 4 years of work on molecular biology of SR-Ca ATPase in David McLenann's laboratory in Toronto. During orientation in a new research project at the Tokyo Metropolitan Institute of Medical Science, Kei sequenced a 4 kb cDNA cloned from a chicken embryo skeletal muscle cDNA using SM1, a monoclonal antibody to an elastic (I-band) portion of connectin (Maruyama *et al.*, 1993). This cDNA had been cloned under the guidance of Dr. T. Endo of Chiba University.

Dr. H. Yajima, who was trained as a molecular biologist, was a member of our laboratory from 1994 to 1997. Starting from the 5'- terminus of the 4 kb cDNA, Yajima and two graduate students (H. Kume and H. Ohtsuka) determined the sequence of 11.5 kb cDNA (3752 amino acids) up to the N terminus of connectin at the Z-disc (Yajima *et al.*, 1996a). Four portions of the cDNA were expressed in *E. coli* and their antibodies were used for immunoelectron microscopy to locate the Z-disc binding portion of connectin consisting of 800 amino acids.

A graduate student, H. Ohtsuka, used the yeast two-hybrid system to search for any Z-disc protein to which the 63 kDa N-terminal fragment (CN63K) bound, and found that it bound to the C terminal portion of α-actinin, the main constituent of the Z-disc (Ohtsuka *et al.*, 1997a).

Further study revealed that the 26 amino acids, 447-472, of the N-terminal region of connectin bind to α-actinin (Ohtsuka *et al.*, 1997b). These 26 amino acids form an N-terminal half of the Z repeat 5 (Zr5) described by Gautel *et al.* (1996). In human cardiac connectin there are several repeat regions called Zr between the third motif II (Ig domain) and the 4th motif II (Gautel *et al.*, 1996). There are two isoforms of chicken breast muscle connectin, i.e., Zr2 Zr5 and Zr2 X Y Zr5.

It was found that the C-terminal 73 amino acids (825-897) of α-actinin bind to connectin. Our work is in good agreement with that of Sorimachi *et al.* (1997).

Giant sarcomeres

In vertebrate skeletal muscle the length of a sarcomere at rest is relatively uniform, 2~3 μm. However, in invertebrate striated muscle, especially in arthropods, the sarcomere length at rest varies greatly from 2.5 (insect indirect flight muscle), 3~4 (crayfish flexor), 5~7 (insect leg muscle), 7~8 (barnacle adductor) to 10 μm (crayfish claw opener and closer). The regular and giant sizes of sarcomeres are shown in Figure 3.

In vertebrate skeletal muscle sarcomeres, the 3000 kDa connectin molecule covers approximately 1.2 μm from the Z-disc to the M line of the thick filament at rest. What kind of elastic protein has the same function as vertebrate connectin in arthropod muscle sarcomeres of various sizes? The answer is somewhat complicated (see Maruyama, 1999). There is an elastic filamentous protein called projectin specific to arthropod striated muscle. Projectin links the Z-disc to the myosin filament in insect flight and crayfish flexor muscle (Nave and Weber, 1990; Hu *et al.*, 1990). However, in larger opener and closer sarcomeres of crayfish claw muscle and of insect leg muscle, projectin is located only on the myosin filament (Hu *et al.*, 1990; Ohtani *et al.*, 1996).

Manabe *et al.* (1993) observed that a 3000 kDa connectin-like protein linked the Z-disc to the myosin filament in giant sarcomeres of crayfish claw closer muscle. The length of a half sarcomere of closer muscle ranges from 5 (rest) to 7 μm (the I-band length is from 2 to 4 μm), as shown in Figure 4. Is it possible for a 3000 kDa protein to cover such a long distance? Theoretically it

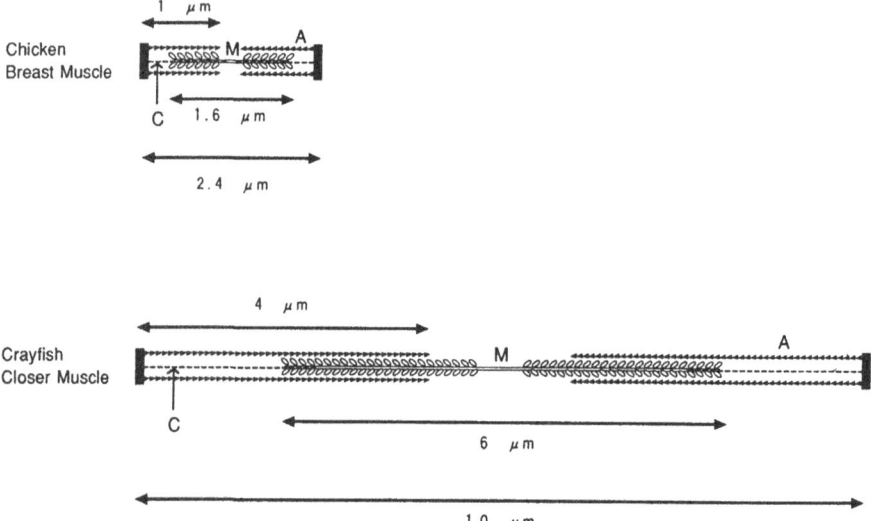

Figure 3. Regular and giant sizes of sarcomeres. C, connectin filament; A, actin filament; M, myosin filament.

is possible: an extended peptide chain of some 25,000 amino acids could be as long as 8 μm, but a straight chain is not extensible at all.

Atsushi Fukuzawa, a graduate student, cloned several cDNAs from a cDNA library of crayfish closer muscle using antiserum to crayfish 3000 kDa protein, and has sequenced the cloned cDNAs.

There have to date been two interesting findings in the sequence work up to approximately 40 kb. First, the so-called PEVK region, known as an elastic

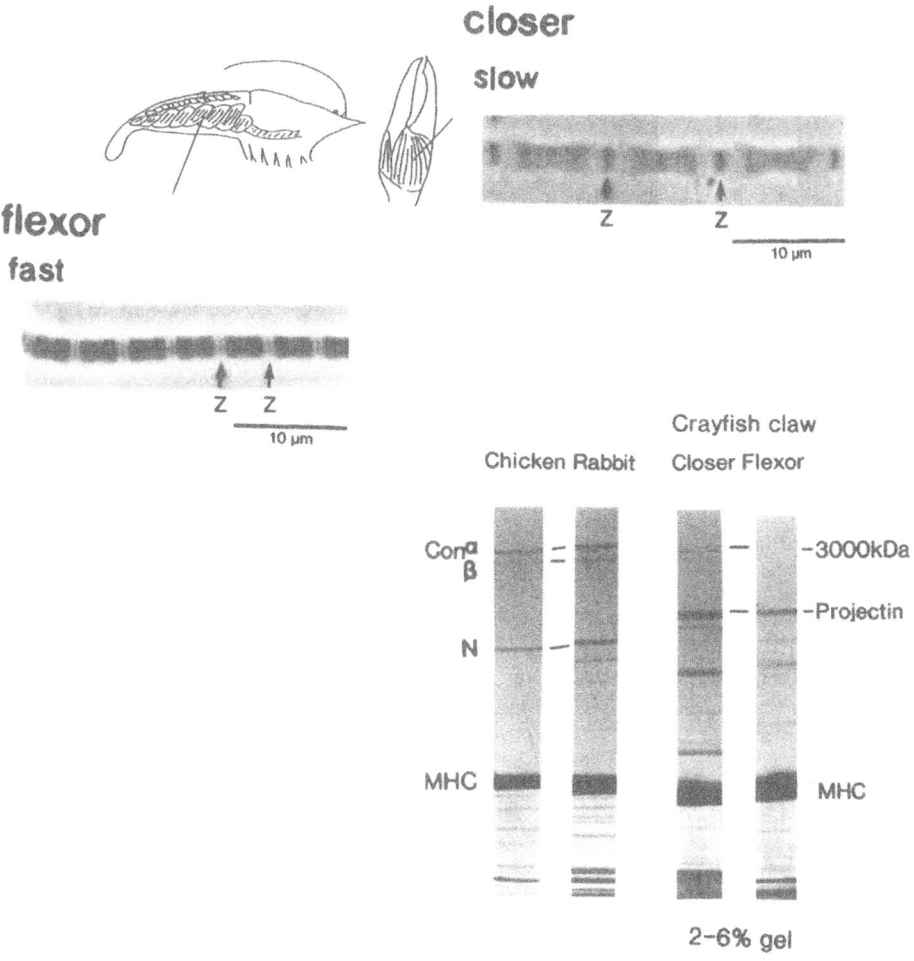

Figure 4. Regular and giant sizes of crayfish muscle sarcomeres and their high molecular weight proteins. Flexor, phase contrast image of flexor (fast) sarcomeres. Sarcomere length at rest, 3 μm. Z, Z-disc. Closer, phase contrast image of closer (slow) sarcomeres. Sarcomere length at rest, 10 μm. Lower right, 2-6% SDS PAGE. Note that 3000 kDa band is missing in crayfish flexor muscle. α, α-connectin; β, β-connectin; N, nebulin; MHC, myosin heavy chain.

region of vertebrate connectin (Labeit and Kolmerer, 1995), is present at two sites. Second, there are region sequences that have not been recorded in the DNA database (Fukuzawa *et al.*, 1999). The properties of such regions are now being examined with bacteria-expressed recombinants.

References

Gautel M, Goulding D, Bullard B, Weber K, Fürst DO. The central Z-disk region of titin is assembled from a novel repeat in variable copy numbers. *J Cell Sci* 1996;109:2747-2754.

Fukuzawa A, Kimura S, Maruyama K. Studies on connectin-like 3000 kDa protein of crayfish giant sarcomeres. *Zool Sci* 1999;16:49.

Hu DH, Kimura S, Maruyama K. Cross-linking of native myosin forms oligomers of myosin heavy chain dimers. *J Biochem* 1988;104:509-511.

Hu DH, Matsuno A, Terakado K, Matsuura T, Kimura S, Maruyama K. Projectin is an invertebrate connectin (titin): Isolation from crayfish claw muscle and localization in crayfish claw muscle and insect flight muscle. *J Muscle Res Cell Motil* 1990;11:497-511.

Huxley HE, Hanson J. Changes in the cross striations of muscle during contraction and stretch and their structural interpretations. *Nature* 1954;173:973-976.

Itoh Y, Suzuki T, Kimura S, Ohashi K, Higuchi H, Sawada H, Shimizu T, Shibata M, Maruyama K. Extensible and less-extensible domains of connectin filaments in stretched vertebrate skeletal muscle sarcomeres as detected by immunofluorescence and immunoelectron microscopy using monoclonal antibodies. *J Biochem* 1988;104:504-508.

Kimura S, Maruyama K. Preparation of native connectin from chicken breast muscle. *J Biochem* 1983;94:2083-2085.

Kimura S, Maruyama K. Isolation of α-connectin, an elastic protein, from rabbit skeletal muscle. *J Biochem* 1989;106:952-954.

Kimura S, Matsuura T, Ohtsuka S, Nakauchi Y, Matsuno A, Maruyama K. Characterization and localization of α-connectin (titin 1): an elastic protein isolated from rabbit skeletal muscle. *J Muscle Res Cell Motil* 1992;13:39-47.

Labeit S, Kolmerer B. Titins: giant proteins in charge of muscle ultrastructure and elasticity. *Science* 1995;270:293-296.

Manabe T, Kawamura Y, Higuchi H, Kimura S, Maruyama K. Connectin, giant elastic protein, in giant sarcomeres of crayfish claw muscle. *J Muscle Res Cell Motil* 1993;14:654-665.

Maruyama K. Comparative aspects of muscle elastic proteins. *Rev Physiol Biochem Pharmacol* 1999;138:1-18.

Maruyama K, Matsubara S, Natori Y, Nonomura Y, Kimura S, Ohashi K, Murakami F, Handa S, Eguchi G. Connectin, an elastic protein of muscle: characterization and function. *J Biochem* 1977;82:317-337.

Maruyama K, Kimura S, Yoshidomi H, Sawada H, Kikuchi M. Molecular size and shape of β-connectin, an elastic protein of striated muscle. *J Biochem* 1984;95:1423-1433.

Maruyama K, Endo T, Kume H, Kawamura Y, Kanzawa N, Nakauchi Y, Kimura S, Kawashima S, Maruyama K. A novel domain sequence of connectin localized at the I-band of skeletal muscle sarcomeres - Homology to neurofilament subunits. *Biochem Biophys Res Commun* 1993;194:1288-1291.

Maruyama K, Ohtani Y, Maki S, Kawamura Y, Benian G, Kagawa H, Kimura S. Connectin-related phenomena and biodiversity of the connectin family. In *Calcium as Cell Signal*, K Maruyama, Y Nonomura, K. Kohama, eds. Tokyo/New York: Igaku-Shoin. 1994;73-79.

Nave R, Weber K. A myofibrillar protein of insect muscle related to vertebrate titin connects Z-band and A-band:purification and molecular characterization of invertebrate mini-titin. *J Cell Sci* 1990;95:535-544.

Obermann WMJ, Gautel M, Steiner F, van der Ven PFM, Weber K, Fürst DO. The structure of the sarcomeric M-band: localization of defined domains of myomesin, M-protein and the 250 kDa carboxyterminal region of titin by immunoelectron microscopy. *J Cell Biol* 1996;134:1441-1453.

Ohtani Y, Maki S, Kimura S, Maruyama K. Localization of connectin-like proteins in leg and flight muscles of insects. *Tissue and Cell* 1996;28:1-8.

Ohtsuka H, Yajima H, Maruyama K, Kimura S. Binding of the N-terminal 63 kDa protein of connectin/titin to α-actinin as revealed by the yeast two-hybrid system. *FEBS Lett* 1997a;401:65-67.

Ohtsuka H, Yajima H, Maruyama K, Kimura S. The N-terminal Z repeat 5 of connectin/titin binds to the C-terminal region of alpha-actinin. *Biochem Biophys Res Commun* 1997b;235:1-3.

Sorimachi H, Freiburg A, Kolmerer B, Ishiura S, Stier G, Gregorio CC, Labeit D, Linke WA, Suzuki K, Labeit S. Tissue-specific expression and α-actinin binding properties of the Z-disc titin: implications for the nature of vertebrate Z-discs. *J Mol Biol* 1997;270:688-695.

Trombitás C, Tigyi-Sebes A. Fine structure and mechanical properties of insect muscle. In *Insect Flight Muscle*, RT Tregear, ed. North Holland: Amsterdam/Oxford.1977;79-90.

Wang K, McClure J, Tu A. Titin: Major myofibrillar components of striated muscle. *Proc Natl Acad Sci USA*, 1979;76:3698-3702.

Yajima H, Ohtsuka H, Kawamura Y, Kume H, Murayama T, Abe H, Kimura S, Maruyama K. A 11.5-kb 5'-terminal cDNA sequence of chicken breast muscle connectin/titin reveals its Z line binding region. *Biochem Biophys Res Commun* 1996a;223:160-164.

Yajima H, Ohtsuka H, Kume H, Endo T, Maruyama K, Kimura S, Maruyama K. Molecular cloning of a partial cDNA clone encoding the C terminal region of chicken breast muscle connectin. *Zool Sci* 1996b;13:119-123.

MOLECULAR TOOLS FOR THE STUDY OF TITIN'S DIFFERENTIAL EXPRESSION

Thomas Centner[1*], Francoise Fougerousse[2*],
Alexandra Freiburg[3], Christian Witt[1], Jacque S.
Beckmann[2], Henk Granzier[4], Karoly Trombitás[4],
Carol C. Gregorio[5], and Siegfried Labeit[1]

[1]*European Molecular Biology Laboratory, Meyerhofstrasse,
Heidelberg, Germany,* [2]*Genethon-URA1922, Evry, France*
[3]*Institut für Anästhesiologie und Operative Intensivmedizin,
Universitätsklinikum Mannheim, Mannheim, Germany*
[4]*Department of Veterinary and Comparative Anatomy, Pharmacology
and Physiology, Washington State University, Pullman, WA*
[5]*Departments of Cell Biology and Anatomy, and Molecular and Cellular
Biology, University of Arizona, Tucson, AZ*

Abstract: Although vertebrate genomes appear to contain only one titin gene, a large variety of quite distinct titin isoforms are expressed in striated muscle tissues. The isoforms appear to be generated by a series of complex, not yet fully characterized differential splicing mechanisms. Here, we provide an overview of the titin-specific antibodies that have been raised by our laboratory to study individual differentially expressed isoforms of titin. The staining patterns obtained in different tissues will contribute to the identification of both the particular titin isoforms that are expressed in the different tissues, as well as their intracellular distributions. In addition, antibodies to titin that are available are rapidly allowing for the refinement of our knowledge of titin's elastic spring properties. Knowledge of the nature and structure of vertebrate titins that may also be expressed in nonmuscle tissues may be broadened using these antibodies.

*Both authors contributed equally to the work.

Elastic Filaments of the Cell, edited by Granzier and Pollack
Kluwer Academic/Plenum Publishers, 2000

Introduction

Striated muscle myofibrils are composed of myosin-based thick filaments and actin-based thin filaments. Interaction of the actin and myosin S1 subunits coupled with ATP hydrolysis allows sliding together of the thick and thin filament under force generation (for a review see Squire, 1997). Passive overstretch of a myofibril would be expected to separate actin filaments from myosin heads so that the capacity for active contraction would be lost. Therefore, it is critical that myofibrils are intrinsically elastic; *i.e.,* myofibrils respond to stretch with a reversible force directed to restore thick and thin filament overlap at rest. It is now firmly established that the presence of a third filament system is responsible for the elastic properties residing in vertebrate myofibrils (Horowits *et al.*, 1986; Linke *et al.*, 1996; 1998; Trombitás *et al.*, 1998; Linke and Granzier, 1998).

This third filament system connecting the thick and thin filaments of vertebrate striated muscles is formed by a giant protein referred to as titin, or connectin (Maruyama *et al.*, 1977; Wang *et al.*, 1979; Fürst *et al.*, 1988). During physiological amounts of stretch, most of the elastic passive tension response of a myofibril can be accounted for by the titin filament system, whereas collagens and intermediate filaments become increasingly important during further stretch and thus may prevent overstretch (Granzier and Irving, 1995). Titin polypeptide chains are about three megadaltons in size. *In situ*, they span from Z- to M-lines, and thus, depending on muscle type, a distance of 1-2 μm (for recent reviews, see Wang, 1996; Trinick, 1996; Maruyama, 1997; Gregorio *et al.*, 1999). The titin filament system provides unique binding sites for numerous other sarcomeric proteins within each half-sarcomere. Therefore, it has been speculated that titin may act as a molecular blueprint in charge of thick filament, as well as overall sarcomeric assembly; thus, allowing for myofibril assembly to be achieved with the precision observed *in vivo* (Whiting *et al.*, 1989; Gregorio *et al.*, 1999 and references therein).

Myofibrils from different striated muscle tissues respond differently to stretch with cardiac myofibrils behaving much more stiffly than skeletal muscle myofibrils (Fabiato and Fabiato, 1978; Allen and Kentish, 1985). Also, significant differences exist in the passive tension properties among the skeletal muscle types; for example, myofibrils from psoas behave stiffer than myofibrils from soleus skeletal muscle. These differences in passive tension properties correlate with the different electrophoretic mobilities of titins from different muscle tissues (Wang *et al.*, 1991; Horowits, 1992).

Since the availability of human cardiac and soleus skeletal muscle titin cDNAs (Labeit and Kolmerer, 1995), molecular approaches to investigate titin isoforms have become feasible. Recent RT-PCR studies have identified the expression of distinct isoforms of titin within it's Z-disk and M-line regions

(Gautel *et al.*, 1996; Kolmerer *et al.*, 1996; Sorimachi *et al.*, 1997). In the case of Z-disk titin isoforms, the differentially expressed isoforms contain different copy numbers of Z-repeats (Gautel *et al.*, 1996; Sorimachi *et al*, 1997), which have α-actinin binding properties (Sorimachi *et al.*, 1997; Gregorio *et al.*, 1998). Interestingly, the differential expression of Z-line titin appears to correlate with the width of the Z-line. In the case of M-line titin, a single differentially expressed exon has been identified that is either included (such as in cardiac titin) or excluded (such as in fast fiber-type muscles). Protein/protein interaction studies, using the yeast two-hybrid approach, have demonstrated that the muscle-specific p94/CAPN3 calpain protease binds to the differentially expressed Mex5 M-line titin domain (Sorimachi *et al.*, 1995). It is likely that the binding properties of these isoform-specific regions contribute to the different physiological properties conferred by different muscle types.

The study of I-band region titin isoforms has been complicated by the large number of potential isoforms, and the large size (~800 kDa) of the I-band region where differential splice decisions may occur (Labeit and Kolmerer, 1995; Gregorio *et al.*, 1999). However, during the last four years, multifaceted approaches have identified three sequence families in the extensible I-band part of titin that are responsible for its elastic properties. The first element is composed of tandemly arranged Ig domains, which extend 3- to 4-fold at low forces. A second distinct sequence element within the central I-band region is rich in proline (P), glutamate (E), valine (V), and lysine (K) residues, and thus is referred to as PEVK titin (Labeit and Kolmerer, 1995). This spring element extends at higher forces (Linke *et al.*, 1996, 1998; Gautel and Goulding, 1996; Trombitás *et al.*, 1998). Together, the tandem Ig and the PEVK segments act as a two-spring system upon stretch, which enables the myofibril to respond with increasing passive tension over a wide range of forces. Recently identified, and specific to heart muscle, is a third spring element formed by the N2B segment (Helmes *et al.*, 1999; Linke *et al.*, 1999). The N2B unique segment behaves stiffer than the tandem-Ig/PEVK springs and therefore may act as a reserve element to limit stretch towards the end of the physiological passive tension curve. In summary, our current knowledge of the elastic properties of the titin spring indicate that skeletal muscle titin acts as a two-spring system, while cardiac titin acts as a three-spring system.

As an initial attempt toward the molecular characterization of the structure of I-band titin isoforms in different muscle tissues, we performed finger-printing studies. For this, sets of cardiac and skeletal I-band titin fragments were probed with cDNAs prepared from different rabbit muscle tissues (Labeit and Kolmerer, 1995). This provided an overview of the differential expression events in the central I-band region. However, the complexity of the titin molecule has become more apparent recently with the discovery that there is significant species variability in titin isoform patterns (Helmes *et al*, 1999; Cazorla *et al*., unpublished). Therefore a fingerprinting approach, probing labeled human titin cDNAs to blotted animal muscle RNAs, may provide us with inaccurate information due to the interspecies variations of the expressed titin isoforms. Therefore, we recently determined the I-band titin cDNA sequence from rabbit soleus muscle and performed RT-PCR studies on a panel of rabbit muscles (Freiburg *et al*., unpublished). In addition, we are working on determining the complete titin gene sequence from human. This will allow us to identify the splice routes leading to the expression of tissue-specific isoforms.

In this study, we provide an overview of the antibodies that have been recently generated (SL) for the characterization of differentially expressed titin isoforms. Additionally, we present examples of the distinct staining patterns obtained using these novel anti-titin antibodies. Titin antibodies are available that recognize differentially and constitutively expressed titin I-band epitopes. In collaboration with the groups of Linke (Universität Heidelberg, Germany) and Granzier (Washington State University, Pullman, WA), the utilization of these antibodies has provided information on how I-band titin behaves elastically in its different subdomains. Also, recent immunoelectron microscopy studies (KT) using the antibodies generated against titin Z-disk epitopes have further refined our knowledge of the layout of the titin filament system within the sarcomere structure. Currently, our laboratories are using the available titin antibodies for the further investigation of the tissue-specific expression of the titin isoform family.

Materials and Methods

Amplification of titin cDNA fragments

We have previously published a set of 54 partial cardiac titin cDNAs that span human cardiac titin, and a set of 19 partial cDNAs that span the central I-band region of human soleus I-band titin. In principle, these fragments can be used as templates to amplify any partial titin cDNAs of interest. In practice, we have found it more convenient to use commercial human cardiac and skeletal cDNA libraries, to amplify them in our laboratory, and use the amplified material as templates for the PCR reaction. This allows for the amplification of any desired fragment/domain arrangement or deletion series without prior knowledge of the fragment boundaries (Labeit and Kolmerer, 1995). In our hands, this method works well for the amplification of 2-3 kb partial titin cDNAs. Only when encountering a low efficiency of amplification do we use single purified cDNA clones as templates.

The following libraries have been amplified and aliquots are being used for titin subclone construction: human cardiac cDNA library in lambda-ZapII vector backbone (Stratagene #936208), human skeletal cDNA library in gt10

(Clontech, Cat. # HL1024) and human skeletal muscle cDNA library in pGAD10 (Clontech, Cat. # HL4010AB). This latter library (with cDNA inserts in the plasmid vector) has become our "default template" because the amplification efficiency of the PCR reaction is higher in comparison with cDNA pools inserted into the 50 kb GC-rich lambda phage vector backbone. For all three library amplifications, about 800,000 clones are plated, and the lambda phage or plasmid DNA is prepared from the clone pools.

For PCR amplification (Saiki *et al.*, 1985), synthetic oligonucleotides, 30-50 bases in length are used. Generally, our PCR primers have about 30 bases of perfect match to titin at their 3' ends (Tm 65°C or higher), and 10-20 bases at their 5' ends for introducing additional features such as cloning sites or epitope-tags. For the amplification we use 20 to 25 cycles with about 1 ng amplified human cDNA per microliter PCR reaction. Nucleotide-triphosphates are used at a concentration of 30 μM. Typical cycling conditions are 10 s at 95°C denaturation, followed by 3-10 min at 68°C for annealing/extension. For each kb of target cDNA, we calculate 2 min at 68°C.

Antibodies to titin

For raising anti-titin domain-specific antibodies, respective cDNA fragments amplified by the PCR method (see above) were precipitated with isopropanol in the presence of glycogen as a carrier (AGS Heidelberg, Germany). Pellets were washed briefly with 70% ethanol and dried. After filling in 3' and 5' overhangs by a brief Klenow treatment (Henikoff, 1984), fragments were gel-purified and excised while visualizing them with a long wave ultraviolet light (366 nm). Fragments were extracted from an agarose matrix by using the Jetsorb kit (Genomed, Bad Oyenhausen, Germany). Extracted DNA was digested with NcoI and KpnI restriction enzymes in 0.8 x 10x Y-Tango buffer (MBI Fermentas, Vilnius, Lithunia). After incubation for 2 hours at 37°C, restriction enzymes were inactivated by a phenol extraction, excess phenol was removed by an isopropanol precipitation, followed by a 70% ethanol wash. Obtained fragments were then ligated into a pET9D vector (Studier and Moffat, 1986) which was modified to express their insert sequences as N-terminal His_6-tags fusions. The recombinant peptides were purified from the soluble fractions by nickel chelate affinity chromatography on NTA (Ni-NTA) resins (LeGrice and Grueninger-Leitch, 1990) as specified by the manufacturer (Quiagen, Chatsworth, CA). Antibodies to the respective peptides were raised in rabbits (Eurogentec. Belgium), or in both rabbits and chickens (Biogenes, Berlin). The specific IgG fraction (rabbit), or IgY fraction (chicken) was isolated by affinity chromatography. Western blot analysis of extracts of cardiac and skeletal muscle was used to verify the specificity of the antibodies. The anti-titin monoclonal antibody T12 (Fürst *et al.*, 1988) was purchased from Boehringer-Mannheim Biochemicals (Indianapolis, IN) and another was pur-

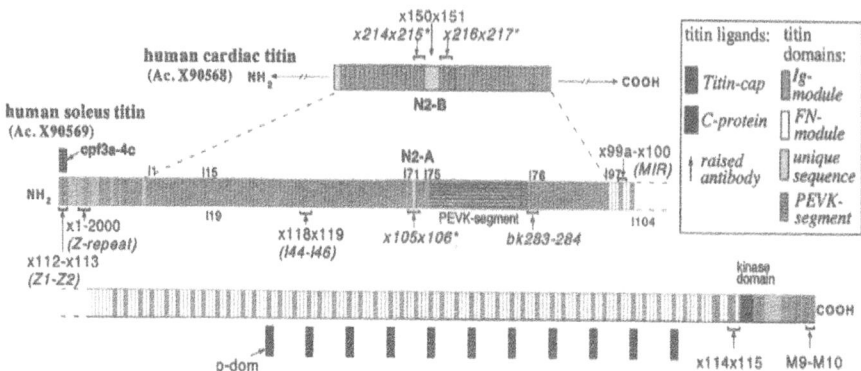

Figure 1. Overview of titin's domain architecture from human soleus muscle (middle and bottom) and the anti-titin domain-specific antibodies generated to date. On top, the domain structure of the cardiac-specific N2B isoform is shown which has a distinct structure within the central I-band region. This isoform results from differential splicing events. The dashed lines connecting the partial cardiac N2B titin (top) and the full-length soleus titin (middle/bottom columns) indicate where splicing events occur. The N- and C-terminal regions of the dashed lines in cardiac and soleus titins are essentially identical in structure. Arrows indicate the sets of 2-3 domain fragments that have been expressed for raising antibodies. The numbers above/below the arrows list the lab code names for the respective antisera which are based upon the primer names used for amplification of the respective cDNA fragments. Specific antisera against C-protein (MyBP-C) or Tcap/telethonin were also generated (p-dom and cpf3a-4c, respectively). Please note that the numbering of the domains within the I-band section differs from previous nomenclature (Labeit and Kolmerer, 1995; Witt et al., 1998). The suggested nomenclature presented here is based upon the titin genomic sequence data, which will be presented in detail elsewhere.

chased from Novocastra (Burlingame, CA). For the titin epitope positions monitored in this study, see Figure 1 above.

Immunohistochemistry

To study the developmental expression pattern of titin, human embryos (aged from 4 to 7 post-ovulatory weeks) were obtained from legal abortions induced by Mifepristone (RU486) at Hospital Broussais in Paris. All procedures were approved by the Ethical Committee of Saint-Vincent de Paul Hospital in Paris. Tissues were collected shortly after delivery and frozen within the first 24 h post mortem in dry ice and stored at −80°C until use. Seven micron-thick cryostat sections were mounted on slides and stained using indirect immunofluorescence techniques.

Sections were fixed in 2% paraformaldehyde for 10 minutes, and permeabilized in 0.1% Triton X-100/PBS for 15 min. To minimize nonspecific binding of antibodies, the coverslips were preincubated in 2% BSA/PBS for 30 min. To analyze the distribution of titin N2B or the proximal tandem Ig segment, anti-x216-x217 or anti-x118-x119 antibodies, respectively, were used at 1:50 overnight followed by Cy3-conjugated goat anti-rabbit antibodies (1:100) (Jackson Immunoresearch Laboratories, Inc.). For the double staining experiments demonstrated, the stained slides were also incubated with monoclonal anti-titin T12 antibodies (Fürst *et al.*, 1988) or with the commercial anti-titin antibodies (Novocastra), followed by FITC-conjugated goat anti-mouse antibodies (1:100) (Silenus).

Results

Polyclonal antibodies directed to specific titin domains

With the availability of the full length titin cDNA and information on its domain boundaries, it has become relatively straightforward to express specific titin Ig/FN3 domains in *E. Coli* using the pET system, and add histidine tags either to the N- or C-terminus of the expressed titin domains. Milligram quantities of pure peptides can be obtained within a week. Due to the repetitive substructure of the titin filament, there is always the concern that the raised antibodies may cross-react with multiple epitopes along the length of the titin filament. In an attempt to minimize the risk of nonspecific cross-reactivities, some laboratories have developed panels of titin domain-specific monoclonal antibodies using expressed titin domains as antigens (Jin, 1995; Obermann *et al.*, 1997).

In our experience, polyclonal antibodies raised to single titin domains are unreliable. We find that it is unpredictable whether the generated antisera will specifically recognize only the domain of interest or will also react with a series of other repeats within the 300-repeats present in the titin molecule. Interestingly, however, if two- or three-domain fragments are expressed and used as an antigen, we find that these multidomain-antigens reproducibly give rise to antisera that specifically recognize only the domains of interest within each half-sarcomere. An example of a two-domain specific antiserum is the anti-titin Z1-Z2 serum, directed against titin's N-terminus. Immunoelectron microscopy revealed that this antibody stains the periphery of the Z-line. Since titin Z1-Z2 domains bind to T-cap/telethonin, antibodies generated against this protein were used to confirm the epitope location of titin Z1-Z2. Immunoelectron microscopy using anti-T-cap/telethonin antibodies also detected staining at the periphery of the Z-line, providing additional independent proof for the localization of titin's N-terminus (Gregorio *et al.*, 1998). Concurrently, this result was also reported by Young and colleagues (Young *et al.*, 1998). In contrast to the staining patterns observed with the polyclonal anti-titin Z1-Z2 antibodies, staining with the monoclonal antibodies T20 and T21 (also directed to the Z-disk region of titin; Fürst *et al.*, 1988) did not distinguish whether the titin N-terminus was located in the periphery of the Z-line or in its center. This led to a contradicting model (Gautel *et al.*, 1996).

Currently, we have raised a total of ten rabbit polyclonal antibodies against two to three titin domain fragments within the Z-disk, I-band, A-band or M-line titin regions (see Fig. 1 for overview). Multiple-labeling staining procedures are often desirable; for example, to localize one particular domain of titin with respect to another domain utilizing two different fluorochromes. Thus, we have also begun to generate antibodies against the ten expressed titin domains in both goats and chickens. Figure 1 also includes information obtained to date using anti-titin domain-specific avian antibodies.

Antibody provision policy. Figure 1 lists the antibodies that we plan to provide (in 10 µg quantities) upon request in the future. Antibody generation in goats is planned so that larger quantities will be obtained. If higher quantities, or a larger number of different antibodies are requested, we will request a contribution to the cost of production. A continually updated version of Figure 1 will be available under http://www.embl-heidelberg.de/ExternalInfo/Titin/.

Study of titin isoforms in cardiac and skeletal muscles

Anti-x216-x217 antibody (N2B titin). All titins sequenced so far share the Ig repeat I19, while the Ig repeat N2B-I18 is expressed only in heart muscle (Fig. 1). Since in our experience, single titin domains are often poor antigens, we did not attempt to express the single repeats N2B-I18 or N2B-I19 for raising antibodies. Rather, we expressed the two-domain fragment N2B-I18/N2B-I19, to which polyclonal antibodies were raised. By immunofluorescence microscopy the obtained antibody (lab code x216-x217) stains the human developing heart in a striated pattern; a much weaker staining is observed in fetal skeletal muscles (Fig. 2). Staining of nonmuscle tissues was not detected using this antibody. These data suggest that the two-domain fragment acts as an antigen, which is not simply the sum of two single domains. Rather, our data indicate that the N2B-I18/I19 two-domain fragment behaves preferentially as a cardiac-specific antigen, a property residing in the N2B-I18, but not in the N2B-I19 repeat. Furthermore, we conclude that the cardiac-specificity of the N2B isoform is determined early during human development.

Anti-x118-x119 antibody (proximal tandem Ig segment). The proximal tandem Ig segment (Witt *et al.*, 1998) is expressed in a variety of length isoforms (Labeit and Kolmerer, 1995), which will be described in detail elsewhere (Freiburg *et al.*, unpublished). For an initial attempt to characterize the differ-

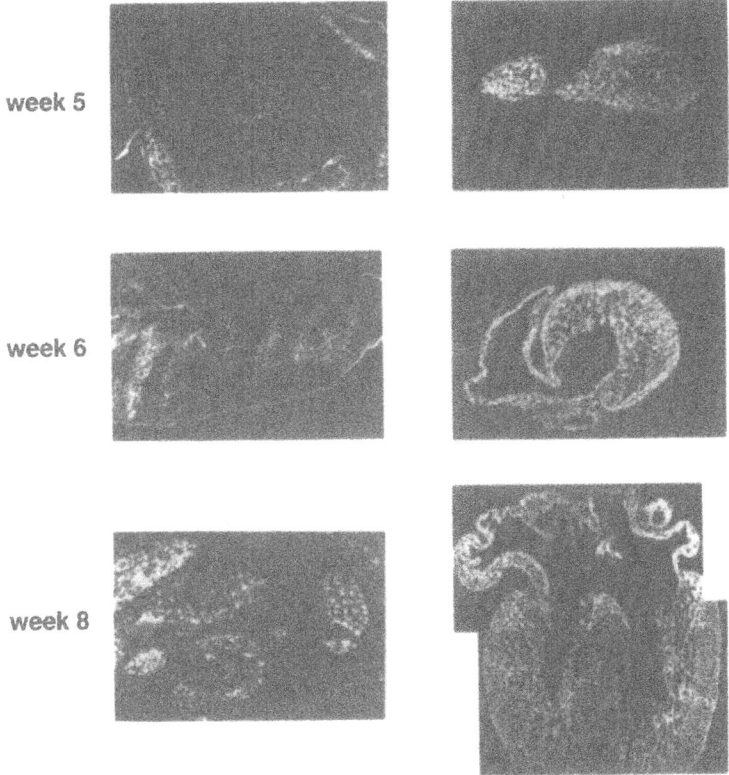

Figure 2. *Staining of human fetuses at 5, 6 and 8 weeks of gestation, using anti-titin x216-x217 (N2B-I18/I19) antibodies. This antibody generated against the titin I18-I19 fragment stains the developing heart, whereas much less staining is observed elsewhere in the developing fetus.*

ential expression of the proximal tandem Ig segment, the three Ig domains I47/I48/I49 were expressed and used as an antigen to raise rabbit polyclonal antibodies. Immunofluorescence microscopy revealed that this antibody stains both human developing limb muscle (Fig. 3A, right panel) and heart muscle (Fig. 3B, right panel), although we previously suggested that the repeats I47/I48/I49 are not expressed in the human heart (Fig. 1 and Labeit and Kolmerer, 1995). To further investigate if this staining is due to a cross-reactivity to Ig repeats from the constitutive titin regions, we compared the x118-x119 staining with the labeling pattern of the anti-titin T12 antibody (Fürst *et al.*, 1988) or with another anti-titin antibody purchased from Novocastra. Interestingly, the x118-x119 antibody does not decorate all fibers that are T12-or Novocastra antibody-positive by immunofluorescence microscopy (both of the latter antibod-

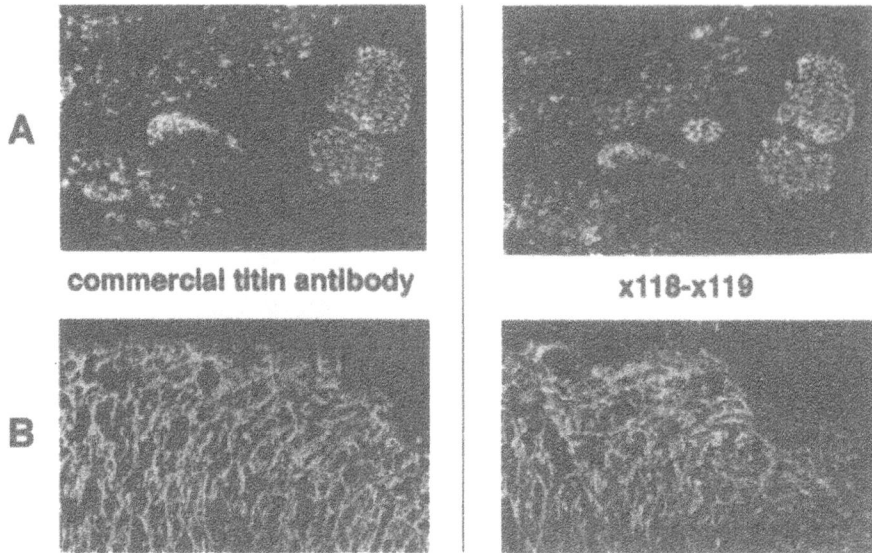

Figure 3. Comparison of the staining patterns obtained using anti-titin antibodies that appears to recognize all striated muscle fiber types and the anti-titin x118-x119 antibodies (anti-I47/I48/I49 antibodies). A: Some structures in developing limb muscle stain differently when costained with both anti-titin antibodies. No staining is detected in a subset of muscle fibers stained with the anti-titin x118-x119 antibodies but is labeled using the commercial antibody. The central oval structure (right panel), corresponding to a peripheral nerve, is stained with the anti-titin x118-x119 antibody but not with the commercial anti-titin monoclonal antibody. The significance/specificity of peripheral nerve labeling with the polyclonal anti-titin x118-x119 antibody will require further studies. B: Developing human hearts stained with the anti-titin x118-x119 antibody and with the commercial titin antibody. Note, the anti-titin x118-x119 antibody does not decorate all fibers that are T12-or Novocastra antibody-positive. 0.9 cm = 75 μm.

ies stain in indistinguishable patterns, data show for the Novocastra antibody; Fig. 3A and 3B, left panels). Therefore, the anti-titin x118-x119 antibody may recognize a distinct titin isoform that is expressed in some, but not all fibers in heart and skeletal muscle. Moreover, the anti-titin x118-x119 antibody stains the peripheral nerves in sections of the developing limb (for example see Fig. 3A, right panel, center). We conclude that further studies are required to 1) investigate the significance of the observed staining pattern in the peripheral nerves, 2) identify the fibers that stain with the anti-x118-x119 antibody, and 3) identify the precise proximal tandem Ig segments that are expressed in the human heart.

Discussion

Specific antibodies generated against titin have played a critical role in the investigation of the properties of the third filament system. A few years after titin's discovery on low percentage SDS gels (Maruyama *et al.*, 1977; Wang *et al.*, 1979), the use of polyclonal antibodies specific to titin demonstrated that the titin molecules form a filament system which apparently connected the Z-lines with the A-band (for a review: Wang, 1996; Maruyama, 1997). A few years later, a set of ten monoclonal antibodies specific to titin were generated recognizing epitopes from the Z-line up to almost the M-lines (Fürst *et al.*, 1988). From these antibody studies it was recognized that single titin molecules are likely to span the entire length of the half sarcomere, from Z-lines to M-lines.

Since the cloning of the human titin cDNA, it has become possible to express any titin domain of interest, and to raise antibodies to it. Using this approach, Obermann and colleagues obtained a set of monoclonal antibodies generated against regions of titin that spanned the M line. Immunoelectron microscopy studies with these anti-titin antibodies demonstrated that the C-terminal region (~ 200 kD) of titin spans and fully overlaps with titins from opposite half-sarcomeres within the M-line region (Obermann *et al.*, 1997). Initially, most workers had reservations about raising anti-titin polyclonal antibodies for immunlocalization studies, because such antibodies could potentially cross-react extensively within the three hundred titin Ig/FN3 repeats. More recently, however, a number of polyclonal antibodies have been generated and successfully used for immunolocalization studies to refine our knowledge of titin's (and also nebulin's) sarcomeric layout in the Z-line region (Millevoi *et al.*, 1998; Gregorio *et al.*, 1998; Young *et al*, 1998). Immunoelectron microscopy studies with specific polyclonal anti-titin Z1-Z2 antibodies (Fig. 1), recognizing the extreme N-terminal end of titin, demonstrated an intense staining pattern at the periphery of the Z-line. Since this finding contradicted earlier localization studies (Gautel *et al.*, 1996) using the anti-titin monoclonal antibodies T20 and T21 (Fürst *et al.*, 1988), additional immunoelectron microscopy experiments were performed using anti-T-cap antibodies. T-cap or telethonin (Valle *et al.*, 1997) is a 19 kDa sarcomeric protein that binds to titin's Z1-Z2 domains (Gregorio *et al.*, 1998; Mues *et al.*, 1998). In agreement with the finding that titin Z1-Z2 is localized at the periphery of the Z-line lattice, immunoelectron microscopy studies revealed that T-cap was also localized in the identical location (Gregorio *et al.*, 1998). We concluded from these studies that the Z-disk titin spans the entire Z-line region, analogous to the layout of the C terminus of titin in the M-line, where titins of opposite polarity from different half-sarcomeres fully overlap (Obermann *et al.*, 1997).

We also conclude that at least in some cases, when titin antibodies are raised to multi-domain fragments containing members of the titin Ig repeat family, the resulting polyclonal antibodies are generated against unique features. An example of this is shown in Figure 2 where the staining patterns obtained suggest that the anti-titin antibodies generated against the titin domains N2BI18 and N2BI19 preferentially recognize the N2BI18 domain (see discussion above). Another example of a bias of an animal's immune system towards selecting unique epitope features when raising polyclonal sera is the response to titin's MIR (main immunogenic region; Gautel *et al.*, 1993). When whole native titin is injected into animals, antibodies are raised preferentially to a specific site in titin that is located close to the A/I junction region. The same immunogenic site appears to be recognized by the human immune system if a Myasthenia gravis autoimmune disorder develops in the thymus, leading to a thymus tumor. Patients having both Myasthenia gravis and a thymus tumor are referred to as MGT patients (**m**yasthenia **g**ravis coupled with **t**hymoma). Patients with MGT almost always generate autoantibodies in their sera directed to the MIR ("main immunogenic region"), located on a two-domain fragment in titin (indicated as "MIR" in Fig. 1).

Furthermore, the anti-titin x118-x119 antibodies (against the proximal Ig segments) stain the peripheral nerve in sections of the developing limb (Fig. 3A). Further work will be required to determine whether this staining is indeed reflecting titin expression or cross-reactivity with Ig-like domains from a non-titin protein. Interestingly though, this finding is consistent with other reports that titins or titin-like proteins have been reported to be expressed in nonstriated muscle cells including the brush border of chicken intestinal epithelial cells, fibroblasts (C-titin; Eilertsen and Keller, 1992; Eilertsen *et al.*, 1994), human HEp-2 epithelial cells and in *Drosophila* embryos as a structural constituent of chromosomes (D-titin; Machado *et al.*, 1998). It was suggested that C-titin regulates the assembly of nonmuscle myosin II bipolar filaments, and based on the predicted functional properties of striated muscle titin, that D-titin confers elasticity and structure to chromosomes by providing an elastic filament system. With regard to nuclear titins, it should be noted that a recent unpublished survey of nuclear titin-like epitopes using 11 different anti-titin antibodies (generated against different regions of the molecule) in a panel of nonmuscle tissues revealed that only two of the antibodies cross-reacted with nuclear antigens (M. Way and S. Labeit, EMBL Heidelberg, unpublished data). Also, a titin protein was not found in the sequenced eucaryotic genome from *S. cerevisae*, an organism with extensive chromosomal structure. Possibly, there are unidentified nuclear proteins with extensive similarity to titin, which are encoded by a distinct gene(s). For the giant proteins from brush border and fibroblasts, further molecular studies will be required to identify if they represent particular splice variants from the single titin gene locus, or if they are derived from different gene loci.

We previously expressed a three-domain fragment within I-band titin (I76/ I77/I78) for raising specific antibodies to be utilized for immunofluorescence and immunoelectron microscopy studies. This antibody in combination with anti-titin N2A antibodies allowed for the monitoring of the behavior of the PEVK segment during myofibrillar stretch (Linke *et al.*, 1998). In the future, it will become increasingly important to monitor the behavior of several specific titin epitopes simultaneously. For biophysical studies on titin's elasticity, this will allow for the analysis of different subsegments of I-band titin's passive tension response. For studies on the differential expression of titin, double/ multiple labeling should reveal information on the distribution of unique iso-forms with respect to one another. To simplify this experimental approach, we have now started to raise anti-titin antibodies using both goat and chicken as hosts. Using a triple labeling approach, Gregorio and colleagues (Linke *et al.*, 1999) recently compared the assembly of the thick, thin and titin filaments within individual myofibrils. Their results revealed that titin N2B is required for thin filament stability and are consistent with the hypothesis that thick and thin filament systems assemble independently from each other (for discussion see Holtzer *et al.*, 1997).

Acknowledgments

This work was supported by the Deutsche Forschungsgemeinschaft La 668/5-1 (SL), the National Institute of Health Heart, Lung and Blood Institute NIH HL57461 (CCG) and HL61497 and HL62881 (KT), the Human Frontier Science Programme (SL, CCG), and by the Association Francaise contre les Myopathies (FF, JB).

References

Allen DG, Kentish JC. The cellular basis of the length-tension relation in cardiac muscle. *J Mol Cell Cardiol* 1985;17:821-840.

Eilertsen KJ, Kazmierski ST, Keller TC III. Cellular titin localization in stress fibers and interaction with myosin II filaments in vitro. *J Cell Biol* 1994;126:1201-1210.

Eilertsen KJ, Keller TC III. Identification and characterization of two huge protein components of the brush border cytoskeleton: Evidence for a cellular isoform of titin. *J Cell Biol* 1992;119: 549-557.

Fabiato A, Fabiato F. Myofilament-generated tension oscillations during partial calcium activation and activation dependence of the sarcomere length-tension relation of skinned cardiac cells. *J Gen Physiol* 1978;72:677-699.

Fürst DO, Osborn M, Nave R, Weber K. The organization of titin filaments in the half-sarcomere revealed by monoclonal antibodies in immunoelectron microscopy: A map of ten nonrepetitive epitopes starting at the Z line extends close to the M line. *J Cell Biol* 1988;106:1563-1572.

Gautel M, Lakey A, Barlow DP, Holmes Z, Scales S, Leonard K, Labeit S, Mygland A, Gilhus NE, Aarli JA. Titin antibodies in myasthenia gravis: Identification of a major auto-immunogenic region of titin. *Neurology* 1993;43:1581-1585.

Gautel M, Goulding D. A molecular map of titin/connectin elasticity reveals two different mechanisms acting in series. *Febs Lett* 1996;385:11-14.

Gautel M, Goulding D, Bullard B, Weber K, Fürst DO. The central Z-disk region of titin is assembled from a novel repeat in variable copy numbers. *J Cell Sci* 1996;109:2747-2754.

Granzier HL, Irving TC. Passive tension in cardiac muscle: contribution of collagen, titin, microtubules, and intermediate filaments. *Biophys J* 1995;68:1027-1044.

Gregorio CC, Trombitás K, Kolmerer B, Stier G, Granzier H, Kunke K, Suzuki K, Obermayr F, Herrmann B, Sorimachi H, Labeit S. The N terminal of titin spans the Z-Disk. Its interaction with a novel 19 kDa ligand (T-cap) is required for sarcomeric integrity. *J Cell Biol* 1998;143:1013-1027.

Gregorio CC, Granzier H, Sorimachi H, Labeit S. Muscle assembly: a titanic achievement? *Curr Opin Cell Biol.* 1999;11:18-25.

Helmes M, Trombitás K, Centner T, Kellermayer M, Labeit S, Linke WA, Granzier H. Mechanically driven contour-length adjustment in rat cardiac titin's unique N2B sequence. *Circ Res* 1999;84:1339-1352.

Henikoff S. Unidirectional digestion with exonuclease III creates targeted breakpoints for DNA sequencing. *Gene* 1984;28:351-359

Holtzer H, Hijikata T, Lin ZX, Zhang ZQ, Holtzer S, Protasi F, Franzini-Armstrong C, Sweeney HL. Independent assembly of 1.6 μm long bipolar MHC filaments and I-Z-I bodies. *Cell Struc Func* 1997;22:83-93.

Horowits R, Kempner ES, Bisher ME, Podolski RJ. A physiological role for titin and nebulin in skeletal muscle. *Nature* 1986;323:160-164.

Horowits, R. Passive force generation and titin isoforms in mammalian skeletal muscle. *Biophys J* 1992;61:392-398.

Jin JP. Cloned rat cardiac titin class I and class II motifs. Expression, purification, characterization, and interaction with F-actin. *J Biol Chem* 1995;270:6908-6916.

Kolmerer B, Olivieri N, Witt CC, Herrmann BG, Labeit S. Genomic organization of the M-line titin and its tissue-specific expression in two distinct isoforms. *J Mol Biol.* 1996;256:556-563.

Labeit S, Kolmerer B. Titins, giant proteins in charge of muscle ultrastructure and elasticity. *Science* 1995;270:293-296.

LeGrice SF, Grueninger-Leitch F. Rapid purification of homodimer and heterodimer HIV-1 reverse transcriptase by metal chelate affinity chromatography. *Eur J Biochem* 1990;187:307-314.

Linke WA, Granzier H. A spring tale: new facts on titin elasticity. *Biophys J* 1998;75:2613-2614.

Linke WA, Ivemeyer M, Olivieri N, Kolmerer B, Rüegg JC, Labeit S. Towards a molecular understanding of the elasticity of titin. *J Mol Biol* 1996;261:62-71.

Linke WA, Ivemeyer M, Mundel P, Stockmeier MR, Kolmerer B. Nature of PEVK-titin elasticity in skeletal muscle. *Proc Natl Acad Sci. USA* 1998;95:8052-8057.

Linke WA, Rudy DE, Centner T, Gautel M, Witt CC, Labeit S, Gregorio CC. I-band titin in cardiac muscle is a three-element molecular spring and is critical for maintaining thin filament structure. *J Cell Biol* 1999;146:631-644.

Machado C, Sunkel CE, Andrew DJ. Human autoantibodies reveal titin as a chromosomal protein. *J Cell Biol* 1998;141:321-333.

Maruyama K, Matsubara S, Natori R, Nonomura Y, Kimura S, Ohashi K, Murakami F, Handa S, Eguchi G. Connectin, an elastic protein of muscle: characterization and function. *J Biochem* (Tokyo) 1977;82:317-337.

Maruyama K. Connectin/titin, giant elastic protein of muscle. *Faseb J* 1997;11:341-345.

Millevoi S, Trombitás K, Kostin S, Schaper J, Pelin K, Kolmerer B, Granzier H, Labeit S. Characterization of nebulette and nebulin and emerging concepts of their roles for vertebrate Z-disks. *J Mol Biol* 1998;282:111-123.

Mues A, van der Ven PF, Young P, Fürst DO, Gautel M. Two immunoglobulin-like domains of the Z-disk portion of titin interact in a conformation-dependent way with telethonin. *FEBS Lett* 1998;428:111-114.

Obermann WM, Gautel M, Weber K, Fürst DO. Molecular structure of the sarcomeric M band: mapping of titin and myosin binding domains in myomesin and the identification of a potential regulatory phosphorylation site in myomesin. *EMBO J* 1997;16:211-220.

Saiki RK, Scharf S, Faloona F, Mullis KB, Horn GT, Erlich HA, Arnheim N. Enzymatic amplification of beta-globin genomic sequences and restriction site analysis for diagnosis of sickle cell anemia. *Science* 1985;230:1350-1354.

Squire JM. Architecture and function in the muscle sarcomere. *Curr Opin Struct Biol* 1997;7:247-257.

Sorimachi H, Kinbara K, Kimura S, Takahashi M, Ishiura S, Sasagawa N, Sorimachi N, Shimada H, Tagawa K, Maruyama K, Suzuki K. Muscle-specific calpain, p94, responsible for limb girdle muscular dystrophy type 2A, associates with connectin through IS2, a p94-specific sequence. *J Biol Chem* 1995;270:31158-31162.

Sorimachi H, Freiburg A, Kolmerer B, Ishiura S, Stier G, Gregorio CC, Labeit D, Linke WA, Suzuki S, Labeit S. Tissue-specific expression and α-actinin binding properties of the Z disk titin. Implications for the nature of vertebrate Z disks. *J Mol Biol* 1997;270:688-695.

Studier FW, Moffat BA. Use of bacteriophage T7 RNA polymerase to direct selective high-level expression of cloned genes. *J Mol Biol* 1986;189:113-130.

Trinick J. Titin as a scaffold and spring. Cytoskeleton. *Curr Biol.* 1996;6:258-260.

Trombitás K, Greaser M, Labeit S, Jin JP, Kellermayer M, Helmes M, Granzier H. Titin extensibility in situ: Entropic elasticity of permanently folded and permanently unfolded molecular segments. *J Cell Biol.* 1998;140:853-859.

Valle G, Faulkner G, De Antoni A, Pacchioni B, Pallavicini A, Pandolfo D, Tiso N, Toppo, S, Trevisan S, Lanfranchi G. Telethonin, a novel sarcomeric protein of heart and skeletal muscle. *FEBS Lett* 1997;415:163-168.

Wang K, McClure J, Tu A. Titin: Major myofibrillar component of striated muscle. *Proc Natl Acad Sci USA* 1979;76:3698-3702.

Wang K. Titin/connectin and nebulin: giant protein rulers of muscle structure and function. *Adv Biophys* 1996;33:123-134.

Wang K, McCarter R, Wright J, Beverly J, Ramirez-Mitchell R. Regulation of skeletal muscle stiffness and elasticity by titin isoforms: A test of the segmental extension model of resting tension. *Proc Natl Acad Sci USA* 88: 1991;7101-7105.

Whiting A, Wardale J, Trinick J. Does titin regulate the length of muscle thick filaments? *J Mol Biol* 1989;205:263-8.

Witt CC, Olivieri N, Centner T, Kolmerer B, Millevoi S, Labeit D, Jockusch H, Pastore A, Labeit S. A survey of the primary structure and the interspecies conservation of I-band titin's elastic elements in vertebrates. *J Struc Biol* 1998;122:1-10.

Young P, Ferguson C, Bañuelos S, Gautel M. Molecular structure of the sarcomeric Z-disk: two types of titin interactions lead to an asymmetrical sorting of alpha-actinin. *EMBO J* 1998;17:1614-1624.

Discussion
(Presented by Siegfried Labeit)

Greaser: I see that you have a different system for numbering Ig domains.

Labeit: This is really a problem. Our initial nomenclature is based upon counting Ig domains consecutively in the first completely sequenced titin, coming from human heart (Labeit and Kolmerer, 1995: *Science* 270, 293). Then, it became apparent that skeletal titins have more Ig domains. We have tried to keep our original nomenclature by specifying that additional Ig repeats are skeletal muscle specific. But now it turns out that some of these additional Ig domains from skeletal titins are expressed in some heart isoforms as well. My preference now would be to rename Ig repeats from the I-band titin by counting them in the gene sequence from 5' prime to 3'. Using such a nomenclature, we would not need to worry anymore about isoforms. We may use this new nomenclature, while keeping the original numberings in brackets. How do you feel about this?

Greaser: I do not care exactly what it is as long as we do not confuse people. Perhaps you can put something on your web site and send a copy to everybody in the field. I don't think there are problems with other people accepting your nomenclature. It will be helpful to straighten this out before we get too far.

Labeit: I think ultimately, a decision on I-band titin's nomenclature will come from the titin elasticity researchers who study the respective isoforms and refer to them in their publications. But my intuitive feeling at present is if we introduce now a systematic nomenclature and we can keep it, it will be an advantage.

Bullard: I have a question about the N-terminal end that is in the Z-disk. The third Ig domain comes and goes.

Labeit: Yes, this again is a nomenclature problem. A 90-residue stretch located within Z-disk titin's large unique sequence has been referred to as Z3 Ig domain (Labeit and Kolmerer, 1995: *Science* 270, 293). Now we have Charles Trombitás' data that indicate that the N-terminal 600 residue span the entire Z-line. Therefore, it is unlikely that the N-terminal residues can fold into a globular structure, they appear to be extended. So we don't believe that this is an Ig domain any more. I think we should throw the original Z3 out.

Bullard: What will you do with the nomenclature?

Labeit: Yes, this is the point. What are we doing with nomenclature? Are we now re-counting and, for example, referring to the original 10th Ig domain now as the 9th? It is a pain. It gets even worse when we consider that the domains Z6 to Z10 appear already to locate somewhat outside the Z-line. Again, we need to decide on titin's nomenclature. The above problems are caused by the rapid advances of the titin field. Therefore, I consider the above problems as positive.

Granzier: Along the same lines, we've spoken about this, what do you suggest that the two main cardiac isoforms should be named?

Labeit: It would be nice to have a short quick comprehensible name that researchers will remember easily. We should continue to call the small cardiac isoform N2B, whereras other larger cardiac isoforms may be referrred to as N2B+N2A, or N2BA. And we may use the Ig repeat nomenclature by specifying which Ig repeats are joined by splice events, for example I19 and I76 in N2B titin.

Granzier: I like Marion's suggestion, that you decide a name and post it on a web site.

Labeit: Good idea. If we reach consensus. There is no point if we don't reach consensus.

Linke: A problem is also the nomenclature of Z-line and M-line titin. For example, we know now that what was named Z-line titin in your '95 *Science* article, actually penetrates into the I-band.

Labeit: We could call this junction domains 5-9, or we would call them Z/I domains.

Greaser: I don't think anybody would argue if you named them.

Labeit: Yes, but you would argue if we do this and in a year or so, we decide we have to rename again.

Pollack: I have a question which relates to the structure and stability of the Ig domains. The question is to what extent are the domains similar?

Labeit: If you compare the whole family of Ig domains within titin, the most distantly related members share about 10 residues, and are therefore merely 10% identical. But if you look at subgroups of titin's Ig repeats such as the A-band super-repeat domains, sequence identities are in the range of 60-70%.

Trinick: Do you now think that the diversity in Z-line region of titin is not sufficient to account for Z-line width. Is that what you said?

Labeit: That's a good point, John, I would agree with that. As far as I understand, all the Z-lines from different tissues and species look different, don't they? If we look at the titin splice pattern, we find only 3 major isoforms.

SEQUENCE AND MECHANICAL IMPLICATIONS OF TITIN'S PEVK REGION

Marion L. Greaser, Seu-Mei Wang[1], Mustapha Berri[2], Paul Mozdziak, and Yashiyuki Kumazawa

Muscle Biology Laboratory, University of Wisconsin-Madison, Madison, WI
Current: [1]Dept. of Anatomy, National Taiwan University, Taiwan; [2]INRA, Tours, France

Abstract: A widely used titin monoclonal antibody (9D10) was epitope mapped to the PEVK region in the I-band portion of titin. Sequence analysis of the titin PEVK region revealed a large number of 28 amino acid modules (termed "PPAK" repeats) alternating with glutamic acid rich segments. Species differences in cardiac rest tension could not be ascribed to differences in the PEVK length of the N2B titin isoform. The low rest tension generated by dog cardiac muscle also does not appear to be explained by the N2 and PEVK segment lengths in the N2A titin isoform.

Introduction

Titin, also referred to as connectin, is a giant protein found in vertebrate cardiac and skeletal muscle (Maruyama *et al.*, 1977; Wang *et al.*, 1979). Titin is a very long protein, and there appear to be two sets of molecules that span from the Z line to the M line in the sarcomere. Monoclonal antibodies against titin have been very useful in mapping different parts of the molecule in the sarcomere and identifying regions that change position during sarcomere extension (Itoh *et al.*, 1988; Furst *et al.*, 1988; Whiting *et al.*, 1989; Gautel and Goulding, 1996; Granzier *et al.*, 1996; Trombitás *et al.*, 1998a; Linke *et al.*, 1998). An

Elastic Filaments of the Cell, edited by Granzier and Pollack
Kluwer Academic/Plenum Publishers, 2000

antibody referred to as 9D10 was developed in our laboratory a number of years ago (Wang and Greaser, 1985) and has been used in a variety of studies (Wang *et al.*, 1988; Handel *et al.*, 1991; Schaart *et al.*, 1991; van der Loop *et al.*, 1995; Trombitás and Granzier, 1997; Trombitás *et al.*, 1997; Trombitás *et al.*, 1998b; Machado *et al.*, 1998). This antibody binds in the I-band at a position suggesting that it may react with the PEVK sequence. The recent demonstration of 9D10 binding over a wide region of the I-band in stretched human soleus muscle fibers (Trombitás *et al.*, 1998b) has prompted us to more accurately map 9D10's epitope binding region and examine the potential repeating structure in the PEVK region.

An understanding of the structure and function of titin has been advanced by cDNA sequencing. It was previously recognized that the titin molecule contained a number of repeating modules of approximately 100 amino acids, which were similar to those found in immunoglobulins and fibronectin domain III (Labeit *et al.*, 1990). The Ig and FNIII modules form a super-repeat pattern in the A-band (Labeit *et al.*, 1992), but only modules of the Ig type were found in the I-band region (Sebestyen *et al.*, 1995; Labeit and Kolmerer, 1995). The Labeit and Kolmerer report also identified an unusual amino acid segment of the titin sequence near the middle of the I-band region (referred to as "PEVK") which contained predominantly the residues proline (P), glutamic acid (E), valine (V), and lysine (K). Different isoforms of titin were found in cardiac and skeletal muscle, and it was postulated that the elasticity and rest tension properties of these tissues may be related to the length of the PEVK (human soleus, which is easily stretched, has a titin PEVK length of 2174 residues while cardiac muscle, which is far less compliant, has a titin isoform with a PEVK as short as 163 residues) (Labeit and Kolmerer, 1995). There are also large variations in rest tension in cardiomyocytes from different species (Fabiato and Fabiato, 1978), leading us to test the hypothesis that these variations are due to changes in expressed PEVK length.

Results and Discussion

9D10 monoclonal antibody binding regions in myofibrils

The binding pattern of the 9D10 on isolated rabbit psoas myofibrils is shown in Figure 1. The strong fluorescence bands appear in the I-band closer to the A-I junction than to the Z line (Fig. 1B). However, many myofibrils show a different staining pattern (Fig. 1D). In spite of virtually identical sarcomere lengths (A - 3.07 μm; C - 3.04 μm; and phase contrast appearance, the spacing between the fluorescence bands within a sarcomere is different (B - 1.89μm; D - 1.63μm). Visually the titin bands appear more evenly spaced and farther from the Z lines in Figure 1D than in 1B. Similar phenomena were observed with chicken breast myofibrils that were stained with 9D10 and observed by high voltage electron microscopy (Fig. 2). A majority of the unstained myofibrils had little or no perpendicular banding in the I-band region (Fig. 2A). Staining with 9D10 (Fig. 2B) produced a fairly broad zone in the mid I-band regions (identified by arrows). However, many unstained myofibrils had appearances like that in Figure 2C. A pair of strong perpendicular stripes were visible (white bars), and the distance of one of these stripes from the Z line varied somewhat with sarcomere length (compare Fig. 2C and 2D). The 9D10 antibody binding

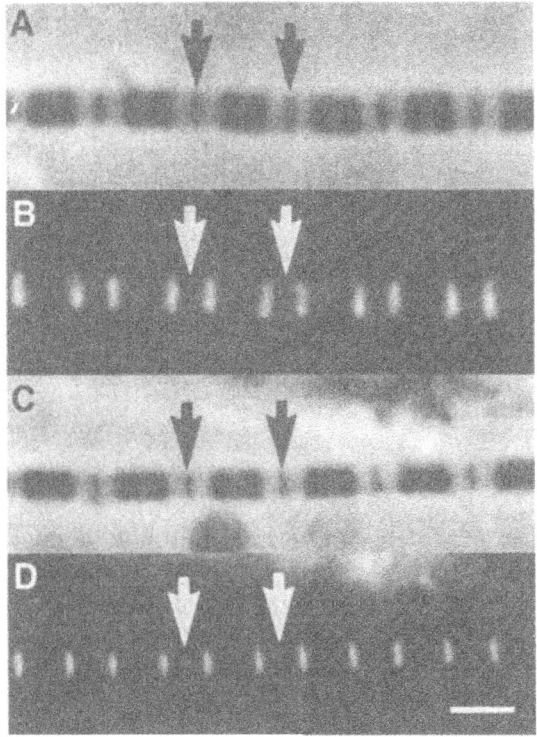

Figure 1. *Different patterns of 9D10 staining of rabbit psoas myofibrils. Myofibrils were fixed with 0.1% glutaraldehyde, incubated with anti-titin 9D10 cell supernatant, and subsequently incubated with fluorescently labeled anti-mouse IgM. A. Phase. B. Fluorescence (same as A). C. Phase. D. Fluorescence (same as C). Arrows - Z line positions. Note that the spacing between the stained bands within a sarcomere is greater in B than in D. Magnification bar = 2μm.*

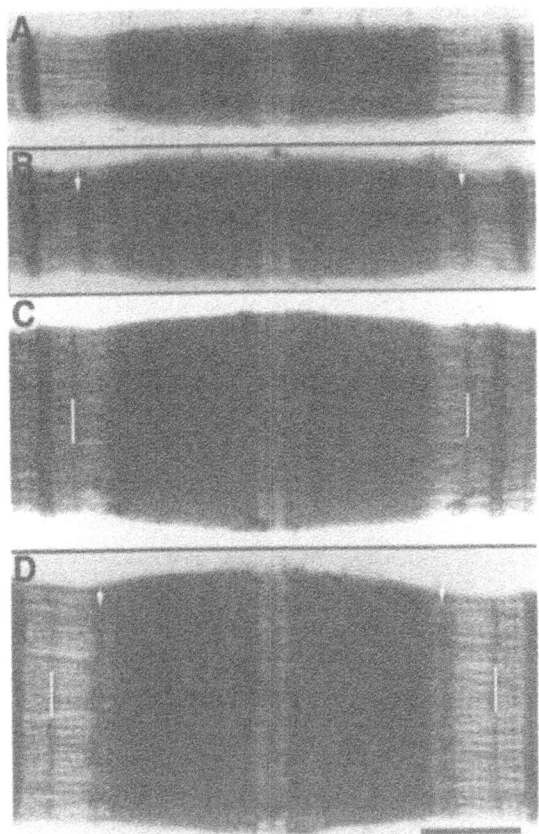

Figure 2. *High voltage electron micrographs of chicken breast myofibrils with and without 9D10 staining. Myofibrils were incubated with antibody, stained with osmic acid and uranyl acetate, critical point dried, and photographed in a 1 million volt electron microscope. A. Unstained. B. 9D10 stained. C. Unstained. An extra perpendicular band in the I band is visible (white bar). D. 9D10 stained. The extra band is visible, but the position of the 9D10 staining (arrows) is closer to the edge of the A band than in B. Magnification bar = 0.5μm.*

positions in all of the myofibrils that showed this extra band were closer to the A-I-band junction (Fig. 2D) than in those where the extra bands were absent. We interpret these results to indicate that part of the myofibrils have broken

Figure 3. *Highly stretched rabbit psoas myofibrils stained with 9D10. **A**. Phase. **B**. Fluorescence. Although the phase image is grossly distorted, the sarcomere periodicity of the 9D10 antibody locations remains clear. Magnification bar = 2 μm.*

titin molecules and the portion of the titin that contains the 9D10 epitope retracts toward the edge of the A band.

Pure mechanical stretching of muscle does not appear to be solely responsible for the breakage. Rabbit psoas muscle strips held under extreme stretch until they are in rigor often yield myofibrils with patterns like those shown in Figure 3. The sarcomere structure is obscured when observed by phase contrast (Fig. 3A), but the 9D10 staining patterns show a fairly regular appearance (Fig. 3B). The distance of the 9D10 epitope from the Z line reaches a constant value at very long sarcomere lengths. The patterns are consistent with the idea that the titin is attached so firmly to the thick filaments that the latter are dislodged from the original A band. The constant distance of the Z to the epitope is also inconsistent with postulations that the Ig domains may unravel under extreme stretch (Erickson, 1994). Similar conclusions can be made from observations of bovine cardiac muscle subjected to extreme stretch. Patterns similar to those in Figure 4 are often observed. In this case the apparent slippage of the thick filaments shows a more consistent pattern than with the skeletal muscle myofibrils. In each case the distance from the edge of the phase dense apparent A-band to the other side of a dense internal stripe is exactly 1.6 μm (see interpretive diagram in Fig. 4, bottom).

The suggested breakage of titin in the PEVK region is consistent with observations made several years ago on the postmortem changes in 9D10 staining pattern (Ringkob *et al.*, 1988). The antibody stains in the usual position in the mid I-band of most bovine psoas myofibrils isolated soon after death, but a four band per sarcomere pattern is observed in myofibrils prepared after cold storage of the muscle for 3 days (Fig. 5B). The myofibril shown in Figure 5 was also stained with the anti-titin 514 (an antibody prepared against a peptide

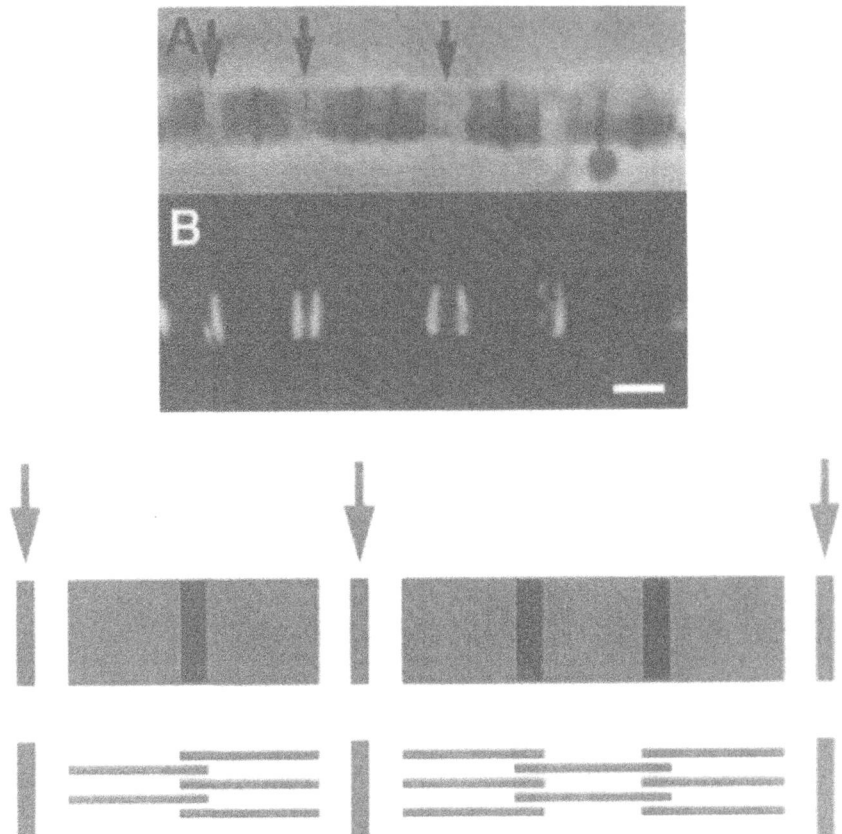

Figure 4. *Highly stretched bovine cardiac myofibrils stained with 9D10.*
A. Phase. B. Fluorescence. Arrows correspond to the positions of the Z lines.
The bottom diagram is a thick filament arrangement that may explain the phase
structure. Magnification bar = 2μm.

near the carboxyl end of the PEVK region; see Trombitás *et al.*, 1998a). The
antibody staining results can be interpreted by assuming that the 9D10 binds
more than one region in the PEVK and that breakage results in titin retraction
in both directions in the half I-band zone. It should be noted that the PEVK
segment that moves toward the Z line does not collapse all the way. The fluorescent stripe closest to the Z line is in a position similar to the extra band
observed in the chicken myofibrils (see Fig. 2B) and may correspond to the N2
line described previously (Franzini-Armstrong, 1970).

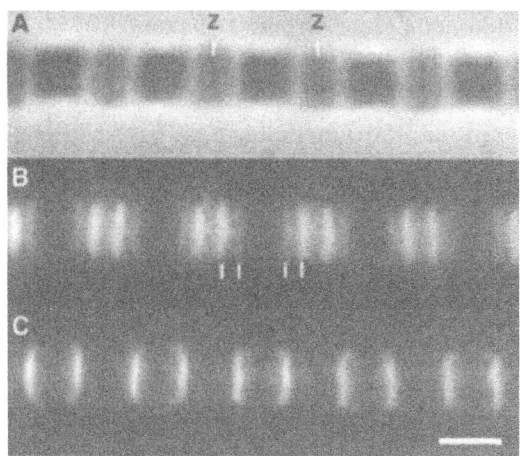

Figure 5. *Postmortem bovine psoas myofibrils stained with 9D10 and 514 titin antibodies. A. Phase. B. 9D10 stained. White bars indicate the 4 antibody staining positions within a sarcomere. C. 514 stained. The patterns are interpreted to indicate that the titin is broken in the PEVK region and that multiple epitope positions are stained with the 9D10 antibody. Magnification bar = 2μm.*

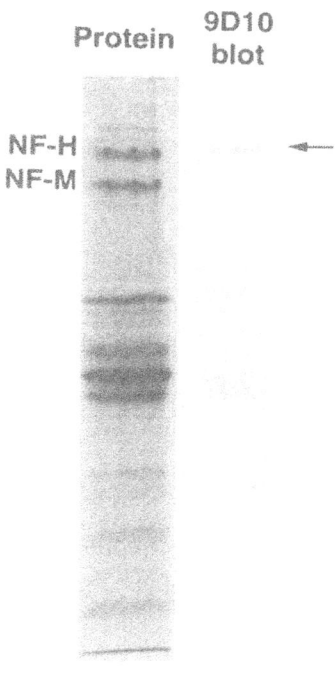

Figure 6. *Immunoblot using a crude neurofilament preparation from pig spinal cord and anti-titin 9D10. Neurofilament H and neurofilament M are identified with the protein lane. The arrow shows the position of the major reaction product at the NF-H position.*

9D10 epitope identification

The 9D10 monoclonal antibody was raised against bovine cardiac titin. It is immunoblot specific for titin in heart and skeletal muscle, and it reacts with only the T1 and T2 bands. Early work indicated antibody staining with some protein(s) in spinal cord (Wang and Greaser, unpublished). Other workers have found that several titin monoclonals cross-react with neurofilament proteins (Shimizu *et al.*, 1988; Matsumura *et al.*, 1989; Mencarelli *et al.*, 1991). Therefore a crude neurofilament preparation from bovine spinal cord was tested by immunoblotting against 9D10 and the results are shown in Figure 6. Reaction product is evident at the position corresponding to the neurofilament H band. In spite of this cross-reactivity, BLAST searches (Altschul *et al.*, 1990) show only weak similarity between titin and the neurofilament amino acid sequences.

We have recently found that an expressed titin fragment that contains the short human cardiac N2B PEVK region gives a strong immunoblot reaction with 9D10. V8 protease digests of the expressed polypeptide have been separated on an HPLC reverse phase column, and the peptides examined by slot-blots. Two positive reacting peptides with overlapping sequence have been obtained; the sequences have been determined by mass spectrometry. We therefore can conclude that at least one of the 9D10 epitopes maps to the PEVK region of titin.

Repeating structure in the PEVK

The broad (over 1 μm) 9D10 staining region in the I-band region of highly stretched human soleus muscle fibers (Trombitás *et al.* 1998b) led us to search for possible repeating structure within the PEVK region. A number of short amino acid repeats have been identified (Trombitás *et al.*, 1998b). However, further sequence analysis has revealed a much more extensive and regular pattern of amino acid repeats than previously recognized. Numerous 26-28 residue repeats (referred to as "PPAK" after the initial four amino acids in a typical module) are found in many parts of the PEVK sequence. Multiple copies of this pattern usually occur sequentially, and segments rich in glutamic acid separate the sequential PPAK copies. The PPAK segments have more lysines than glutamic acid residues, and this leads to a cycling of the isoelectric points from near 10 for the PPAK regions to near 4 in the polyglutamic acid sections. The role of the alternating charge and the highly regular repeating structures in titin's physiological function remains to be determined.

Relation of PEVK length to resting tension

It has been previously postulated that the PEVK segment length may be responsible for differences in elasticity between various muscle tissues (Labeit and Kolmerer, 1995). Observations on the passive tension of skinned single muscle cells have shown that slow skeletal muscle is lowest, fast skeletal muscle is intermediate, and cardiac muscle is highest. Previously published work has also shown that there are species differences in rest tension with rat ventricular myocytes having greater rest tension than similar cells isolated from dogs (Fabiato and Fabiato, 1978). If titin is largely responsible for rest tension, then we might expect that dog cardiac titin would have a longer PEVK segment length than that found in rat. Also, since the cardiac N2B isoform has the shortest PEVK region, we hypothesized that changes in this isoform would be more likely to influence the rest tension. Experiments were therefore conducted to determine if there were either differences in the amino acid sequence or differences in length of the cardiac N2B PEVK. Polymerase chain reaction (PCR) primers were designed (based on the human cardiac N2B titin isoform cDNA sequence; Labeit and Kolmerer, 1995) that would amplify the region between Ig19 and Ig20 (Labeit and Kolmerer 1995 nomenclature). PCR amplification was achieved using RNA isolated from rabbit, pig, rat and dog ventricular tissue. The rabbit, pig, and rat all gave a single band with a size of approximately 760 bp, the same size as expected from the human sequence. Two bands were obtained with dog tissue; a major band with a size of 615 bp and a minor band of 760 bp. Each of these bands has been sequenced, and all the PCR products are similar to the human N2B PEVK. The sequence identity among the various PEVK cDNAs was more than 90%. Thus any species difference in passive tension can not be ascribed to major differences in protein sequence character. In addition, the fact that the major dog cDNA was shorter than that from the other species means that the length change is opposite from expected. Therefore the N2B PEVK segment length can not explain differences in passive tension between the dog and rat.

Recent efforts have been directed towards obtaining sequence from the N2 and PEVK region of the dog cardiac N2A isoform. We initially tried to amplify the full N2 plus PEVK region but without success. However, a 2.3 kb fragment from the N2 region was obtained and sequenced. Subsequently a forward primer (prepared from near the 3' end of this sequence) and a reverse primer (based on the unique N2A PEVK sequence expressed in human cardiac muscle) were used to generate new dog sequence moving in the 3' direction (see Fig. 7). After three rounds using this strategy, the full N2A PEVK cDNA

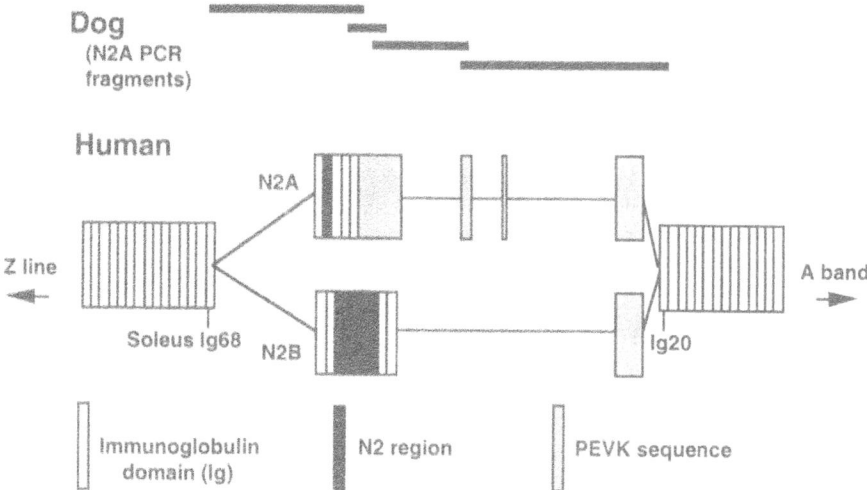

Figure 7. *Diagram of the titin module arrangement in the N2 and PEVK regions of titin. The wide solid lines indicate the different dog cardiac titin N2A isoform PCR fragments. Note that the lengths of these lines do not reflect their actual base pair size. The narrow horizontal lines indicate connecting or splicing regions in the isoform sequences. The Ig domain numbering is that of Labeit and Kolmerer (1995).*

sequence has been completed. Although there are a few minor splice variations when compared to the human N2A PEVK, the lengths of the dog and
human sequence are fairly similar. Therefore it appears that the low rest tension of dog cardiac muscle may come from additional Ig modules in the tandem Ig regions of this species compared to others. Further work is in progress
to determine which additional regions are expressed.

Acknowledgments

This work was supported by the College of Agricultural and Life Sciences of
the University of Wisconsin-Madison (Hatch) and grants from the National
Institutes of Health (HL47053, HL62466).

References

Altschul SF, Gish W, Miller W, Meyers EW, Lipman DJ. Basic local alignment search tool. *J
    Mol Biol* 1990;215:403-10.
Erickson H. Reversible unfolding of fibronectin type III and immunoglobulin domains provides
    the structural basis for stretch and elasticity of titin and fibronectin. *Proc Natl Acad Sci
    USA* 1994;91:10114-18.
Fabiato A, Fabiato F. Myofilament-generated tension oscillations during partial calcium activation
    and activation dependence of the sarcomere length-tension relation of skinned cardiac cells.
    J Gen Physiol 1978;72:667-99.
Franzini-Armstrong C. Details of the I-band structure as revealed by the localization of ferritin.
    Tissue & Cell 1970;2:327-38.
Furst DO, Osborn M, Nave R, Weber K. The organization of the titin filaments in the half-
    sarcomere revealed by monoclonal antibodies in immunoelectron microscopy: a map of
    ten nonrepetitive epitopes starting at the Z-line extends close to the M-line. *J Cell Biol*
    1988;106:1563-72.
Gautel M, Goulding D. A molecular map of titin/connectin elasticity reveals two different
    mechanisms acting in series. *FEBS Lett* 1996;385:11-4.
Granzier H, Helmes M, Trombitás K. Nonuniform elasticity of titin in cardiac myocytes: a study
    using immunoelectron microscopy and cellular mechanics. *Biophys J* 1996;70:430-42.
Handel SE, Greaser ML, Schultz E, Wang S-M, Bulinski JC, Lin JJ-C, Lessard JL. Chicken
    cardiac myofibrillogenesis studied with antibodies specific for titin and the muscle and
    non-muscle isoforms of actin and tropomyosin. *Cell & Tissue Res* 1991;263:419-430.
Itoh Y, Suzuki T, Kimura S, Ohashi K, Higuchi H, Sawada H, Shimizu T, Shibata M, Maruyama
    K. Extensible and less-extensible domains of connectin filaments in stretched vertebrate
    skeletal muscle sarcomeres as detected by immunofluorescence and immunoelectron
    microscopy using monoclonal antibodies. *J Biochem (Tokyo)* 1988;104:504-8.
Labeit S, Barlow DP, Gautel M, Gibson T, Holt J, Hsieh CL, Francke U, Leonard K, Wardale K,
    Whiting A, Trinick J. A regular pattern of two types of 100-residue motif in the sequence of
    titin. *Nature* 1990;345:273-6
Labeit S, Gautel M, Lakey A, Trinick J. Towards a molecular understanding of titin. *EMBO J*
    1992;11:171-6.
Labeit S, Kolmerer B. Titins: giant proteins in charge of muscle ultrastructure and elasticity.
    Science 1995;270:293-6.
Linke WA, Stockmeier MR, Ivemeyer M, Hosser H, Mundel P. Characterizing titin's I-band Ig
    domain region as an entropic spring. *J Cell Sci* 1998;111:1567-74.

Machado C, Sunkel CE, Andrew DJ. Human autoantibodies reveal titin as a chromosomal protein. *J Cell Biol* 1998;141:321-33.

Maruyama K, Matsubara S, Natori R, Nonomura Y, Kimura S. Connectin, an elastic protein of muscle. Characterization and function. *J Biochem (Tokyo)* 1977;82:317-37.

Matsumura K, Shimizu T, Mannen T, Maruyama K. The immunological homology between two filamentous cross-linker phosphoproteins, connectin and cross-bridge region of neurofilament-H, is not affected by the phosphorylation state. *J Biochem (Tokyo)* 1989;105:226-30.

Mencarelli C, Magi B, Marzocchi B, Armellini D, Pallini V. Evolution of the "titin epitope" in neurofilament proteins. *Comp Biochem Physiol* 1991;100B:741-4.

Ringkob TP, Marsh BB, Greaser ML. Change in titin position in postmortem bovine muscle. *J Food Sci* 1988;53:276-7.

Schaart G, Pieper FR, Kuijpers HJ, Bloemendal H, Ramaekers FC. Baby hamster kidney (BHK-21/C13) cells can express striated muscle type proteins. *Differentiation* 1991;46:105-15.

Sebestyen MG, Wolff JA, Greaser ML. Characterization of a 5.4 kb cDNA fragment from the Z-line region of rabbit cardiac titin reveals phosphorylation sites for proline-directed kinases. *J Cell Sci* 1995;108:3029-37.

Shimizu T, Matsumura K, Itoh Y, Mannen T, Maruyama Y. An immunological homology between neurofilament and muscle elastic filament: a monoclonal antibody cross-reacts with neurofilament subunits and connectin. *Biomed Res* 1988;9:227-33.

Trombitás K, Granzier H. Actin removal from cardiac myocytes shows that near Z line titin attaches to actin while under tension. *Am J Physiol* 1997;273:C662-70.

Trombitás K, Greaser ML, Pollack GH. Interaction between titin and thin filaments in intact cardiac muscle. *J Muscle Res Cell Motil* 1997;18:345-51.

Trombitás K, Greaser M, Labeit S, Kellermayer M, Helmes M, Granzier H. Titin extensibility in situ: entrophic elasticity of both native and permanently unfolded molecular segments. *J Cell Biol* 1998a;140:853-9.

Trombitás K, Greaser M, French G, Granzier H. PEVK extension of human soleus muscle titin revealed by immunolabeling with the anti-titin antibody 9D10. *J Struct Biol* 1998b;122:188-96.

van der Loop FT, Schaart G, Langmann H, Ramaekers FC, Viebahn C. Rearrangement of intercellular junctions and cytoskeletal proteins during rabbit myocardium development. *Eur J Cell Biol* 1995;68:62-9.

Wang K, McClure J, Tu A. Titin: major myofibrillar components of striated muscle. *Proc Natl Acad Sci USA* 1979;76:3698-702.

Wang S-M, Greaser ML. Immunocytochemical studies using a monoclonal antibody to bovine cardiac titin on intact and extracted myofibrils. *J Muscle Res Cell Motil* 1985;6:293-312.

Wang S-M, Greaser ML, Schultz E, Bulinski JC, Lin JJ-C, Lessard JL. Studies on cardiac myofibrillogenesis with antibodies to titin, actin, tropomyosin, and myosin. *J Cell Biol* 1988;107:1075-83.

Whiting A, Wardale J, Trinick J. Does titin regulate the length of muscle thick filaments? *J Mol Biol* 1989;205:263-8.

Discussion

Granzier: You raised the possibility that the dog may express a high level of N2A titin. Our results that will be presented later this week are consistent with this.

Greaser: The question I think still remains is what is happening to N2B titins in those species when you put them under a large degree of stretch. Are they in fact being torn loose, and are you only seeing what is left in terms of the N2A titins, or what is happening?

Bullard: You mentioned that the PEVK might bind to actin. Could the difference in stiffness be due to a difference in actin binding? Do you know which part of the PEVK sequence binds to actin?

Greaser: We don't know specifically. There's some collaborative work between Henk and Rob Yamasaki, who is in the back row over here. We've shown that this I19 PEVK I20 does in fact bind to actin. But the specific regions of the PEVK have not yet been defined. We expect that it is the PEVK part but we haven't absolutely proven that as of yet.

Pollack: Marion, in our experience, the apparent stiffness can differ among different species, conceivably because of differences in the degree of damage at the ends of the preparation. Could you say in just a few words how you made the measurements of resting tension versus length in the two different species?

Greaser: This data was generated in Rick Moss' lab and I can't give you the exact details at this point about how they were attached. But they are single cells or small numbers of cells. The data that was presented is corrected for the maximal active tension. So it is a ratio between resting tension and maximal active tension so that usually in ones that are badly damaged, they won't give as much active tension either. But it certainly is a concern. Just in terms of simple things, it is very easy to make myofibrils with long sarcomere lengths from dog muscle compared to rat or rabbit for example. So there is something quite different about the stretch properties of these tissues.

Pollack: It would be interesting to measure the resting length tension in single myofibrils.

Granzier: Just a comment. Differences in resting tension of cardiac cells or small bundles of cardiac myofibrils have also been found by Fabiato and Fabiato. These authors reported that dog myofibrils were most compliant.

Greaser: Our findings are consistent with Fabiato's work.

Vigoreaux: First a comment. Insect flight muscle has very high resting stiffness but does not contain a PEVK domain. How does this fit with your model? My question is, if you look at your charge distribution it reminds me of the myosin rod with alternating positive and negative charges which has been implicated in rod-rod interactin.

Greaser: Well, first of all, the charge distribution of myosin is on a much narrower sort of scale than we are talking about here. We're talking about groups of 28 amino acids. There are fairly long regions in titin that all have a positive charge; then there are fairly long segments that have a negative charge. And it's much more than the 28 amino acid repeat that one finds in myosin. What we make of it, I think, is open to question. Whether or not these positively charged regions interact with the negative charged ones that are very close to and adjacent to it, or whether or not they interact with actin, we don't know. You have an idea? Throw it out.

Trombitás: In my last lecture I will address how insect flight muscle may be able to develop such high resting stiffness.

TerKeurs: When you say dog is different from pig, one thing always strikes me about dog experiments and that is that brand is ill defined. I would expect a large degree of variability among dogs themselves. Even if we look at the passive elastic properties of rat trabeculae, you find from specimen to specimen there is a noticeable difference. My question to this forum would be, is there not a solitaire ability in N2A and N2AB, N2BA?

Labeit: These P-pack repeats, are they contained in the Drosophila titin? Did you look at it?

Greaser: No. I haven't looked carefully at the Drosophila titin.

Linke: It is of course puzzling that with the 514 antibody you see a different distance across the A-band. I would like to comment on this. I have done experiments with the MIR antibody which labels right at the A/I-junction. I also see a difference measured across the A-band at high stretch, which should not be, because there is a highly conserved region. So I think one explanation could be that because you see the difference as far as I understood, at very high stretch of cardiac myofibrils, you already stretch out the A-band titin so that you basically pulled it out already once you come to about 2.5 micron sarcomeres.

Greaser: That begs the question about where the association points are of the titin with the thick filament. How far away from the tip of the thick filament is the titin bound, and how tightly is it bound? And we don't know much about how much slack there is in the A-band but based on what we know about some antibodies that bind in say the C-protein zone, etc., it is unlikely that the A-band titin has much in terms of looping out or much slack. We're talking about so much extension here that I am not sure that we can account for it by just straightening the distance between domains.

PROBING THE FUNCTIONAL ROLES OF TITIN LIGANDS IN CARDIAC MYOFIBRIL ASSEMBLY AND MAINTENANCE

Abigail S. McElhinny[1], Siegfried Labeit[2], and Carol C. Gregorio[1,3]

[1]Department of Cell Biology and Anatomy
University of Arizona, Tucson, AZ
[2]European Molecular Biology Laboratory
Heidelberg, Germany
[3]Department of Molecular and Cellular Biology
University of Arizona, Tucson, AZ

Abstract: Sarcomeres of cardiac muscle are comprised of numerous proteins organized in an elegantly precise order. The exact mechanism of how these proteins are assembled into myofibrils during heart development is not yet understood, although existing *in vitro* and *in vivo* model systems have provided great insight into this complex process. It has been proposed by several groups that the giant elastic protein titin acts as a "molecular template" to orchestrate sarcomeric organization during myofibrillogenesis. Titin's highly modular structure, composed of both repeating and unique domains that interact with a wide spectrum of contractile and regulatory ligands, supports this hypothesis. Recent functional studies have provided clues to the physiological significance of the interaction of titin with several titin-binding proteins in the context of live cardiac cells. Improved models of cardiac myofibril assembly, along with the application of powerful functional studies in live cells, as well as the characterization of additional titin ligands, is likely to reveal surprising new functions for the titin third filament system.

Elastic Filaments of the Cell, edited by Granzier and Pollack
Kluwer Academic/Plenum Publishers, 2000

Introduction

Efficient muscle contraction is dependent upon the precise assembly and uniform organization of contractile proteins into sarcomeric units to form myofibrils. The process of myofibrillogenesis cumulates in the construction of one of the most highly ordered macromolecular structures found in eukaryotic cells. As a result of myofibril assembly, many classes of proteins function together to convert the molecular–level movements of actin and myosin into macroscopic movements of contractile activity.

The detailed mechanisms controlling the assembly of myofibrils have yet to be elucidated. Historically, cardiac myofibrillogenesis was difficult to study since the heart is the first functional organ in the developing vertebrate: that is, its rapid development makes it difficult to identify distinct stages in the process. As a result, delineation of regulatory pathways required for the expression of contractile proteins and the events contributing to their assembly into myofibrils has mainly been derived from studies using skeletal muscle. Here, however, we will focus on what is currently known about myofibril assembly in cardiac tissue. This process requires coordinated synthesis of the constituent proteins, polymerization of actin and myosin (and their associated proteins) into thin and thick filaments, respectively, and association of the two filament systems into the sarcomere. These newly assembled sarcomeres consist of parallel arrays of approximately 1.0 μm-long thin filaments that interdigitate with laterally aligned 1.6 μm-long thick filaments. A third filament system is formed by titin (also known as connectin), the largest protein yet discovered (approximately 3,000-3,700 kDa; Labeit and Kolmerer, 1995). Single molecules of titin span an entire half-sarcomere (Maruyama *et al.*, 1977; Wang *et al.*, 1979; Maruyama *et al.*, 1985; Fürst *et al.*, 1988). The N-terminal regions of titin from opposite sarcomeres overlap in the Z-lines, while the C-terminal regions of titin from opposite half-sarcomeres overlap in the M-lines (Obermann *et al.*, 1997; Gregorio *et al.*, 1998; Young *et al.*, 1998). Therefore, titin filaments (via their specific ligands; see section *Titin Ligands* this chapter) form a continuous filament system from one sarcomere to another (for reviews see Trinick, 1996; Wang, 1996; Keller, 1997; Labeit *et al.*, 1997; Maruyama, 1997; Gregorio *et al.*, 1999). In skeletal (but not cardiac) muscle another giant molecule, nebulin (770-900 kDa), forms a fourth filament system and has been proposed to regulate thin filament length. A single molecule of nebulin inserts its C-terminal region into the Z-disc, associates with the thin filaments along their entire length, and terminates at or near the pointed ends of the thin filaments (for reviews see Trinick, 1994; Horowits *et al.*, 1996; Fowler, 1996; Wang *et al.*, 1996; Shimada *et al.*, 1996; Gregorio *et al.*, 1999). To date, nebulin has not been found in cardiac muscle, although smaller, nebulin-like proteins have been identified recently (Moncman and Wang, 1995; Luo *et al.*, 1997; Millevoi *et al.*, 1998).

Although titin was discovered two decades ago (Maruyama *et al.*, 1977; Wang *et al.*, 1979), its large size prohibited detailed studies concerning its structure and function. However, recent complete cDNA sequencing (Labeit and Kolmerer, 1995), immunoelectron microscopy studies, and advances in molecular, structural and cell biology techniques have yielded important clues concerning the roles that titin has in the sarcomere. Interestingly, this giant protein appears to have many functions. Titin centers thick filaments in the middle of the sarcomere and has been proposed to act as an efficient elastic spring element during muscle contraction (Horowits *et al.*, 1986; Funatsu *et al.*, 1990; Granzier and Irving, 1995; Linke *et al.*, 1996; Erickson, 1997; Linke and Granzier, 1998). In addition, the C-terminal region of titin (located at the M-line region) contains a serine/threonine (SP) kinase domain (Labeit and Kolmerer, 1995), the atomic 3D structure of which was solved recently (Mayans *et al.*, 1998). Therefore, titin may also have a role in signal transduction pathways (Labeit *et al.*, 1997; Mayans *et al.*, 1998). Finally, titin has been proposed by many groups to be the "molecular template" that orchestrates sarcomeric organization.

Results and Discussion

The role of titin in cardiac myofibril assembly: is it a molecular template?

To date, most investigations of cardiac myofibril assembly have focused on analyzing the localization of contractile proteins in isolated cardiac myocytes from embryonic chicken or fetal, newborn or adult rats. These model systems have proven to be remarkable resources for studying cytoskeletal protein assembly and interactions due to the precise organization and length of the sarcomeric units. For instance, actin filaments (~1 µm in length) can be resolved by light microscopy; thus, perturbations of thin filament length can be detected easily (e.g., Gregorio *et al.*, 1995). Additionally, primary cultures of cardiac myocytes are accessible to experimental manipulation (see *Deciphering the functional roles of titin and its ligands in myofibrillogenesis* in this chapter). Finally, it is possible to manipulate the timing of myofibril assembly (which occurs over the course of several days in this system) by modulating cell culture conditions.

Using primary cultures of chick cardiac myocytes, two prominent models of myofibrillogenesis have been proposed. In one model, microfilament bundles resembling the stress fibers of nonmuscle cells (stress fiber-like structures: SFLS) function as a scaffold during sarcomere assembly (Dlugosz *et al.*, 1984). I-Z-I complexes, containing α-actinin, sarcomeric actin and titin then accumulate early on these filamentous structures to form nonstriated (stress-fiber-like) myofibrils (NSMF) in the absence of muscle-myosin containing thick filaments. The thick filaments assemble independently and incorporate later into these preformed structures (Wang *et al.*, 1988; Schultheiss *et al.*, 1990; Epstein and Fischman, 1991). Interestingly, the integration of titin into sarcomeres proceeds with the assembly of titin's N-terminus (associated with the Z-line) and continues toward the C-terminal region (associated with the M-line region) (Fürst *et al.*, 1989). These studies suggest that titin has a vital role in the assembly of the I-Z-I bodies and in "docking" these structures to the thick filaments (e.g., Holtzer *et al.*, 1997).

A different model of myofibril assembly proposes the importance of "premyofibrils" as precursor structures (Rhee *et al.*, 1994; Dabiri *et al.*, 1997). In this model, premyofibrils are composed of mini-sarcomeres (with the distance between Z-bodies ranging from 0.3-1.4 μm), consisting of thin filaments that grow in length as the assembly process proceeds. The premyofibrils also contain nonmuscle myosin molecules but do not appear to contain titin. In this model, as sarcomeres mature, the distance between individual Z-discs increases as the thin filaments grow in length. Titin assembles and nonmuscle myosin is replaced by its muscle isoforms to form the mature sarcomeres (with Z-line spacings ranging from 1.8-2.5 μm).

Although primary cultures of cardiac myocytes are valuable systems for studying myofibril assembly and maintenance, they are not perfect models. This is because some aspects of myofibril assembly in cultured cells may not reflect true myofibrillogenesis *in vivo*. For instance, cultures of cardiac myocytes are obtained from differentiated heart tissue when the myocytes already contain functional sarcomeres. Therefore, cultured cells appear to assemble their myofibrils by reutilizing preexisting material; hence, it is difficult to determine whether myofibrillar structures are undergoing "assembly" or "disassembly" (Lin *et al.*, 1989). Furthermore, while it is known that proper cell-cell contact and shape are critical factors that affect myofibrillogenesis, cells in culture modify their cytoarchitecture in response to isolation procedures and the artificial tissue culture environment. That is, cardiac myocytes in culture spread out on the substrate and are largely two-dimensional. On the other hand, myocytes *in vivo* have a three-dimensional cytoarchitecture and are interconnected through intercellular junctions and intercalated disks. Therefore, it is advantageous to compare the results obtained from *in vitro* studies with those from *in vivo* studies.

The elegant and detailed immunolocalization studies in the chick heart by Tokuyasu and Maher (1987a,b) and more recently by Perriard and colleagues (Ehler *et al.*, 1999) provide a detailed description of myofibril assembly *in situ*, and demonstrate that titin appears early in myofibrillogenesis during heart development. At the earliest stages of cardiac myocyte differentiation, all sarcomeric proteins studied accumulate diffusely within the cytoplasm as early as the 4 somite stage (stage 8, 26-29 hrs; Tokuyasu and Maher 1987a), with the exception of titin, which is present both diffusely and in a punctate (dot-like) pattern in the cytoplasm. (These dots may be actual aggregates of several titin molecules.) Within a few hours, the titin dots appear at the sites of α-actinin localization at the plasma membrane, forming nascent Z-bodies. These accumulations are components of the "dense Z-bodies/plaques" observed by ultra-structural analysis (Manasek, 1968; Legato, 1972; Markwald, 1973; Hill and Lemanski, 1985; Tokuyasu and Maher, 1987a,b). Thus, the early appearance of the titin and α-actinin containing Z-bodies suggests that they may function as fundamental organizing structures during myofibrillogenesis and may provide the cue for the organization of filaments in nascent sarcomeres. Filamentous actin then appears to accumulate near the cell membrane and is observed to associate with the Z-bodies, forming I-Z-I complexes consisting of (at least) actin filaments, the N terminus of titin, and α-actinin (See Fig. 1a) (Tokuyasu and Maher 1987a; Ehler *et al.*, 1999).

Like other myofibrillar constituents, myosin first appears diffusely in the cytoplasm by immunofluorescence at the 4 somite stage, and slightly later, filaments that are not associated with I-Z-I complexes can be observed in the cytoplasm (Manasek, 1968; Hiruma and Hirakow, 1985; Tokuyasu and Maher, 1987a) (Fig. 1a). Between the 8-9 somite stage (stage 9-10; ~30 hours), a pattern for M-line epitopes of titin is discerned (Ehler *et al.*, 1999) and myosin thick filaments are observed in a sarcomeric arrangement organized together with I-Z-I structures. The assembly of the myosin-binding protein myomesin correlates with this event, suggesting that this molecule may serve to anchor thick filaments to titin filaments, analogous to the crosslinking role of α-actinin within Z-lines (Ehler *et al.*, 1999). Interestingly, myosin thick filaments are observed to be aligned in their sarcomeric pattern before actin thin filaments are organized into their mature pattern (Tokyasu and Maher, 1987a; Ehler *et al.*, unpublished). Ultrastructural analysis revealed that the shortest sarcomere lengths (Z-band spacings) seen in nascent myofibrils are approximately 1.5 microns with no visible I-bands, suggesting that the A-band width determines the initial sarcomeric length of the first myofibrils (Markwald, 1973; Tokuyasu and Maher, 1987b; Bishop *et al.*, 1990) (See Fig. 1b). The final alignment of the thick filaments and thin filament in the mature sarcomere is likely to be regulated, in part, by factors such as cell adhesion and mechanical forces (i.e., contractile activity) (e.g., Bishop *et al.*, 1990; Simpson *et al.*, 1993; Littlefield and Fowler, 1998). By the 12 or 13 somite stage (sarcomere length is now 1.8-2.3 μm), extensive intercalated disks and branching of myofibrils are apparent, and individual myofibrils are properly aligned in register, forming a three-dimensional network (Tokuyasu and Maher, 1989a,b; Shiraishi *et al.*, 1993; 1995; 1997) (See Fig. 1c). The circumferential alignment and stability of myofibrils

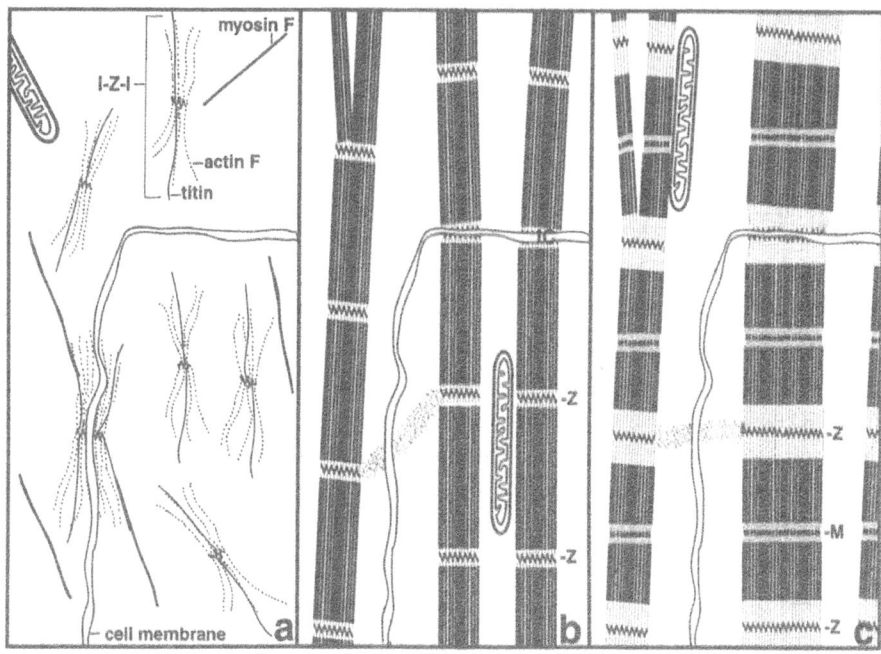

Figure 1. *Model of proposed temporal steps involved in cardiac myofibrillogenesis in chick heart* in vivo. *This schematic was modified from Figure 1 presented in Markwald, 1973.* **A:** *At the 8-9 somite stage, I-Z-I complexes are composed of Z bodies that contain α-actinin and titin linked to filaments composed of actin (actin F) (and other thin filament proteins), often in close proximity to the cell membrane. Myosin thick filaments (myosin F) are scattered diffusely in the cytoplasm and become associated with I-Z-I complexes, perhaps via interaction with the giant molecule titin. Note, in this model the actin filaments are at variable lengths at this stage (for discussion see Gregorio, 1997).* **B:** *At the 10-11 somite stage, myosin thick filaments are organized together with I-Z-I structures forming nascent sarcomeres. The shortest sarcomere lengths (Z-band spacings) seen in nascent myofibrils are approximately 1.5 microns with no visible I bands, suggesting that the A band width determines the initial sarcomeric length of the first myofibrils.* **C:** *At the 12 or 13 somite stage, thin filaments can be resolved at their mature lengths, sarcomere lengths (distance between Z lines) range from 1.8-2.5 mm and individual myofibrils are properly aligned in register, forming a three dimensional network. Note continuity of myofibrils of neighboring cells at intercalated disks. (IC).*

is the last step during myofibrillogenesis and seems to be important for optimal load management.

What is apparent from most studies to date is that a molecular template, or molecular "blueprint", is required to specify the precise position of contractile proteins within each half-sarcomere. The titin third filament system has emerged as the primary candidate for this role, based on its early appearance in myofibril assembly (as observed in most, but not all, *in vitro* studies and in all *in vivo* studies; see above), its ability to associate with both the thin and the thick filaments, its huge size that spans the half sarcomere, and its modular structure of repeating and unique domains that contain numerous binding sites for many different classes of ligands. Additionally, many distinct titin isoforms have been identified that are generated by differential splicing within its Z-line, I-band and/or M-line regions (Labeit and Kolmerer, 1995; Gautel *et al.*, 1996; Kolmerer *et al.*, 1996; Sorimachi *et al.*, 1997; see Centner *et al.*, this issue for discussion; Freiburg *et al.*, unpublished). As a result, sarcomeres in different muscle types could be created with varying ultrastructures and consequently, different elastic and mechanical properties (e.g., Vigoreaux, 1994; Squire, 1997). For example, the variability of Z-line length and protein composition seems to correlate with the presence of different Z-line titin isoforms (Gautel *et al.*, 1996; Yajima *et al.*, 1996; Sorimachi *et al.*, 1997; Gregorio *et al.*, 1998). Furthermore, several titin isoforms possessing different structural elements within their I-band region are expressed in human heart tissue. This is likely due to differing physiological demands of the distinct regions of the myocardium (Helmes *et al.*, 1999; Freiburg *et al.*, unpublished; Centner *et al.*, this issue). It is also possible that various titin isoforms may interact uniquely with (different or the same) titin ligands, which also could impart distinct elastic and mechanical properties to specific muscle types. If titin is the template for sarcomere assembly, one would predict that these muscle-specific titin splice pathways would emerge early during myogenesis (Gregorio *et al.*, 1999).

One puzzle concerning the potential role of titin as a molecular blueprint is that based on its structural architecture, it is expected that newly synthesized titin would behave as a spring. Thus, it could be somewhat difficult to imagine how this protein remains linearized in order to organize other contractile proteins in the sarcomere. One possible explanation is that the linearizing of titin might occur from the periodic organization of titin's mRNA with a large fraction of titin protein being assembled and linearized upon translation (Fulton and Alftine, 1997). It is also possible that ligands of titin might function to straighten the otherwise coiled protein at their specific binding sites, thereby allowing it to function as a ruler. In this respect, specific titin-binding proteins (such as T-cap/telethonin, see *Titin ligands* section below) may act as "bolts" to anchor titin in the sarcomere, providing a framework upon which sarcomeres, as well as myofibrils, could be constructed (Gregorio *et al.*, 1998).

Titin ligands

Studies to identify various ligands of titin became feasible after the epitope-mapping of titin-specific antibodies (e.g., Fürst *et al.*, 1988, 1989; Obermann

et al., 1996, 1997; Gregorio *et al.*, 1998; Young *et al.*, 1998) and the sequencing of human titin cDNA (Labeit and Kolmerer, 1995). The deduced protein
sequence of titin indicates that 90% of the molecule is comprised of two classes
of repeating domains that belong to the fibronectin (FN) type III superfamily
and to the immunoglobulin (Ig) superfamily. The remaining 10% of titin's
mass is composed of non-repetitive sequences situated between the Ig and FNIII
repeats (Labeit and Kolmerer, 1995). Seventeen of these interdomain inser-
tions are present in titin, sixteen of which have no clear homology to each
other, nor to any other known proteins. One interdomain insertion is the "titin
kinase domain," which is located near the M-line region of titin and has sequence homology to the serine/threonine kinase domain of myosin light chain
kinase, twitchin and projectin (Benian *et al.*, 1989; Ayme-Southgate *et al.*, 1991;
Labeit *et al.*, 1992; Mayans *et al.*, 1998). The unique regions of titin have been
the focus of many functional studies, including recent investigation as potential binding sites for novel titin ligands.

An obvious limitation in the study of titin-ligand interactions is the gigantic size of titin (i.e., it is very difficult to isolate intact titin). Fortunately, however, techniques such as the yeast two-hybrid system and various *in vitro* binding assays (using recombinant fragments of titin) have proven to be valuable
tools in the search for titin-binding proteins. To date, three distinct protein
interactions have been identified in titin's N-terminal Z-line region as a result
of yeast two-hybrid studies: titin/T-cap (telethonin), titin/α-actinin's C-terminal region, and titin/α-actinin's spectrin-like repeats. The most extreme titin
N-terminal residues 1-200 (Ig repeats Z1 and Z2), bind to the novel 19 kDa Z-
line protein, titin-cap (T-cap/telethonin) (Gregorio *et al.*, 1998; Mues *et al.*,
1998). [The sequence of T-cap is identical to the cDNA sequence from telethonin
(Valle *et al.*, 1997), which was originally identified as a putative thick filament-associated protein. Based on its actual location in the Z-disc and its association with titin's extreme N-terminal repeats, it was suggested to rename this
protein titin-cap (T-cap) (Gregorio *et al.*, 1998).] Both titin Z1 and Z2 repeats
were shown to be necessary for T-cap/telethonin binding. Titin residues 450-
750 encode up to seven unique Z repeats that bind to the C-terminal region of
α-actinin (Ohtsuka *et al.*, 1997; Sorimachi *et al.*, 1997; Young *et al.*, 1998).
These yeast two-hybrid studies were complemented with an *in vitro* fusion-
protein binding assay, which demonstrated that titin Z repeats provide at least
four potential binding sites for α-actinin (Gregorio *et al.*, 1998). Additionally,
the Z-line residues 750-826 may also bind α-actinin in its spectrin-like repeat
domains (Young *et al.*, 1998). These investigations reveal that α-actinin dimers
may attach at both ends to titin filaments, while their N-terminal ends crosslink
them to the thin filaments. It has been hypothesized that the differential expression of titin Z-repeats and therefore, the number of α-actinin binding sites,
could regulate the width of the Z-line (Gautel *et al.*, 1996; Sorimachi *et al.*,

1997; Gregorio *et al.*, 1998) and affect the mechanical properties of different muscle types.

To date, less is known concerning potential ligands of titin's I-band region. Biochemical data suggest that titin interacts with the thin filaments near the Z-disc region, since following thin filament removal, titin becomes more extensible (Granzier *et al.*, 1997; Linke *et al.*, 1997; Trombitás and Granzier, 1997). However, the specific titin domains responsible (directly or indirectly) for this interaction have not yet been identified. Whether or not titin–thin filament interactions occur elsewhere in the I-band is unclear. A deep etch electron microscopy study suggested that the I-band region of titin may be laterally associated with the thin filaments (Funatsu *et al.*, 1993). A relatively weak interaction that affects I-band titin Ig-domain unfolding in cardiac muscle is indicated by one study (Granzier *et al.*, 1997), but not by another (Linke *et al.*, 1997). Additionally, both native titin and recombinant titin fragments interact with actin in *in vitro* assays (Soteriou *et al.*, 1993; Jin, 1995). Finally, functional studies on the differentially expressed I-band N2-B region of titin in cardiac myocytes indicate that this domain contributes either directly or indirectly to the stabilization of the thin filaments (Linke *et al.*, 1999) (See *Deciphering the functional roles of titin in myofibrillogenesis* in this chapter). Clearly, additional studies are needed to examine further the potential interaction of titin with thin filament components in the I-band.

A ligand that has been shown to bind to the central I-band region and also within the M-line of titin is the muscle-specific calpain protease p94 (Sorimachi *et al.*, 1995). Interestingly, the M-line p94 binding motif is present in cardiac muscle but is absent in some fast fiber-type muscles (Kolmerer *et al.*, 1996). The biological significance of the binding of p94 to titin and differential expression of the p94 binding sites is presently unknown, but loss of function of calpains results in the activation of other proteases (Kinbara *et al.*, 1998). This suggests the exciting possibility that p94 calpain may function as a regulatory mechanism in muscle types where its titin-binding domains are expressed.

In contrast to the I-band region of titin, more is known about the specific proteins that bind to its A-band region. The central region of A-band titin (also called the C-zone) consists of two types of superrepeats (containing Ig and FNIII domains) (Labeit and Kolmerer, 1995). The central A-band region of titin contains seven FNIII and four Ig domains that are repeated 11 times. The 11-domain superrepeats contain multiple binding sites for the tail portion of myosin, as well as for the thick filament-associated protein, C-protein (MyBP-C) (Labeit *et al.*, 1992; Houmeida *et al.*, 1995). The C-terminal region of C-protein also binds to the myosin tail (Okagaki *et al.*, 1993). Thus, within the C-zone of the A-band, titin may act as a molecular ruler to regulate the number and location of myosin and C-protein molecules that assemble into thick filaments (Whiting *et al.*, 1989; Houmeida *et al.*, 1995).

M5, an Ig repeat found at the M-line region of titin, has been shown to bind myomesin *in vitro* (Obermann *et al.*, 1996). M5 is next to tandem copies of K (lysine), S (serine) and P (proline) (KSP) repeats. Titin's KSP motifs are potential substrates of a yet unidentified kinase (Gautel *et al.*, 1993; Obermann *et al.*, 1997), which could serve a regulatory role for titin-myomesin interactions.

Finally, titin contains one catalytic serine/threonine kinase domain in its M-line region, referred to as the titin kinase domain (Labeit *et al.*, 1992). This domain is an autoregulated SP kinase that binds calcium/calmodulin (Mayans *et al.*, 1998). From *in vitro* assays, the Z-disc protein (and ligand of titin's Z1 and Z2 repeats), T-cap/telethonin, (see above) was predicted to be the substrate of this kinase (Mayans *et al.*, 1998). It has therefore been proposed that the titin kinase may coordinate both M-line and Z-disc assembly during myofibrillogenesis. The feasibility of how the Z-disc and M-line regions of a sarcomere interact during the early stages of myofibrillogenesis is unknown at this time.

Deciphering the functional roles of titin and its ligands in myofibrillogenesis

As mentioned previously, most of the titin ligand studies are based on *in vitro* binding data. Clearly, *in vivo* studies to investigate the physiological significance of titin's interactions are warranted. In this respect, the functional roles of titin and some of its known binding partners have been investigated recently using dominant-negative transient expression studies in cultured myocytes (See Fig. 2 for targets of disruption). These studies, to date, utilized microinjection or transfection techniques to overexpress truncated forms of the protein of interest, which can inhibit the function of the endogenous protein (Herskowitz, 1987). The recombinant expressed protein can be distinguished from the endogenous protein by fusion with epitope tags (e.g., myc) that can be detected by immunofluorescence, or by fusion with a reporter tag (e.g., green fluorescence protein, GFP) that can be monitored in real time. This approach has allowed for the study of titin-ligand interactions in the context of live cells and has led to an exciting new direction in the field of myofibrillogenesis.

Overexpression of the first 362 residues of titin in living myocytes resulted in myofibril disassembly (Turnacioglu *et al.*, 1997). Similarly, overexpression of the Z-disc titin Z1-Z2 domain (residues 1-200), its ligand T-cap (Gregorio *et al.*, 1998), the entire Z-disc region of titin (Peckham *et al.*, 1997), or a C-terminal fragment of α-actinin that was missing its titin-binding domain (Lin *et al.*, 1998), all resulted in severe, or "explosive," disruption of myofibrils (See Fig. 2). It is surprising that overexpression of full-length T-cap disrupted myofibril integrity, suggesting that an excess of this protein also prevented endogenous titin from functioning correctly. The results from these

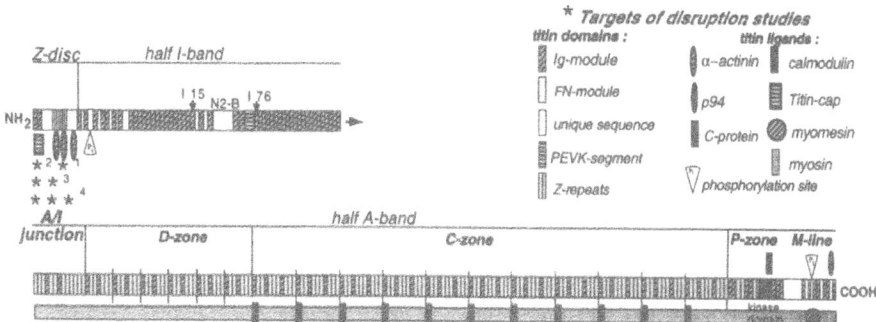

Figure 2. *Molecular layout of cardiac titin filament and its putative ligand-binding sites. Functional overexpression study sites are indicated by *. [1]Lin et al., 1998; [2]Gregorio et al., 1998; [3]Turnacioglu et al., 1997; [4]Peckham et al., 1997; and [5]Linke et al., 1999. Ig-molecules 15 and 76 from N2-A titin are marked for reference (Linke et al., 1999).*

functional studies indicate that the association of titin with Z-discs (i.e., via its interaction with T-cap and α-actinin) is critical for the assembly and maintenance of sarcomeric integrity.

Another study that illustrated the power of the dominant negative approach in living myocytes was performed to probe the functional significance of titin cardiac muscle's N2-B region (Linke *et al.*, 1999); this is the third region identified within titin that has elastic properties (Helmes *et al.*, 1999; Linke *et al.*, 1999). Interestingly, the resulting phenotype was not as "explosive" as the titin Z-disc component overexpression phenotypes (described above), since expression of the N2-B domain alone resulted in disruption of thin filaments, but not of thick filaments or titin filaments. The specific region of N2-B responsible for the disruption of the thin filaments was mapped by using subfragments of the domain. It was found that expression of the middle region or the C-terminal region of N2-B did not affect thin filament structure in cardiac myocytes, but that only the expression of the N-terminal end of N2-B caused severe disruption of the thin filaments. Thus, the N-terminal end of cardiac N2-B titin is either directly or indirectly critical for maintaining the stability of the thin filaments, providing the first direct evidence for a structural role of I-band titin in cardiac muscle (Linke *et al.*, 1999).

It is apparent from all of these studies that the dominant-negative approach for investigating the physiological roles of titin and its ligands should be extended to other regions of the titin filaments. However, it is likely that other, complementary approaches will also provide important clues to determine the functional significance of titin-ligand interactions in the context of living myocytes. For example, microinjection of domain-specific, monovalent (Fab) immunoglobulin (Ig) fragments has been used to block the function of both barbed and pointed end thin-filament capping proteins in skeletal and cardiac myocytes (Schafer *et al.*, 1995; Gregorio *et al.*, 1995; Gregorio and Fowler, 1996). Advantages of using this technique are that the investigator can control the amount of protein that is injected into the myocyte as well as the timing of the injection; for example, during particular stages of myofibrillogenesis, the antibodies generally exert their effects soon after microinjection (within minutes to hours). This rapid effect is most likely due to the dynamic structure of the sarcomere, where newly synthesized contractile proteins are continually turning over (Sanger *et al.*, 1984, 1986; McKenna *et al.*, 1985a, b; Mittal *et al.*, 1987; Bouche *et al.*, 1988; Simpson *et al.*, 1993; Gregorio and Fowler, 1995; Littlefield and Fowler, 1998; Michele *et al.*, 1999). Finally, recovery of the sarcomeric structure can be monitored since the cells tend to degrade the injected antibodies over time (the half-life is ~15 hours) (i.e., Gregorio *et al.*, 1995). Thus, microinjection of function-blocking anti-titin antibodies could prove to be a valuable approach for analyzing the potential roles of titin-ligand interactions during myofibrillogenesis.

Antisense technology also could be a useful tool for investigating the role of titin and its ligands. Oligonucleotides against the specific mRNA sequence can be introduced into cells by microinjection or transfection techniques, or by bulk application. Additionally, transfection of expression vectors containing cDNAs of the protein of interest, inserted in the antisense direction, allow for sustained high level expression of antisense mRNA in cells (for reviews, see Wagner, 1995; Bennett, 1996). These approaches have been used successfully to block the expression of several muscle proteins, including dystrophin (Dunckley *et al.*, 1998), troponin C (Ojala *et al.*, 1997; Choudhury and Bag, 1998), desmin (Li *et al.*, 1994), tropomodulin (Sussman *et al.*, 1998), and myosin (e.g., Martin *et al.*, 1996).

Future directions
Substantial progress has been made during the last few years in elucidating the structure and function of the giant sarcomeric protein, titin. Although advances in molecular and cell biology have led to a greater understanding of titin's structural roles in the sarcomere, there is still much work to be done to determine its actual roles in the process of myofibril assembly and maintenance. This challenge undoubtedly will be aided by the further characterization of the specific ligands and functional domains of the third filament system.

To meet this goal, improvements in existing experimental models of cardiac myofibril assembly and technical approaches are required. For instance, cDNA transfection methods used to date to study titin-ligand interactions allow for ~50% transfection efficiency in primary cultures of cardiac myocytes, at best. The use of more efficient transfection methods would allow, for example, the application of biochemical studies on transfected cells (e.g., to perform binding studies *in situ*), as well as the application of electron microscopy studies (that is, to observe ultrastructural changes in transfected cells). In this respect, adenoviral vectors have been used successfully to introduce certain sarcomeric proteins into cardiac myocytes (reviewed recently in Kass-Eisler and Leinwand, 1998). Additionally, the use of inducible gene promoters would allow for control over the amount and the timing of exogenous protein expression in cultured myocytes. Improved *in vitro* culture systems for studying the mechanisms of cardiac myofibril assembly are also required (for discussion of the problems with existing systems, see *The role of titin in myofibril assembly: is it a molecular template?* in this chapter). In this respect, one *in vitro* system that may prove useful for investigating the function of various proteins during cardiac myofibril assembly is explants of precardiac regions from late gastrula (stage 5-6) avian embryos. These explants can be maintained in culture as they differentiate, assemble contractile myofibrils and begin beating in a temporal sequence that correlates with myofibrillogenesis *in vivo* (DeHaan, 1963; Lough *et al.*, 1990; Antin *et al.*, 1994; Imanaka-Yoshida, 1997; Yatskievych *et al.*, 1997; Rudy *et al.*, 1999). Excitingly, mouse embryonic stem (ES) cells is another cell culture system that seems promising for studying myofibril assembly in differentiating cardiac myocytes. Although ES cells have been used to study many aspects of cardiac myocyte development, including contractile protein gene expression (Robbins *et al.*, 1990; Ng *et al.*, 1997), the function of signaling molecules in the regulation of heart muscle cell development (Wobus *et al.*, 1997) and various aspects of myocyte ultrastructure and cell-cell coupling (Westfall *et al.*, 1997), little if any work, to date, has addressed the issue of myofibril assembly. ES cells are especially appealing to investigate the physiological significance of titin-ligand interactions because of the ability to knockout genes or stably introduce mutated genes in this system. This would allow for a more thorough analysis of the protein interactions involved in myofibril assembly, as well as the mechanisms regulating titin's potential role as a molecular template in this process.

Although improved *in vitro* culture models of cardiac myofibril assembly are necessary, genetic models would be invaluable for analyzing the functional significance of titin-ligand interactions during this process. This remains quite a challenge, however, because of titin's large size. Additionally, severe mutations in proteins involved in myofibril assembly are expected to be lethal to the animal since proper heart function is critical for embryogenesis. In this respect, most mice homozygous for mutations in cardiac α-actin (Kumar *et al.*, 1997), α-tropomyosin (Rethinasamy *et al.*, 1998), and α-myosin heavy chain (Jones *et al.*, 1996), for example, die during gestation or soon after birth. However, there has been recent success in creating genetic models to study familial hypertrophic cardiomyopathy (FHC), an autosomal dominant disease characterized by ventricular hypertrophy, myocellular disarray, arrhythmias, and sudden death (reviewed in Bonne *et al.*, 1998). This disease is caused by mutations in one of at least seven different contractile genes, and is therefore a disease of the sarcomere (Thierfelder *et al.*, 1994). Hence, an understanding of the molecular basis for FHC is expected to yield insight into sarcomeric interactions during myofibril assembly and maintenance. Mice with cardiac-specific expression of defined mutations in myosin heavy chains (Vikstrom *et al.*, 1996; Geisterfer-Lowrance *et al.*, 1996), MyBP-C (Yang *et al.*, 1998), and troponin T (Tardiff *et al.*, 1998) all exhibit myofibril degeneration and histopathological features characteristic of the human disease. These mouse models, along with approaches to create more subtle mutations (e.g., point mutations) in specific domains of contractile proteins, are likely to be useful for investigating cardiac myofibrillogenesis.

In conclusion, studies from existing *in vitro* and *in vivo* systems have established a solid foundation for further investigation into the molecular mechanisms involved in cardiac myofibril assembly and maintenance. It is evident that the giant protein titin plays a key role in this complex process, although many questions regarding its exact functions remain unanswered. The future of the field promises to provide great insight into the significance of titin's modular and unique domain structure and its potential interactions with various ligands. It is expected that these studies will also yield significant clues regarding the function of recently identified titin-like molecules in non-striated muscle cells (Eilertsen and Keller, 1992; Eilertsen *et al.*, 1997; Machado *et al.*, 1998). All evidence to date indicates that the elastic titin protein is multifunctional, with many potential roles in organizing and maintaining the sarcomere. The stage is now set for dissecting further titin's physiological significance in myofibril assembly and maintenance.

Acknowledgments

We gratefully acknowledge Thomas Centner and David Carroll for help with the illustrations. Work is supported by the National Institute of Health Heart, Lung and Blood Institute NIH HL57461 and HL03985 (CCG), the National Institute of Health Heart, Lung and Blood Training Grant NIH HL07249 (ASM), the Human Frontier Science Programme (SL, CCG), and by the Deutsche Forschungsgemeinschaft La 668/5-1 (SL).

References

Antin PB, Taylor RG, Yatskievych TA. Precardiac mesoderm is specified during gastrulation in quail. *Dev Dyn* 1994;200:144-153.
Ayme-Southgate A, Vigoreaux J, Benian G, Pardue ML. Drosophila has a twitchin/titin-related gene that appears to encode projectin. *Proc Natl Acad Sci USA* 1991;88:7973-7977.

Benian GM, Kiff JE, Neckelmann N, Moerman DG, Waterston RH. Sequence of an unusually large protein implicated in regulation of myosin activity in C. elegans. *Nature* 1989;342:45-50.

Bennett CF. Antisense research. *Science* 1996;271:434.

Bishop SP, Anderson PG, Tucker DC. Morphological development of the rat heart growing in oculo in the absence of hemodynamic work load. *Circ Res* 1990;66:84-102.

Bonne G, Carrier L, Richard P, Hainque B, Schwartz K. Familial hypertrophic cardiomyopathy: from mutations to functional defects. *Circ Res* 1998;83:580-593.

Bouche M, Goldfine SM, Fischman DA. Posttranslational incorporation of contractile proteins into myofibrils in a cell-free system. *J Cell Biol* 1988;107:587-596.

Choudhury M, Bag J. Stabilization of slow troponin C polypeptide compensates for its reduced synthesis in antisense oligodeoxynucleotide-treated cells. *Nucleic Acids Res* 1998;26:4765-4770.

Dabiri GA, Turnacioglu KK, Sanger JM, Sanger JW. Myofibrillogenesis visualized in living embryonic cardiomyocytes. *Proc. Natl. Acad. Sci USA* 1997;19:9493-9498.

DeHaan RL. Migration patterns of the precardiac mesoderm in the early chick embryo. *Exptl Cell Res* 1963;29:544-560.

Dlugosz AA, Antin PB, Nachmias VT, Holtzer H. The relation between stress fiber-like structures and nascent myofibrils in cultured cardiac myocytes. *J Cell Biol* 1984;99:2268-2278.

Dunckley MG, Manoharan M, Villiet P, Eperon IC, Dickson G. Modification of splicing in the dystrophin gene in cultured Mdx muscle cells by antisense oligoribonucleotides. *Hum Mol Genet* 1998;7:1083:90.

Ehler E, Rothen BM, Hämmerle SP, Komiyama M, Perriard J-C. Myofibrillogenesis in the developing chicken heart: assembly of Z-disc, M-line and the thick filaments. *J Cell Sci* 1999;112:1529-1539.

Eilertsen KJ, Keller TC 3rd. Identification and characterization of two huge protein components of the brush border cytoskeleton: evidence for a cellular isoform of titin. *J Cell Biol* 1992;119:549-557.

Eilertsen KJ, Kazmierski S, Keller TC 3rd. Interaction of alpha-actinin with cellular titin. *Eur J Cell Biol* 1997;74:361-364.

Epstein HF, Fischman DA. Molecular analysis of protein assembly in muscle development. *Science* 1991;251:1039-1044.

Erickson HP. Stretching single protein molecules: titin is a weird spring. *Science* 1997;276:1090-1092.

Fowler VM. Regulation of actin filament length in erythrocytes and striated muscle. *Curr Opin Cell Biol* 1996;8:86-96.

Fulton AB, Alftine C. Organization of protein and mRNA for titin and other myofibril components during myofibrillogenesis in cultured chicken skeletal muscle. *Cell Struc Func* 1997;22:51-58.

Funatsu T, Higuchi H, Ishiwata, S. Elastic filaments in skeletal muscle revealed by selective removal of thin filaments with plasma gelsolin. *J Cell Biol* 1990;110:53-62.

Funatsu T, Kono E, Higuchi H, Kimura S, Ishiwata S, Yoshioka T, Maruyama K, Tsukita S. Elastic filaments in situ in cardiac muscle: deep-etch replica analysis in combination with selective removal of actin and myosin filaments. *J Cell Biol* 1993;120:711-724.

Fürst DO, Osborn M, Nave R, Weber K. The organization of titin filaments in the half-sarcomere revealed by monoclonal antibodies in immunoelectron microscopy: A map of ten nonrepetitive epitopes starting at the Z-line extends close to the M-line. *J Cell Biol* 1988;106:1563-1572.

Fürst DO, Nave R, Osborn M, Weber K. Repetitive titin epitopes with a 42 nm spacing coincide in relative position with known A-band striations also identified by major myosin-associated proteins. An immunoelectron-microscopical study on myofibrils. *J Cell Sci* 1989;94:119-125.

Gautel M, Lakey A, Barlow DP, Holmes Z, Scales S, Leonard K, Labeit S, Mygland A, Gilhus NE, Aarli JA. Titin antibodies in myasthenia gravis: Identification of a major auto-immunogenic region of titin. *Neurology* 1993;43:1581-1585.

Gautel M, Goulding D, Bullard B, Weber K, Fürst DO. The central Z-disc region of titin is assembled from a novel repeat in variable copy numbers. *J Cell Sci* 1996;109:2747-2754.

Geisterfer-Lowrance AA, Christie M, Conner DA, Ingwal JS, Schoen FJ, Seidman CE, Seidman JG. A mouse model of familial hypertrophic cardiomyopathy. *Science* 1996;272:731-734.

Granzier HL, Irving TC. Passive tension in cardiac muscle: contribution of collagen, titin, microtubules, and intermediate filaments. *Biophys J* 1995;68:1027-1044.

Granzier H, Kellermayer M, Trombitás, K. Titin elasticity and mechanism of passive force development in rat cardiac myocytes probed by thin-filament extraction. *Biophys J* 1997;73:2043-2053.

Gregorio CC. Models of striated muscle thin filament assembly. *Cell Struc Func* 1997;22:191-195.

Gregorio CC, Fowler VM. Mechanisms of thin filament assembly in embryonic chick cardiac myocytes: tropomodulin requires tropomyosin for assembly. *J Cell Biol* 1995;129:683-695.

Gregorio CC, Fowler VM. Tropomodulin function and thin filament assembly in cardiac myocytes. *Trends Cardiovas Med* 1996;6:136-141.

Gregorio CC, Granzier H, Sorimachi H, Labeit S. Muscle assembly: a titanic achievement? *Curr Opin Cell Biol* 1999;11:18-25.

Gregorio CC, Trombitás K, Centner T, Kolmerer B, Stier G, Kunke K, Suzuki K, Obermayr F, Herrmann B, Granzier H, Sorimachi H, Labeit S. The NH2 terminus of titin spans the Z-disc; Its interaction with a novel 19 kD ligand (T-cap) is required for sarcomeric integrity. *J Cell Biol* 1998;143:1013-1027.

Gregorio CC, Weber A, Bondad M, Pennise CR, Fowler VM. Requirement of pointed-end capping by tropomodulin to maintain actin filament length in embryonic chick cardiac myocytes. *Nature* 1995;377:83-86.

Helmes M, Trombitás K, Centner T, Kellermayer M, Labeit S, Linke WA, Granzier H. Mechanically driven contour-length adjustment in rat cardiac titin's unique N2B sequence: titin is an adjustable spring. *Circ Res* 1999;84:1339-1352.

Herskowitz I. Functional inactivation of genes by dominant negative mutations. *Nature* 1987;329:219-222.

Hill CS, Lemanski LF. Immunoelectron microscopic localization of alpha actinin and actin in embryonic hamster heart cells. *Euro J Cell Biol* 1985;39:300-312.

Hiruma T, Hirakow R. An ultrastructural topographical study on myofibrillogenesis in the heart of the chick embryo during pulsation onset period. *Dev Dyn* 1985;196:291-299.

Holtzer H, Hijikata T, Lin ZX, Zhang, ZQ, Holtzer S, Protasi F, Franzini-Armstrong C, Sweeney HL. Independent assembly of 1.6μm long bipolar MHC filaments and I-Z-I bodies. *Cell Struc Func* 1997;22:83-93.

Horowits R, Kempner ES, Bisher ME, and Podolski RJ. A physiological role for titin and nebulin in skeletal muscle. *Nature* 1986;323:160-164.

Horowits R, Luo G, Zhang JZ, Herrera AH. Nebulin and nebulin-related proteins in striated muscle. *Adv Biophys* 1996;33:143-150.

Houmeida A, Holt J, Tskhovrebova L, Trinick J. Studies of the interaction between titin and myosin. *J Cell Biol* 1995;131:1471-1481.

Imanaka-Yoshida K. Myofibrillogenesis in precardiac mesoderm explant culture. *Cell Struct Func* 1997;22:45-49.

Jin JP. Cloned rat cardiac titin class I and class II motifs. Expression, purification, characterization, and interaction with F-actin. *J Biol Chem* 1995;270:6908-6916.

Jones WK, Grupp IL, Doetschman T, Grupp G, Osinska H, Hewett TE, Boivin G, Gulick J, Ng WA, Robbins J. Ablation of the murine alpha myosin heavy chain gene leads to dosage effects and functional deficits in the heart. *J Clin Invest* 1996;98:1906-1917.

Kass-Eisler A, Leinwand L. DNA and Adenovirus-mediated gene transfer into cardiac muscle. *Methods Cell Biol* 1998;52:423-437.

Keller TCS. Molecular bungees. *Nature* 1997;387:233-235.

Kinbara K, Ishiura S, Tomioka S, Sorimachi H, Jeong S, Amano S, Kawasaki H, Kolmerer B, Kimura S, Labeit S, Suzuki K. Purification of native p94, a muscle-specific calpain, and characterization of its autolysis. *Biochem J* 1998;335:589-596.

Kolmerer B, Olivieri N, Witt CC, Herrmann BG, Labeit S. Genomic organization of the M-line titin and its tissue-specific expression in two distinct isoforms. *J Mol Biol* 1996;256:556-563.

Kumar A, Crawford K, Close L, Madison M, Lorenz J, Doetschman T, Pawlowski S, Duffy J, Neumann J, Robbins J, Boivin G P, O'Toole BA, Lessard JL. Rescue of cardiac alpha-actin deficient mice by enteric smooth muscle gamma-actin. *Proc Natl Acad Sci USA* 1997;94:4406-4411.

Labeit S, Kolmerer B. Titins, giant proteins in charge of muscle ultrastructure and elasticity. *Science* 1995;270:293-296.

Labeit S, Kolmerer B, Linke W. The giant protein titin: emerging roles in physiology and pathophysiology. *Circ Res* 1997;80:290-294.

Labeit S, Gautel M, Lakey A, Trinick J. Towards a molecular understanding of titin. *EMBO J* 1992;11:1711-1716.

Li H, Choudhary SK, Milner DJ, Munir ML, Kuisk IR, Capetanaki Y. Inhibition of desmin expression blocks myoblast fusion and interferes with the myogenic regulators MyoD and myogenin. *J Cell Biol* 1994;124:827:841.

Lin Z, Kijikata T, Zhang Z, Choi J, Holtzer S, Sweeney HS, Holtzer H. Dispensability of the actin-binding site and spectrin repeats for targeting sarcomeric α-actinin into maturing Z-bands in vivo: implications for in vitro binding studies. *Development.* 1998;199:291-308.

Lin Z, Holtzer S, Schultheiss T, Murray J, Masaki T, Fischman DA, Holtzer H. Polygons and adhesion plaques and the disassembly and assembly of myofibrils in cardiac myocytes. *J Cell Biol* 1989;10:2355-2367.

Linke WA, Granzier H. A spring tale: new facts on titin elasticity. *Biophys J* 1998;75:2613-2614.

Linke WA, Ivemeyer M, Labeit S, Hinssen H, Ruegg JC, Gautel M. Actin-titin interaction in cardiac myofibrils: probing a physiological role. *Biophys J* 1997;73:905-919.

Linke WA, Ivemeyer M, Olivieri N, Kolmerer B, Rüegg JC, Labeit S. Towards a molecular understanding of the elasticity of titin. *J Mol Biol* 1996;261:62-71.

Linke WA, Rudy DE, Centner T, Witt C, Labeit S, Gregorio CC. I-band titin in cardiac muscle is a three-element molecular spring and is critical for maintaining thin filament structure. *J Cell Biol* 1999;146:631-644.

Littlefield R, Fowler VM. Defining actin filament length in striated muscle: rulers and caps or dynamic stability? *Annu Rev Dev Biol* 1998;14:487-525.

Lough JW, Markwald RR. A culture model for cardiac morphogenesis. *Ann NY Acad Sci* 1990;588:421-424.

Luo G, Zhang JQ, Nguyen TP, Herrera AH, Paterson B, Horowits R. Complete cDNA sequence and tissue localization of N-RAP, a novel nebulin-related protein of striated muscle. *Cell Motil Cytoskel* 1997;38:75-90.

Machado C, Sunkel CE, Andrew DJ. Human autoantibodies reveal titin as a chromosomal protein. *J Cell Biol* 1998;141:321-333.

Manasek FJ. Embryonic development of the heart. I. A light and electron microscopic study of myocardial development in the early chick embryo. *J Morphol* 1968;125:329-365.

Markwald RR. Distribution and relationship of precursor Z material to organizing myofibrillar bundles in embryonic rat and hamster ventricular myocytes. *J Mol Cell Cardiol* 1973;5:341-350.

Martin XJ, Wynne DG, Glennon PE, Moorman A, Boheler KR. Regulation of expression of contractile proteins with cardiac hypertrophy and failure. *Mol Cell Biochem* 1996;157:181-189.

Maruyama K. Connectin/titin, giant elastic protein of muscle. *Faseb J* 1997;11:341-345.

Maruyama K, Matsubara R, Natori Y, Nonomura S, Kimura S, Ohashi K, Murakami F, Handa S, Eguchi G. Connectin, an elastic protein of muscle. *J Biochem* 1977;82:317-337.

Maruyama K, Yoshioka T, Higuchi H, Ohashi K, Kimura S, Natori R. Connectin filaments link thick filaments and Z-lines in frog skeletal muscle as revealed by immunoelectron microscopy. *J Cell Biol* 1985;101:2167-2172.

Mayans O, van der Ven P, Wilm M, Mues A, Young P, Fürst DO, Wilmanns M, Gautel M. Structural basis for activation of the titin kinase domain during myofibrillogenesis. *Nature* 1998;395:863-869.

McKenna N, Meigs JB, Wang YL. Identical distribution of fluorescently labeled brain and muscle actins in living cardiac fibroblasts and myocytes. *J Cell Biol* 1985a;100:292-296.

McKenna N, Meigs JB, Wang YL. Exchangeability of alpha-actin in living cardiac fibroblasts and muscle cells. *J Cell Biol* 1985b;101:2223-2232.

Michele DE, Albayya P, Metzger JM. Thin filament protein dynamics in fully differentiated adult cardiac myocytes: toward a model of sarcomere maintenance. *J Cell Biol* 1999;145:1483-1495.

Millevoi S, Trombitás K, Kostin S, Schaper J, Pelin K, Kolmerer B, Granzier H, Labeit S. Characterization of Nebulette and Nebulin and emerging concepts of their roles for vertebrate Z-discs. *J Mol Biol* 1998;282:111-123.

Mittal B, Sanger JM, Sanger JW. Visualization of myosin in living cells. *J Cell Biol* 1987;105:1753-1760.

Moncman CL, Wang K. Nebulette: a 107 kD nebulin-like protein in cardiac muscle. *Cell Motil Cyto* 1995;32:205-225.

Mues A, van der Ven PF, Young P, Fürst DO, Gautel M. Two immunoglobulin-like domains of the Z-disc portion of titin interact in a conformation-dependent way with telethonin. *FEBS Lett* 1998;428:111-114.

Ng WA, Doetschman T, Lessard JL. Muscle isoactin expression during in vitro differentiation of murine embryonic stem cells. *Pediatr Res* 1997;41:285-292.

Obermann WM, Gautel M, Weber K, Fürst DO. Molecular structure of the sarcomeric M band: mapping of titin and myosin binding domains in myomesin and the identification of a potential regulatory phosphorylation site in myomesin. *EMBO J* 1997;16:211-220.

Obermann WMJ, Gautel M, Steiner F, van der Ven PFM, Weber K, Fürst DO. The structure of the sarcomeric M band: localization of defined domains of myomesin, M protein, and the 250 kD carboxy terminal region of titin by immunoelectron microscopy. *J Cell Biol* 1996;134:1441-1453.

Ohtsuka H, Yajima H, Kimura S, Maruyama K. Binding of the N terminal fragment of connectin/titin to alpha-actinin as revealed by yeast two-hybrid systems. *FEBS Lett* 1997;401:65-67.

Ojala J, Choudhury M, Bag J. Inhibition of troponin C production without affecting other muscle protein synthesis by the antisense oligonucleotide. *Antisense Nucleic Acid Drug Dev* 1997;7:31-8.

Okagaki T, Weber FE, Fischman DA, Vaughan KT, Mikawa T, Reinach FC. The major myosin-binding domain of skeletal muscle MyBP-C (Cprotein) resides in the COOH-terminal, immunoglobulin C2 motif. *J Cell Biol* 1993;123:619-626.

Peckham M, Young P, Gautel M. Constitutive and variable regions of Z-disc titin/connectin in myofibril formation: a dominant-negative screen. *Cell Struc Func* 1997;22:95-101.

Rethinasamy P, Muthuchamy M, Hewett T, Boivin G, Wolska BM, Evans C, Solaro RJ, Wieczorek DF. Molecular and physiological effects of alpha-tropomyosin ablation in the mouse. *Circ Res* 1998;82:116-123.

Rhee D, Sanger JM, Sanger JW. The premyofibril: evidence for its role in myofibrillogenesis. *Cell Motil Cytoskel* 1994;28:1-24.

Robbins J, Gulick J, Sanchez A, Howles P, Doetschman T. Mouse embryonic stem cells express the cardiac myosin heavy chain genes during development in vitro. *J Biol Chem* 1990;265:11905-11909.

Rudy DE, Yatskievych TA, Antin PB, Gregorio CC. Investigation of the assembly of thick, thin and titin filaments in chick precardiac explants. *Mol Biol Cell* 1999;10:980.

Sanger JW, Mittal B, Sanger JW. Analysis of myofiber structure and assembly using fluorescently labeled contractile proteins. *J Cell Biol* 1984;98:825-833.

Sanger JM, Mittal B, Pochapin MB, Sanger JW. Myofibrillogeneis in living cells microinjected with fluorescently labeled alpha-actinin. *J Cell Biol* 1986;102:2053-2066.

Schafer DA, Hug C, Cooper JA. Inhibition of CapZ during myofibrillogenesis alters assembly of actin filaments. *J Cell Biol* 1995;128:61-70.

Schultheiss T, Lin Z, Lu M-H, Murray J, Fischman DA, Weber K, Masaki M, Imamura M, Holtzer H. Differential distribution of subsets of myofibrillar proteins in cardiac nonstriated and striated myofibrils. *J Cell Biol* 1990;110:1159-1172.

Shimada Y, Komiyama M, Begum S, Maruyama K. Development of connectin/titin and nebulin in striated muscles of chicken. *Adv Biophys* 1996;33:223-234.

Shiraishi I, Takamatsu T, Fujita S. Three-dimensional observation with a confocal scanning laser microscope of fibronectin immunolabeling during cardiac looping in the chick embryo. *Anat Embryol* 1995;191:183-189.

Shiraishi I, Simpson DG, Carver W, Price R, Hirozane T, Terracio L, Borg TK. Vinculin is an essential component for normal myofibrillar arrangement in fetal mouse cardiac myocytes. *J Mol Cell Cardiol* 1997;29:2041-2052.

Shiraishi I, Takamatsu T, Fujita S. 3-D observation of N-cadherin expression during cardiac myofibrillogenesis of the chick embryo using a confocal laser scanning microscope. *Anat Embryol* 1993;187:115-120.

Simpson DG, Decker ML, Clark WA, Decker RS. Contractile activity and cell-cell contact regulate myofibrillar organization in cultured cardiac myocytes. *J Cell Biol* 1993;123:323-336.

Sorimachi H, Kinbara K, Kimura S, Takahashi M, Ishiura S, Sasagawa N, Sorimachi N, Shimada H, Tagawa K, Maruyama K, Suzuki K. Muscle-specific calpain, p94, responsible for limb girdle muscular dystrophy type 2A, associates with connectin through IS2, a p94-specific sequence. *J Biol Chem* 1995;270:31158-31162.

Sorimachi H, Freiburg A, Kolmerer B, Ishiura S, Stier G, Gregorio CC, Linke WA, Suzuki K, Labeit SL. Tissue-specific expression and alpha-actinin binding properties of the Z-disc titin: implications for the nature of vertebrate Z-discs. *J Mol Biol* 1997;270:688-695.

Soteriou A, Gamage M, Trinick J. A survey of the interactions made by titin. *J Cell Sci* 1993;104:119-123.

Squire JM. Architecture and function in the muscle sarcomere. *Curr Opin Struct Biol* 1997;7:247-257.

Sussman MA, Baquè UC-S, Daniels MP, Price RL, Simpson D, Terracio L, Kedes L. Altered expression of tropomodulin in cardiomyocytes disrupts the sarcomeric structure of myofibrils. *Circ Res* 1998;82:94-105.

Tardiff JC, Factor SM, Tompkins BD, Hewett TE, Palmer BM, Moore RL, Schwartz S, Robbins J, Leinwand LA. A truncated cardiac troponin T molecule in transgenic mice suggests multiple cellular mechanisms for familial hypertrophic cardiomyopathy. *J Clin Invest* 1998;101:2800-2811.

Thierfelder L, Watkins H, MacRae C, Lamas R, McKenna W, Vosberg H-P, Seidman JG, Seidman C. Beta-tropomyosin and cardiac troponin T mutations cause familial hypertrophic cardiomyopathy: a disease of the sarcomere. *Cell* 1994;77:701-712.

Tokuyasu KT, Maher PA. Immunocytochemical studies of cardiac myofibrillogenesis in early chick embryos. I. Presence of immunofluorescent titin spots in premyofibrillar stages. *J Cell Biol* 1987a;105:2781-2793.

Tokuyasu KT, Maher PA. Immunocytochemical studies of cardiac myofibrillogenesis in early chick embryos. II. Generation of alpha-actinin dots within titin spots at the time of the first myofibril formation. *J Cell Biol* 1987b;105:2795-2801.

Trinick J. Titin and nebulin protein rulers in muscle? *Trends Biochem Sci* 1994;19:405-408.

Trinick J. Cytoskeleton: titin as a scaffold and spring. *Curr Biol* 1996;6:258-260.

Trombitás K, Granzier H. Actin removal from cardiac myocytes shows that near the Z-line titin attaches to actin while under tension. *Am J Physiol* 1997;273:C662-C670.

Turnacioglu KK, Mittal B, Dabiri GA, Sanger JM, Sanger, JW. An N-terminal fragment of titin coupled to green fluorescent protein localizes to the Z-bands in living muscle cells: overexpression leads to myofibril disassembly. *Mol Biol Cell* 1997;8:705-717.

Valle G, Faulkner G, De Antoni A, Pacchioni B, Pallavicini A, Pandolfo D, Tiso N, Toppo S, Trevisan S, Lanfranchi G. Telethonin, a novel sarcomeric protein of heart and skeletal muscle. *FEBS Lett* 1997;415:163-168.

Vigoreaux JO. The muscle Z-band: lessons in stress management. *J Muscle Res Cell Motil* 1994;15:237-255.

Vikstrom KL, Factor SM, Leinwand LA. Mice expressing mutant myosing heavy chains are a model for familial hyptertrophic cardiomyopathy. *Mol Med* 1996;2:556-567.

Wagner RW. The state of the art in antisense research. *Nat Med* 1995;1:1116-1118.

Wang K, McClure J, Tu A. Titin: Major myofibrillar component of striated muscle. *Proc Natl Acad Sci USA* 1979;76:3698-3702.

Wang K. Titin/connectin and nebulin: giant protein rulers of muscle structure and function. *Adv Biophys* 1996;33:123-134.

Wang K, Knipfer M, Huang Q, van Heerden A, Hsu LC, Gutierrez G, Quian X, Stedman H. Human skeletal muscle nebulin sequence encodes a blueprint for thin filament architecture. Sequence motifs and affinity profiles of tandem repeats and terminal SH3. *J Biol Chem* 1996;271:4304-4314.

Wang SM, Greaser ML, Schultz E, Bulinski JC, Lin JJ, Lessard JL. Studies on cardiac myofibrillogenesis with antibodies to titin, actin, tropomyosin, and myosin. *J Cell Biol* 1988;107:1075-1083.

Westfall MV, Pasyk KA, Yule DI, Samuelson LC, Metzger JM. Ultrastructure and cell-cell coupling of cardiac myocytes differentiating in embryonic stem cell cultures. *Cell Motil Cytoskel* 1997;36:43-54.

Whiting A, Wardale J, Trinick J. Does titin regulate the length of muscle thick filaments? *J Mol Biol* 1989;205:263-268.

Wobus AM, Kaomei G, Shan J, Wellner MC, Rohwedel J, Guanju J, Fleischmann B, Katus HA, Hescheler J, Franz WM. Retinoic acid accelerates embryonic stem cell-derived cardiac differentiation and enhances development of ventricular cardiomyocytes. *J Mol Cell Cardiol* 1997;29:1525-1539.

Yajima H, Ohtsuka H, Kawamura Y, Kume H, Maruyama T, Abe H, Kimura S, Maruyama K. A 11.5 kb 5'-terminal cDNA sequence of chicken breast muscle connectin/titin reveals its Z-line binding region. *Biochem Biophys Res Commun* 1996;223:160-164.

Yang Q, Sanbe A, Osinska H, Hewett TE, Klevitsky R, Robbins J. A mouse model of myosin binding protein C human familial hypertrophic cardiomyopathy. *J Clin Invest* 1998;102:1292-300.

Yatskievych TA, Ladd A, Antin PB. Induction of cardiac myogenesis in avian pregastrula epiblast: the role of the hypoblast and activin. *Development* 1997;124:2561-2570.

Young P, Ferguson C, Bañuelos S, Gautel M. Molecular structure of the sarcomeric Z-disc: two types of titin interactions lead to an asymmetrical sorting of alpha-actinin. *EMBO J* 1998;17:1614-1624.

Discussion

Sanger: With your T-cap over expression, what do you think is happening? You presumably have T-cap at the end of titin. Do you feel that T-cap is coming dynamically on and off? And if you have titin with native T-cap in place, why would over expressing T-cap loosen titin?

Gregorio: One idea would be that you are adding too much/overexpressing the protein perturbing the equilibrium level of T-cap in the cell. Another idea is that T-cap also binds to other sarcomeric proteins. Thus, overexpressing T-cap results in the disruption of more than one T-cap binding interaction, preventing it from functioning correctly within the cell.

Sanger: Were you able to see how quickly the individual myofibril or cell fell apart?

Gregorio: In the microinjection experiments I was able to see the phenotype very quickly, within approximately a half an hour. So basically as soon as I injected cells on a coverslip and fixed them, I could observe the phenotype. In the transfection experiments using GFP-fusion proteins I could observe the phenotypes in the T-cap experiments in live cells. I observed just by the morphology of the cells, that the cells that were overexpressing the titin Z1-Z2 fragment or T-cap were detaching from the coverslips. This was not unexpected since the myofibrils/cytoskeleton were severely disrupted in these cells and, as a result they did not attach very well. Using both techniques, the phenotype occurred quickly.

Sanger: My second question relates to the second part of your talk. Did you use anti-actin antibodies, in case that the titin fragment may be blocking the phalloidin binding site on actin?

Gregorio: Not only did I use anti-actin antibodies, I used anti-tropomyosin antibodies. Staining with either of the antibodies also indicated that the thin filaments were disrupted.

TerKeurs: In your model in which T-cap functions as a bolt that holds titin in place, don't you need more than that model to explain your results? In your T-cap expression experiments you simply add more bolts. Why would the myofibril fall apart as a result?

Gregorio: Absolutely more than one model can account for our results. Today I chose to present the model that best fits our observations. It is difficult to understand why adding more T-cap/bolts results in such a dramatic phenotype, especially with the extensive α-actinin network in the Z-disc crosslinking the thin filaments and titin. Our phenotype really was unexpected and that is why I was so concerned that it was an artifact of the transfection procedure. Thus, I continued to pursue complementary approaches to reproduce the phenotype. We performed many different over-expression studies, using different epitope tags etc., and also did microinjection experiments. The reproducible severity of the phenotype observed using all the methods I described is not easily explained, but clearly suggests that the interaction of the N terminus of titin and T-cap is critical for maintaining sarcomeric integrity in cardiac myocytes.

TerKeurs: How rapid is the turn-over of proteins in the sarcomere?

Gregorio: The turnover of most, if not all, sarcomeric proteins is incredibly high. This is one of the reasons the types of experiments that I described work so fast, and at all. For example, I keep talking about all actin filaments being at fixed lengths. But, if you microinject actin into cardiac myocytes it will incorporate within seconds into the filaments; however, they do not get longer. This phenomenon has been termed by Littlefield and Fowler as "dynamic stability" as opposed to the term "dynamic instability" that is used when referring to the intracellular dynamics of microtubules. Therefore, although tropomodulin has strong capping activity, it still has to be dynamic. It has to be coming off and on to allow for exchange of actin monomers to occur while maintaining the length of the filaments. This is just one example.

TerKeurs: When you removed the capping protein and the actin filament started to grow, the cells stopping beating. How did you measure this?

Gregorio: I designed a beating assay. It is really was simply "do they beat or not beat." Basically, I plated the cells on grided cover slips and I recorded their beating activity. I observed about 400 cells per experimental group at similar densities, and I looked if they were beating or not beating. I recorded these data by video microscopy. Next, I injected all the cells that I observed and went back 4 hours later to again monitor their beating activity. Now I certainly missed some, but what I did find is that the injection procedure itself caused about a 10% decrease in the beating activity of the cells, 4 hours later. In contrast, there was an 80% reduction in the percentage of beating cells that were microinjected with the anti-tropomodulin antibody that inhibited tropomodulin's capping activity. It really was a very crude experiment from a physiological perspective, but yet it was very informative and reproducible.

TerKeurs: Did you measure action potentials?

Gregorio: No, we did not. However, it surprised us that the new actin filament extensions had no detectable tropomyosin as observed by immunofluorescence microscopy. So we had these naked filaments that were unregulated; that is, they did not have tropomyosin on them. Also they were of opposite polarity with respect to the thick filaments.

ASSEMBLY OF MYOFIBRILS IN CARDIAC MUSCLE CELLS

Joseph W. Sanger, Joseph C. Ayoob,
Prokash Chowrashi, Daniel Zurawski,
and Jean M. Sanger

*Department of Cell and Developmental Biology,
University of Pennsylvania, School of Medicine, Philadelphia, PA*

Abstract: How do myofibrils assemble in cardiac muscle cells? When does titin first assemble into myofibrils? What is the role of titin in the formation of myofibrils in cardiac muscle cells? This chapter reviews when titin is first detected in cultured cardiomyocytes that have been freshly isolated from embryonic avian hearts. Our results support a model for myofibrillogenesis that involves three stages of assembly: premyofibrils, nascent myofibrils and mature myofibrils. Titin and muscle thick filaments were first detected associated with the nascent myofibrils. The Z-band targeting site for titin is localized in the N-terminus of titin. This region of titin binds alpha-actinin and less avidly vinculin. Thus the N-terminus of titin via its binding to alpha-actinin, and vinculin could also help mediate the costameric attachment of the Z-bands of mature myofibrils to the nearest cell surfaces.

Introduction

Central to the form and function of the heart are the myofibrils that provide the cytoskeletal structure and the contractile force that characterize cardiomyocytes (Fig. 1). Myofibrillar formation requires the precise ordering of multiple subunits, the sarcomeres, into linear arrays connected at the Z-bands. The peripheral Z-bands, in turn, are linked to the membranes of the cell surfaces via focal adhesion proteins (i.e., costameres; Pardo *et al.*, 1983; Danowski *et al.*, 1992). Intermediate filaments, composed of desmin, connect the neighboring Z-bands

Elastic Filaments of the Cell, edited by Granzier and Pollack
Kluwer Academic/Plenum Publishers, 2000

to form aligned myofibrils across the width of the cells (Granger and Lazarides, 1979). Despite all that has been learned about myofibrillogenesis over the last 100 years, the challenges of identifying the essential steps involved in this process and of understanding how the process is initiated and controlled are still unmet. Myofibril formation in embryonic muscle cells has been examined in a number of different ways. Biochemical data have shown that there is coordinate synthesis of actin, myosin, tropomyosin and alpha-actinin during myofibrillogenesis (Devlin and Emerson, 1978). Immunocytochemical and electron microscopic data suggest that thick (18 nm in diameter), thin (6 nm in diameter) and titin filaments (3 nm in diameter) are formed at the same time in the cytoplasm, and subsequently organize into sarcomeric arrays (Epstein and Fischman, 1991). The biochemical appearance of a protein does not always indicate the order of its assembly or its attachment to myofibrils. Desmin, for example, is one of the first detected proteins in the muscle cells, but it is also one of the last of the proteins to attach to the myofibrils (Hill *et al.*, 1986). Embryonic cardiomyocytes migrate in culture, assembling and disassembling their myofibrils (Sanger *et al.*, 1984a; 1989b). These cells disassemble their myofibrils as they undergo cell division and reassemble their myofibrils as they enter interphase. The ability to clone and manipulate the genes involved in myofibril formation and structure is yielding fundamental insights into the interrelationships between the requisite proteins; yet beyond this powerful genetic information the additional aspects of positional information or self-templating in the cell needs to be considered in order to decipher the assembly process.

A major approach in the study of myofibrillogenesis has been to culture cells from embryonic, neonatal or adult cardiomyocytes (embryonic avian cells reviewed in Dabiri *et al.*, 1999a; rat neonatal cardiomyocytes reviewed in Van Bilsen and Chien, 1993; adult rat cardiomyocytes, Claycomb and Palazzo, 1980), and as myofibrils form, to fix and stain the cells with specific antibodies to correlate structural changes with protein localization. This has generated important information (Holtzer *et al.*, 1957; Wang *et al.*, 1988; Rhee *et al.*, 1994), but it does not yield an unambiguous view of the actual process of myofibrillogenesis. Since the actin/myosin cytoskeleton of muscle cells is a dynamic structure that assembles and disassembles reversibly in response to a variety of cellular signals, it can be difficult to know if these static images represent myofibril assembly or myofibril disassembly. Imaging of the myofibrils and their component subunits inside the same living cell allows us to determine their provenance and fate.

Probes, which are visible with light microscopy and specific for individual cytoskeletal proteins, allow myofibrillogenesis to be studied in live cells, documenting if a new myofibril is being formed or an existing one is being disassembled. Time-lapse observations of the proteins in live muscle cells also

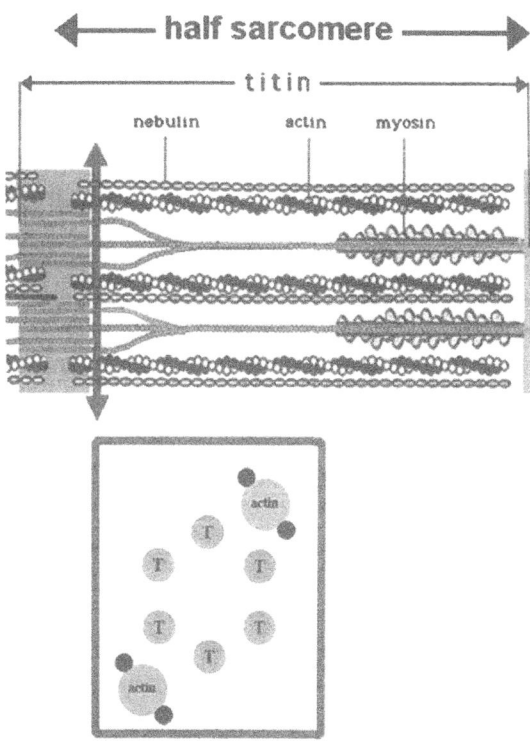

Figure 1. *Diagram of a half sarcomere of skeletal muscle illustrating the distribution of the three types of filaments present. Each end of the thick filament in a sarcomere would be associated with six titins and two actins. Since there are six titin filaments to every two actin filaments, the Z-band contains more titin filamentous ends than actin filaments. The actin filaments would be associated with nebulin isoforms. In cardiac muscle, the nebulin isoform is nebulette and while its C-terminus is in the Z-band, the N-terminus would extend about 0.1 micron along the thin filaments. The cross sectional diagram illustrates the possible array of actin and titin filaments as they exit the Z-band and before the thick filament interacts with the orbit of six titin filaments.*

provide a framework for analyzing the process of myofibril assembly. To generate suitable probes, fluorescent dyes have been coupled to purified cytoskeletal proteins, followed by microinjection of trace amounts of the labeled proteins into live myocytes (Sanger *et al.*, 1984a; 1984b; 1986a; 1986b; 1990; Dome *et al.*, 1988). The development of expression plasmids for linking the cDNA encoding a protein of interest, or a fragment of one, to a cDNA coding for Green Fluorescent Protein (GFP; Chalfie *et al.*, 1994) has supplied the potential for examining the activity of a much wider range of proteins and their domains, particularly those proteins that are difficult to prepare or are only soluble in solution conditions that would be injurious to living cells (Dabiri *et al.*, 1999a; 1999b). Single cells expressing the fluorescent proteins produced by GFP plasmids can be observed over many hours or days in a simple culture chamber on the microscope stage (Dabiri *et al.*, 1999a). Expression of fragments of a sarcomeric protein linked to GFP can specify the region of the protein required for its proper localization in the myofibril (Ayoob *et al.*, 2000). The future development of brighter Blue and Red Fluorescent Proteins along with the existence of the current enhanced GFP presents the possibility of following two or three labeled proteins in the same cardiomyocyte.

Results and Discussion

Model for myofibrillogenesis

We have proposed a model (Rhee *et al.*, 1994; LoRusso *et al.*, 1997) for the assembly of myofibrils based on antibody stainings of embryonic chick cardiomyocytes fixed at different times after spreading in culture. An example

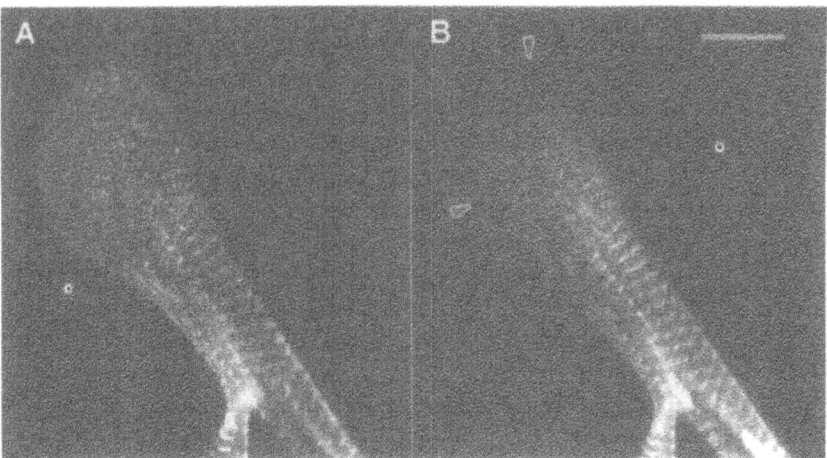

Figure 2. Spreading cardiomyocyte doubly stained with (A) alpha-actinin and (B) zeugmatin (anti-titin) antibodies. Note the absence of titin (arrowheads) from the region containing the premyofibrils (B). There is some co-localization of the anti-titin antibody with the short spacings of alpha-actinin away from the leading edge of the lamella and the fully formed myofibrils. Scale = 5 microns. (Reproduced by permission of Cell Motility and the Cytoskeleton).

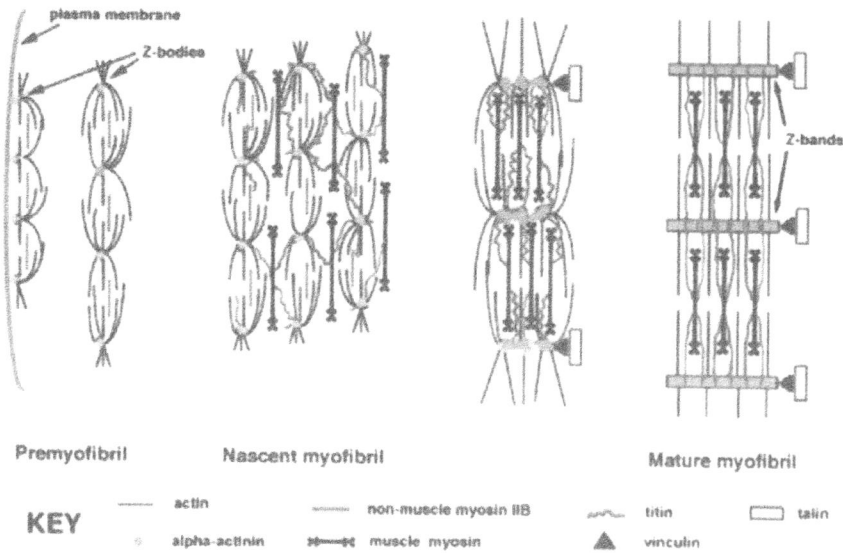

Figure 3. *Model for the assembly of a mature myofibril. The position of alpha-actinin (shaded circles) relative to actin, nonmuscle myosin IIB, muscle myosin II and titin is shown for each stage of myofibril development. Near the spreading edge of the cell (far left), alpha-actinin is found in mini-sarcomeric arrays, the premyofibrils. In the nascent myofibril, the alpha-actinin-rich Z-bodies have begun to laterally associate, possibly through interactions with titin. Overlapping muscle myosin II filaments bound to titin are present at this stage. In the mature myofibril, the alpha-actinin containing Z-bodies fuse to form the wide lateral arrays of mature Z-bands; vinculin and talin are detected in costameric Z-bands of the mature myofibrils; nonmuscle myosin IIB is lost, whereas muscle myosin filaments are now aligned into A-bands. The Z-bodies grow in size in the progression from premyofibrils to mature myofibril. The shape of the Z-band changes after the Z-bodies fuse to form a linear structure.*

of this type of double immunofluorescent staining of cardiomyocytes is presented in Figure 2. This spreading cardiomyocyte was stained with two different antibodies: anti-alpha-actinin and anti-zeugmatin. Zeugmatin (Maher *et al.*, 1985) was determined by our laboratory to be an N-terminal proteolytic fragment of titin (Turnacioglu *et al.* 1996, 1997a, 1997b). The doubly stained image in Figure 2 reveals that alpha-actinin staining extends further into the spreading edge of the cardiomyocyte than the titin (zeugmatin) staining. A number of other combinations of antibody stainings led to the proposed model illustrated in Figure 3. This model postulates that premyofibrils, characterized by banded patterns of alpha-actinin-rich Z-bodies and nonmuscle myosin IIB, form at the edges of spreading cardiomyocytes and develop into mature myofibrils (Fig.

3). Immunofluorescence studies show that the premyofibrils are composed of periodic distributions of most of the sarcomeric isoforms of the proteins present in the mature contractile myofibrils, i.e., alpha-actinin, tropomyosin and the troponins (Schultheiss *et al.*, 1992; Sanger *et al.*, 1994b; LoRusso *et al.*, 1997). During the transition from premyofibril to mature myofibril, it is postulated that there is an exchange of nonmuscle myosin IIB filaments for muscle myosin II filaments and a growth and fusion of Z-bodies into Z-bands (Rhee *et al.*, 1994; LoRusso *et al.*, 1997). The Z-bodies appear initially as discrete aggregates of alpha-actinin along the premyofibrils. As the myofibrils increase in width, the Z-bands appear to be composed of laterally aligned Z-bodies and finally continuous bands of alpha-actinin (Fig. 3). Titin and muscle-specific myosin II are not present in premyofibrils and are first detected in the nascent myofibrils (Figs. 2 and 3).

Immunofluorescent data has indicated that a number of proteins localized in focal adhesions are also associated with Z-Bands. It appears from ongoing experiments in our laboratory that these proteins (vinculin, tensin, talin, paxillin and zyxin; Sanger *et al.*, 1997) become associated with the fusing Z-bodies of the nascent myofibrils and are fully organized as costameres only in the mature myofibrils (Fig. 3).

Role of nonmuscle myosin II in myofibrillogenesis

Tullio *et al.* (1997) have demonstrated that when nonmuscle myosin IIB was knocked out in mice, normal heart and brain development were impaired; both control cardiomyocytes and brain cells possess only one isoform of nonmuscle myosin II, i.e., the IIB isoform. Furthermore, there were no signs of cytokinesis defects in the cardiomyocytes and brain cells (i.e., no signs of multinucleated cells in the heart or brain). This was surprising, because when they divide, cardiac myocytes use the nonmuscle isoform of myosin IIB (Conrad *et al.*, 1991). In half (4/8) of the few animals that survived for only one day after birth, the other isoform of nonmuscle myosin (IIA) was upregulated in the heart, and the authors speculated that this isoform could substitute for the IIB isoform. In the other 4 surviving embryos, no nonmuscle myosin IIA was detected in immunoblots from the heart tissue, although the nonmuscle cells in the tissue must have had nonmuscle myosin IIA. The more sensitive immunofluorescence staining technique by which the IIA isoform is detected in hearts (Murakami *et al.*, 1993) was not used on the knock-out mice. Without further data it is not clear from the Tullio *et al.* study (1997) whether the cardiomyocytes from the IIB knock-out mice lacked nonmuscle myosin IIA.

Support for a role of nonmuscle myosin II in myofibrillogenesis has come from two different laboratories. Ferrari *et al.* (1998) have indicated that a nonmuscle myosin II may be involved in myofibrillogenesis in Xenopus skeletal myocytes. The light chains of nonmuscle myosin II must be phosphorylated to form filaments. Ferrari *et al.* (1998) used an inhibitor of the phosphory-lation of the light chains of nonmuscle myosin II to prevent the assembly of nonmuscle myosin II filaments. This treatment prevented the formation of mature myofibrils, which lack nonmuscle myosin II (Fig. 3). If nonmuscle myosin II had not been involved in the assembly of myofibrils, exposure of the myocytes to this inhibitor should not have affected myofibrillogenesis. Recently, Bloor and Kiehart (1998) reported preliminary data that nonmuscle myosin II is essential for muscle development in *Drosophila*. The premyofibril theory (Fig. 3) postulates a positional information role for nonmuscle myosin II isoforms in myofibrillogenesis (Rhee *et al.*, 1994; Dabiri *et al.*, 1997).

How are the two types on myosin IIs prevented from polymerizing with one another? Pollard (1975) has demonstrated that nonmuscle myosin IIA and skeletal muscle myosin II will co-polymerize to form hybrid filaments. We assume that future work will demonstrate that nonmuscle myosin IIB will also co-polymerize with cardiac muscle myosin II molecules. Tom Keller and his collaborators have presented immunofluorescence and gel electrophoresis data that there are two isoforms of titin: a nonmuscle or cellular isoform and the usual muscle titin isoform (Eilertsen *et al.*, 1994). The cellular titin is also about one micron in length and some three million molecular weight. Sequencing work is still needed on the cellular titin isoform to determine its relationship to muscle titin. Biochemical data was presented to indicate that muscle titin will bind to muscle myosin II filaments but not to the minifilaments formed by nonmuscle myosin IIA molecules (Eilersten *et al.*, 1994). The latter investigators have also demonstrated that cellular titin will not bind muscle myosin thick filaments. Thus, if the two isoforms of titin are present in muscle cells, they may play an important role in preventing the two types of myosin IIs from being present in premyofibrils and mature myofibrils. Since both cellular and muscle titins are capable of binding alpha-actinin (Eilersten *et al.*, 1997; Turnacioglu *et al.*, 1996), one would expect to find both isoforms of titins in the Z-bodies and Z-bands. Specific antibodies are needed to determine if this is the case. If the particular isoforms of titin determine which types of myosin IIs are in the different stages of myofibrils, some other molecules are needed to regulate the binding properties of the N-terminal regions of the cellular titins to prevent them from being incorporated into the Z-bands of the mature myofibrils.

Myofibrillogenesis in living cardiomyocytes

Time-lapse observations of single myocytes in a culture offer a way of following the stepwise assembly of myofibrils under different conditions. Cardiomyocytes spread and locomote in culture (also in the forming heart) assembling myofibrils (Sanger *et al.*, 1989a); embryonic and neonatal cardiomyocytes disassemble their myofibrils in mitosis and then reassemble their myofibrils after cytokinesis (Chacko, 1973; Li *et al.*, 1997) and finally cardiomyocytes accumulate myofibrils during hypertrophy (van Bilsen and Chien, 1993). Myofibrillar proteins can be coupled to fluorescent dyes or Green/ Blue Fluorescent Proteins and introduced into the cardiomyocytes to determine the localization and dynamics of different sarcomeric proteins. Time-lapse observations of GFP-alpha-actinin in single cardiomyocytes undergoing myofibrillogenesis allowed several aspects of the model illustrated in Figure 3 to be tested, supporting the proposal that premyofibrils, with their mix of sarcomeric isoforms of alpha-actinin, troponins and tropomyosin and a nonmuscle isoform of myosin II, provide the blueprint for myofibril formation (Dabiri *et al.*, 1997). We demonstrated that the premyofibrils are indeed formed at the spreading edges of the cell, and that three to four premyofibrils fuse at the level of the Z-Bodies to form mature myofibrils (Fig. 4). These observations have also demonstrated that premyofibrils are not, as some have suggested, the breakdown products of mature myofibrils. The observations of these living transfected cardiomyocytes yielded images supporting the model illustrated in Figure 3. A number of questions remain: Are there alternate pathways for the assembly or reassembly of myofibrils during hypertrophy and following cell division? What are the functions of the filaments of nonmuscle myosin IIB in the premyofibrils? What are the roles of sarcomeric alpha-actinin in the formation of the premyofibrils? What are the roles of alpha-actinin and vinculin in the fusion of the premyofibrillar Z-bodies to form the Z-bands of the mature myofibrils? How does titin connect to the forming Z-bands? Do the three different domains of the Z-band region of titin bind some of the proteins involved in the costameric attachments of the Z-bands to the cell surfaces? Does titin play a role in binding the intermediate filaments to the Z-bands?

Cytokinesis of cardiomyocytes and the reformation of myofibrils
When nonmuscle cells enter prophase, the stress fibers in the living cell are induced to disassemble (Sanger *et al.*, 1989b; 1994a). In an unknown manner, the anaphase mitotic spindle induces the assembly of circular sarcomeric contractile fibers near the cell surfaces between the two sets of separating chromosomes. The contraction of this circular ring of microfilaments leads to the formation of two daughter cells. During the contraction of this cleavage furrow, the filaments are disassembled. At the end of cytokinesis, the stress fibers are observed to reform throughout the two nonmuscle cells (Sanger *et al.*, 1989b; 1994b). In several animal models of cardiomyocytes undergoing cell division, most or almost all of the myofibrils are disassembled at the start of mitosis and the myofibrils reassembled at the end of cytokinesis (newts: Kaneko *et al.*, 1984; 1- to 3-day-old rat neonatal cardiomyocytes: Clubb and Bishop, 1984; Li *et al.*, 1997; embryonic chick cardiomyocytes: Chacko, 1973; Sanger *et al.*, 1994b). Our immunofluorescence staining of these cells reveals that the cleav-

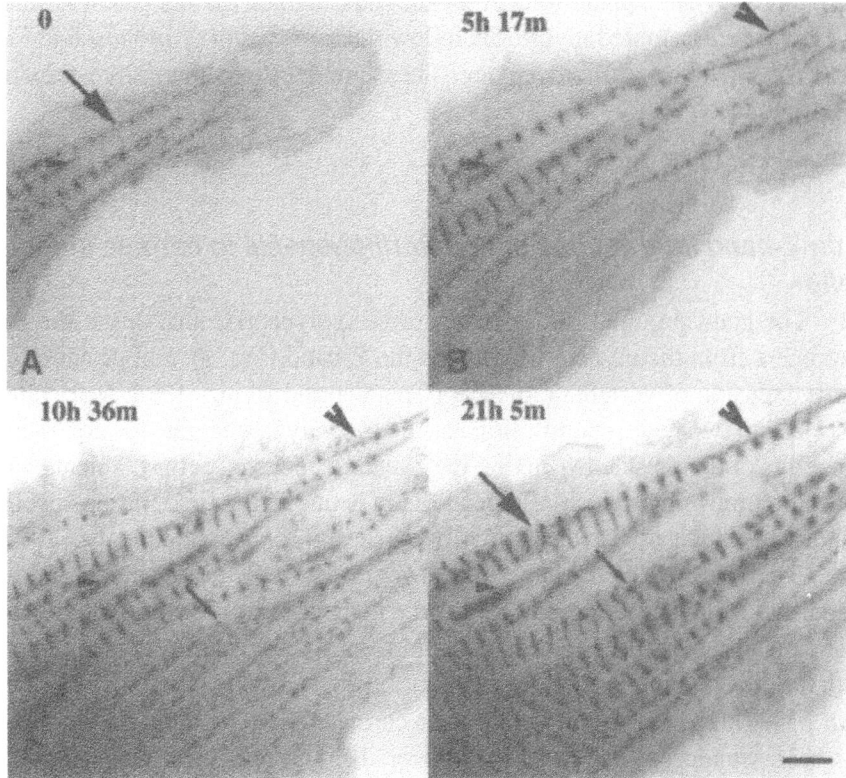

Figure 4. *Four images from a time-lapse sequence of a spreading embryonic chick cardiomyocyte transfected with GFP-alpha-actinin. This cell was contracting as it spread in culture. Premyofibrils appeared in the lower right spreading margin of cytoplasm, and at 21h 5m, myofibrils with solid Z-bands had formed. The image is shown in reverse contrast to render the Z-bands and Z-bodies more visible. The same prominent adhesion plaque is marked in each section of the Figure as a reference mark (A to D, horizontal arrowheads). Note the amount of spreading with respect to this adhesion plaque. During the spreading of the cell towards the lower part of the image, punctate Z-bodies appear and assemble into linear arrays that fuse laterally into Z-bands of myofibrils. A myofibril indicated by a large arrow in (A) doubles in length over the next 21 hours (D). This appears to happen by the lateral fusion of myofibrils (B to D), large oblique arrowheads). Scale = 5 microns. (Images reproduced with permission from the* Proceedings of the National Academy of Sciences, *from Dabiri et al., 1997;94:9493-9498.)*

age furrow is composed of a contractile structure identical to all the components of the premyofibrils (Sanger *et al.*, 1993; 1994 b). Therefore, the cleavage is lacking both muscle myosin II and muscle titin filaments. As in nonmuscle cells, the contractile elements of the muscle cells are reassembled at the end of cytokinesis. What happens to the titin and muscle thick filaments during mitosis? Do the reassembled myofibrils follow the same pathway proposed in Figure 3 for spreading cardiomyocytes? These are questions that future research will try to resolve.

Titin-Z-band interactions in myofibrillogenesis in cardiac muscle cells

The giant protein, titin, forms a single polypeptide that spans the half sarcomere from the myosin filament to the Z-band (Fig. 1), and provides the passive tension of vertebrate muscle (Granzier and Irving, 1995; Maruyama, 1994). With two reactive ends, titin has the capacity to integrate the contractile apparatus of the muscle by linking the Z-band-connected actin filaments with the force-producing myosin filaments. Titin is one of the first muscle-specific proteins to exhibit a striated pattern in myogenesis (Tokuyasu and Maher, 1987). The 43-nm spacing of the super-repeat region at the C-terminal end and the associated M-line proteins may play a role in the assembly of the thick myosin filaments (Trinick, 1994) and allow titin to pull these filaments into register with the developing Z-bands during myofibrillogenesis (Hill *et al.*, 1986; Rhee *et al.*, 1994). The entire length of human cardiac titin, over three million bases has been sequenced (Labeit and Kolmerer, 1995). We have sequenced the Z-band region of chicken cardiac titin (Turnacioglu *et al.*, 1996; 1997c), and demonstrated that this region binds alpha-actinin and less strongly, vinculin, Figure 5 (Turnacioglu *et al.*, 1996, 1997 a, b). There are a number of interesting motifs in the Z-band region of titin including small repeating units of 40 to 50 amino acids each, termed z-repeats (Gautel *et al.*, 1996). In the adult human and rabbit ventricle (Gautel *et al.*, 1996; Sorimachi *et al.*, 1997) there are seven z-repeats, and we have found that in 19 day-old chick ventricles, there are six z-repeats that, when coupled to GFP, can target to the Z-bands of living embryonic chick cardiomyocytes (Ayoob *et al.*, 2000). Several pieces of evidence indicate that the z-repeat region binds alpha-actinin (Turnacioglu *et al.*, 1996; Ohtsuka *et al.*, 1997; Young *et al.*, 1998), but other than the possible interaction with the 19 kDa protein telethonin (Gregorio *et al.*, 1998; Mues *et al.*, 1998) and vinculin, there is no evidence of other interactions of the Z-band region of titin with additional Z-bands proteins. Vinculin, talin and tensin are known to be in costameric associations with Z-bands (Prado *et al.*, 1983; Danowski *et al.*, 1992; Mondello *et al.*, 1996). Future experiments will be needed to determine what Z-band proteins interact with the different domains of the Z-band region of titin and when they become associated with the assembling myofibrils to form costameric attachment to the cell surface (Fig. 6).

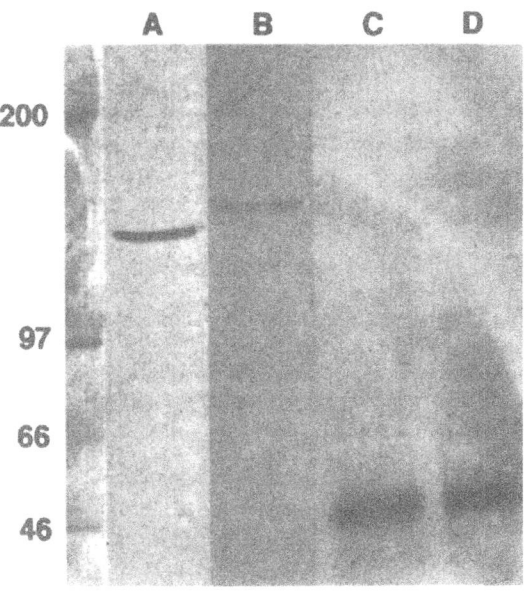

Figure 5. Evidence for the binding of a 46 kD fusion protein with alpha-actinin and vinculin. This 46 kD bacterial fusion protein contains four and half z-repeats of the six present in embryonic chick hearts. The 46 kD purified fusion protein was incubated with alpha-actinin or vinculin. McAb20 (anti-zeugmatin) and a secondary antibody conjugated to agarose beads were added to each of the mixtures and the resultant pellets were collected. The solubilized pellets were run on a 10 % SDS-PAGE gel, transferred to a nitrocellulose sheet and then incubated with either (A) an alpha-actinin antibody, (B) a vinculin antibody, or the McAb20 antibody (C, D). (A) and (C) derive from the alpha-actinin pellet and (B) and (D) from the vinculin pellet. Note the presence of alpha-actinin ((A) and vinculin (B) that have bound the titin fusion protein. (Reproduced by permission of Cell Motility and the Cytoskeleton).

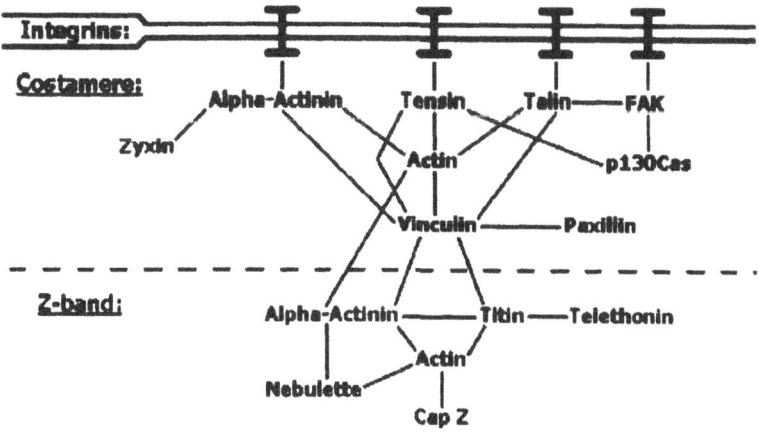

*Figure 6. Word chart indicating the possible roles of alpha-actinin and titin present in the Z-bands for the binding of costameric proteins such as vinculin. This diagram indicates **some** of the multiple binding proteins that are found in costameres (actin, vinculin, talin, tensin, paxillin, zyxin and alpha-actinin).*

References

Ayoob JC, Turnecioglu KK, Mittal B, Sanger JM, Sanger JW. Targeting of cardiac muscle titin fragments to the Z-bands and dense bodies of living muscle and non-muscle cells. *Cell Motil Cytoskel* 2000;45:67-82.

Bloor JW, Kiehart D. Nonsarcomeric myosin II, PS2 integrin and myogenesis. *Mol Biol Cell* 1998;9:146a.

Chacko S. DNA synthesis, mitosis and differentiation in cardiac myogenesis. *Develop Biol* 1973;35:1-18.

Chalfie M, Tu Y, Euskirchen G, Ward WW, Prasher DC. Green fluorescent protein as a marker for gene expression. *Science* 1994;263:802-805.

Claycomb WC, Palazzo MC. Culture of the terminally differentiated adult cardiac muscle cell: A light and scanning electron microscope study. *Dev Bio* 1980;80:466-482.

Clubb FJ, Bishop SP. Formation of binucleated myocardial cells in the neonatal rat. *Lab Invest* 1984;50:571-577.

Conrad AH, Clark WA, Conrad GW. Subcellular compartmentalization of myosin isoforms in embryonic chick ventricle myocytes during cytokinesis. *Cell Motil Cytoskel* 1991;19:189-206.

Dabiri GA, Turnacioglu KK, Sanger JM, Sanger JW. Myofibrillogenesis visualized in living embryonic cardiomyocytes. *Proc Natl Acad Sci USA* 1997;94:9493-9498.

Dabiri GA, Turnacioglu KK, Ayoob JC, Sanger JM, Sanger JW. Transfections of primary muscle cell cultures with plasmids coding for GFP linked to full length and truncated muscle proteins. In *Green Fluorescent Proteins, Methods in Cell Biology*, KF Sullivan and SA Kay, eds. New York: Academic Press 1999a;58:239-260.

Dabiri GA, Ayoob JC, Turnacioglu KK, Sanger JM, Sanger JW. Use of Green Fluorescent Proteins linked to cytoskeletal proteins to analyze myofibrillogenesis in living cells. *Methods in Enzymology (Optical Imaging and Green Fluorescent Proteins)*, PM Coon, ed. New York: Academic Press 1999b;302:171-186.

Danowski BA, Inimaka-Yoshida K, Sanger JM, Sanger JW. Costameres are sites of force transmission to the substratum in adult rat cardiomyocytes. *J Cell Biol* 1992;118:1411-1420.

Devlin RB, Emerson CP. Coordinate regulation of contractile protein synthesis during myoblast differentiation. *Cell* 1978;13:599-611.

Dome JS, Mittal B, Pochapin MB, Sanger JM, Sanger JW. Incorporation of fluorescently labeled actin and tropomyosin into muscle cells. *Cell Differentiation* 1988;23:37-52.

Eilersten KJ, Kazmierski ST, Keller CS. Cellular titin localization in stress fibers and interaction with myosin II filaments in vitro. *J Cell Biol* 1994;26:1201-1210.

Eilersten KJ, Kazmierski ST, Keller CS. Interaction of alpha-actinin with cellular titin. *Eur J Cell Biol* 1997;74: 361-354.

Epstein HF, Fischman DA. Molecular analysis of protein assembly in muscle development. *Science* 1991;251:1039-1044.

Ferrari MB, Ribbeck K, Hagler DJ, Spitzer NC. A calcium signaling cascade essential for myosin thick filament assembly in Xenopus myocytes. *J Cell Biol* 1998;141:1349-1356.

Gautel M, Goulding D, Bullard B, Weber K, Furst DO. The central Z-disk region of titin is assembled from a novel repeat in variable copy numbers. *J Cell Sci* 1996;109:2747-2754.

Granger BL, Lazarides E. Desmin and vimentin coexist at the periphery of the myofibril Z disc. *Cell* 1979;18:1059-1063.

Granzier HL, Irving TC. Passive tension in cardiac muscle: contribution of collagen, titin, microtubules and intermediate filaments. *Biophys J* 1995;68:1027-1044.

Gregorio CC, Trombitás K, Center T, Kolmerer B, Stier G, Kunke K, Sizuki K, Obermayr F, Herrmann B, Granzier H, Sorimachi H, Labeit S. The NH$_2$ terminus of titin spans the Z-disc: its interaction with a novel 19-kD ligand (T-cap) is required for sarcomeric integrity. *J Cell Biol* 1998;143:1013-1027.

Hill CS, Duran S, Lin Z, Weber K, Holtzer H. Titin and myosin, but not desmin, are linked during myofibrillogenesis in postmitotic mononucleated myoblasts. *J Cell Biol* 1986;103:2185-2196.

Holtzer H, Marshall JM, Finck H. An analysis of myogenesis by the use of fluorescent antimyosin. *J Biochem Biochem Cytol* 1957;3:705-724.

Kaneko H, Okamoto M, Goshima K. Structural changes of myofibrils during mitosis of newt embryonic myocardial cells in culture. *Exp Cell Res* 1984;153:483-498.

Labeit S, Kolmerer B. Titins: giant proteins in charge of muscle untrastructure and elasticity. *Science* 1995;270:293-296.

Li F, Wang X, Bunger PC, Gerdes AM. Formation of binucleated cardiac myocytes in rat heart: I. Role of actin-myosin contractile ring. *J Mol Cell Cardiol* 1997;29:1541-1551.

LoRusso SM, Rhee D, Sanger JM, Sanger JW. Premyofibrils in spreading adult cardiomyocytes in tissue culture: evidence for reexpression of the embryonic program for myofibrillogenesis in adult cells. *Cell Motil Cytoskel* 1997;37:363-377.

Maher PA, Cox GF, Singer SJ. Zeugmatin: a high molecular weight protein associated with Z lines in adult and early embryonic striated muscle. *J Cell Biol* 1985;101:1871-1883.

Maruyama K. Connectin/titin, giant elastic protein of muscle. *Biophys Chem* 1994;50:73-85.

Mondello MR, Bramanti P, Cutroneo G, Santoro G, DiMauro D, Anastasi G. Immunolocalization of the costameres in human skeletal muscle: confocal scanning laser microscope investigations. *Anat Rec* 1996;245:481-487.

Mues A, van der Ven PFM, Young P, Fürst DO, Gautel M. Two immunoglobulin-like domains of the Z-disc portion of titin interact in a conformational-dependent way with telethonin. *FEBS Letts* 1998;428:111-114.

Murakami N, Trenkner E, Elzinka M. Changes in expression of nonmuscle myosin heavy chain isoforms during muscle and nonmuscle tissue development. *Devel Biol* 1993;157:19-27.

Ohtsuka H, Yajima H, Maruyama K, Kimura S. The N-terminal z repeat of connectin/titin binds to the C-terminal region of alpha-actinin. *Biochem Biophys Res Comm* 1997;235:1-3.

Pardo JV, Siliciano JD, Craig SW. Vinculin is a component of an extensive myofibril-sarcolemma attachment regions in cardiac muscle fibers. *J Cell Biol* 1983;97:1081-1088.

Pollard TD. Electron microscopy of synthetic myosin filaments. Evidence for cross-bridge flexibility and copolymer formation. *J Cell Biol* 1975;67:93-104.

Rhee D, Sanger JM, Sanger JW. The premyofibril: evidence for its role in myofibrillogenesis *Cell Motil Cytoskel* 1994;28:1-24.

Sanger JW, Mittal B, Sanger JM. Analysis of myofibrillar structure and assembly using fluorescently labeled contractile proteins. *J Cell Biol* 1984a;98:825-833.

Sanger JW, Mittal B, Sanger JM. Formation of myofibrils in spreading chick cardiac myocytes *Cell Motil Cytoskel* 1984b;4:405-416.

Sanger JM, Mittal B, Pochapin MB, Sanger JW. Myofibrillogenesis in living cells microinjected with fluorescently labeled alpha-actinin. *J Cell Biol* 1986a;102:2053-2066.

Sanger JM, Mittal B, Pochapin MB, Sanger JW. Observations of microfilament bundles in living cells microinjected with fluorescently labeled alpha-actinin. *J Cell Sci* 1986b;Suppl 5:17-44.

Sanger JM, Mittal B, Meyer TW, Sanger JW. Use of fluorescent probes to study myofibrillogenesis. In *Cellular and Molecular Biology of Muscle Development,* L Kedes and F Stockdale, eds. New York: Alan R Liss Inc 1989a;221-235.

Sanger JM, Mittal B, Dome JS, Sanger JW. Analysis of cell division using fluorescently labeled actin and myosin in living PtK2 cells. *Cell Motil Cytoskel* 1989b;14:201-219.

Sanger JM, Dabiri G, Mittal B, Kowalski MA, Haddad JG, Sanger JW. Disruption of microfilament organization in living nonmuscle cells by microinjection of plasma vitamin D-binding protein or DNase I. *Proc Natl Acad Sci USA* 1990;87:5474-5478.

Sanger JM, Rhee D, Sanger JW. Cleavage furrows and premyofibrils in embryonic cardiomyocytes. *Mol Biol Cell* 1993;4:53a.

Sanger JM, Dome JS, Hock RS, Mittal B, Sanger JW, Occurrence of fibers and their association with talin in the cleavage furrows of PtK2 cells. *Cell Motil Cytoskel* 1994a;27:26-40.

Sanger JM, Rhee D, Leonard M, Price M, Zhukarev V, Shuman H, Sanger JW. Assembly of myofibrils and cleavage furrows in cardiomyocytes. *Mol Biol Cell* 1994b;5:165a.

Sanger JW, Zhukarev V, Sanger JM. Myofibrillogenesis and the surface attachments of Z-bands. *Mol Biol Cell* 1997;8:375a.

Schultheiss T, Lin ZX, Lu MH, Murray J, Fischman DA, Holtzer H. Differential distribution of myofibrillar proteins in cardiac nonstriated and striated myofibrils. *J Cell Biol* 1992;117:1023-1029.

Sorimachi H, Freiburg A, Kolmerer B, Ishiura S, Stier G, Gregario CC, Labeit D, Suzuki K, Labeit S. Tissue-specific expression and alpha-actinin binding properties of the Z-disc titin: Implications for the nature of vertebrate Z-discs. *J Mol Biol* 1997;207:688-695.

Tokuyasu KT, Maher PA. Immunocytochemical studies of cardiac myofibrillogenesis in early chick embryos. I. Presence of immunofluorescent titin spots in premyofibril stages. *J Cell Biol* 1987;105:2781-2793.

Trinick J. Titin and nebulin: protein rulers in muscle? *Trends Cell Biol* 1994;19:405-408.

Tullio AN, Accili D, Ferrans VJ, Yu Z-X, Takeda KA, Grinberg A, Westphal H, Preston YA, Adelstein RS. Nonmuscle myosin II-B is required for normal development of the mouse heart. *Proc Natl Acad Sci USA* 1997;94:12407-12412.

Turnacioglu K, Mittal B, Sanger JM, Sanger JW. Partial characterization and DNA sequence of zeugmatin. *Cell Motil Cytoske* 1996;34:108-121.

Turnacioglu KK, Mittal B, Dabiri GA, Sanger JM, Sanger JW. Zeugmatin is part of the Z-band region of titin. *Cell Struct Funct* 1997a;22:73-82.

Turnacioglu KK, Mittal B, Dabiri GA, Sanger JM, Sanger JW. An N-terminal fragment of titin coupled to Green Fluorescent Protein localizes to Z-bands in living muscle cells: Overexpression leads to myofibril disassembly. *Mol Biol Cell* 1997b;8:705-717.

Turnacioglu KK, Mittal B, Ayoob JC, Sanger JM, Sanger JW. Targeting of titin to the Z-bands of cardiomyocytes. *Mol Biol Cell* 1997c;8:357a.

Van Bilsen M, Chien KR. Growth and hypertrophy of the heart: towards an understanding of cardiac specific gene expression. *Cardiovasc Res* 1993;27:1140-1149.

Wang S, Greaser ML, Schulty E, Bulinski JC, Lin JJ, Lessard JL. Studies on cardiac myofibrillobenesis with antibodies to titin, actin, tropomyosin and myosin. *J Cell Biol* 1988;107:1075-1083.

Young PY, Ferguson C, Banuelos S, Gautel M. Molecular structure of the sarcomeric Z-disk: two types of titin interactions lead to an asymmetrical sorting of alpha-actinin. *EMBO J* 1998;17:1614-1624.

Discussion

Baatsen: This was very nice work. But I didn't hear any mention of the proteins dystrophin and desmin.

Sanger: I think all of these are important proteins. We have no idea how desmin attaches to the Z-band. That is one reason why I am using the yeast - two hybrid system. I just suspect that titin has to be close to interact. Desmin is interacting with something in the Z-band to stabilize and integrate the Z-band. But in the end, desmin is not an important protein for the assembly of the myofibrils. A number of labs have shown that you can dispense with it and get nice fibrils. It is more the elasticity that is interfered with. Dystrophin, that slide for the proteins (Figure 6), was done 2 days before we left. We had to decide which ones could we put in, and the ones we put in were the ones we had done some work on. I think it is much more complicated than I have shown. It is going to be exciting to sort out the different interactions of these proteins in the Z-band.

Keller: Joe, I think you know we predict that there will be a cellular and non-muscle type titin in the premyofibrils. Now that there are better antibodies available, we should definitely talk about getting together. The other issue is the segregation of the two myosins, the muscle and the nonmuscle myosin. We have some biochemical evidence in the test tube where cellular titin will not interact with muscle myosin and the reverse, muscle titin will not interact with cellular myosin. So I think that is potentially a way that the two systems are segregated. The question then is how do they get integrated into the myofibril? My own personal bias is that maybe that α-actinin has some considerable role in integrating the two systems. I guess the question should be whether muscle α-actinin can interact with cellular titin. We know that smooth muscle α-actinin interacts with cellular titin.

Sanger: Well, we know that the fragments of the muscle region of titin will interact with actinin. We like to find a cellular titin so that it can be sequenced and compared to muscle titin.

Pollack: From your data based on the temporal appearance of myosin and titin and also the spatial distribution, could you say something about the issue of the possibility of titin as a template for the laying down of myosin, i.e., the distinct length of myosin filaments?

Sanger: No, I do not really have any data on titin being a template. In invertebrate muscles, A-band lengths up to 20 microns exist and they certainly do not have titin molecules that stretch all the way from Z-bands to the middle of A-bands. John Aronson in 1961 (*J Biophys Biochem Cytol* 1961;11:147-156) studied living mite muscles on the polarizing microscope. He could see that their A-bands increased in length over the period of observation. In one myofibril, Aronson (1961) followed the growth of the same A-band as it grew from 1.2 to 4.5 microns in length over 25 hours. The sarcomere grew from 2.2 to 10 microns over the same time period. How does titin regulate the lengths of A-bands in this case? I don't know.

Greaser: We have done a bit of the same stuff about 10-15 years ago. We observed that in fact the thick filament was 1.6 microns in length. We didn't see a growing sort of thing. And whenever we saw even single sarcomeres forming on these bundles of filaments, they were always 1.6 microns. I think the template idea is reasonable when we are talking about vertebrate muscle and we are not talking about these other systems. Clearly titin cannot make horseshoe crab thick filaments and so on.

Sanger: I've never said that we see the A-bands of mature vertebrate myofibrils change their length. The only thing we have seen is where they go from an overlapping pattern (nascent myofibril) to where they sort out into A-bands (mature myofibril). I would like to believe that cross-striated muscle cells assemble their myofibrils using similar proteins. If invertebrates can assemble A-bands without titin, why not vertebrates?

Greaser: The other thing is that if you do an actin stain on these premyofibrils, it is fairly continuous and it seems that it takes a while to sort out the orientation. Longer than it does for myosin.

Sanger: No doubt, and that is why I made a point that when we are looking at myofibril forming, we distinguish the three different stages of myofibril formation by the type of myosins that are present, not by the phalloidin staining. There were some groups who asked me at prior meetings whether I am sure that I could see short spacings. I kept thinking, well, yes. I use α-actinin probes, which one should use if you want to see sarcomeres or repeating units as opposed to using actin probes, because it is clear that actin filaments can overlap in sarcomeres. Since cardiac mature myofibrils have overlapping thin filaments, does that indicate that they are not sarcomeric? I think you have to look always at the furthest point in the sarcomere, namely the Z-band. That is why we have always concentrated on that. I think actin-staining patterns are misleading. John Cooper proposed that Cap-Z came in and landed on titin and then organized all the actin into non-overlapping units. There is absolutely no evidence that Cap-Z interacts with titin. Cap-Z in our staining studies of cardiac muscles is actually a late-appearing protein, i.e., present in mature myofibrils. We never see it in premyofibrils.

GENERAL SEMINAR DISCUSSION
(Led by Siegfried Labeit)

Labeit: I am still intrigued by this T2, T1 - T2 business. It is really 100% certain that T2 is a proteolytic breakdown product? Couldn't it be that four copies of T1-like titin molecules span a half sarcomere, whereas in addition, two T2-like titins expressed by an internal titin gene promoter, which span from the N2 line region up to the M-line? So my question is whether T2 titin is a genuine sarcomeric structural protein of the myofibril and not just a breakdown product.

Greaser: At Wisconsin we "muck around" a little bit with post mortem muscle, and clearly if you are very careful initially to prepare the titin you only have T1 present. You have to be very careful, freeze the muscle, pulverize it, dissolve it in hot extraction buffer and use inhibitors and so on. But the picture that I showed with the 9D10 and the 514 antibodies in postmortem muscle clearly indicates that all the titin molecules are in T2 under those conditions. So I don't think that T2 is a different isoform, even though there are different isoforms. I believe that Henk will probably talk a little bit more about different T1 isoforms.

Granzier: Yes.

Greaser: Just a question I'd like to come back to, and that is whether or not the segment between the PEVK and the edge of the A-band is stiff or collapsible. If the end filaments are really 6 titin molecules that are all lined side by side with each other, you'd expect it to be very difficult to collapse that structure. On the other hand, the antibody studies with our 9D10 antibody show that the epitope clearly collapses back closer to the edge of the A-band than the expected 1/10 micron. John, would you like to comment on that?

Trinick: I'm not sure I can really answer your question, but I have sort of a counter question that I'd like to ask the people doing antibodies, and the question I asked Wolfgang in Baltimore a couple of months ago was that I think a lot of the antibody work is done with divalent antibodies. One of the things that struck me was that using divalent molecule to study elasticity doesn't have to have a problem, but I could imagine that it could have a problem, if you could imagine that a divalent antibody could cross link a structure somehow at either, say, inside the PEVK region or between filaments. So the question I would like to ask the people doing antibody work is whether they have ever looked with monovalent antibodies to see that the extensibility they see in different parts of the sarcomere are the same.

Greaser: I'll start off. We've not done it with monovalent antibodies, but there is a certain amount of position uncertainty with divalent antibodies because they could interact with one or the other arms to widen the visible binding. And that's a bigger problem with the 9D10, which is an IgM by the way. So some of the widths of the 9D10 we thought originally were just due to the fact that it was an IgM and you could catch it flipped one way or the other. But it certainly doesn't explain the one-micron-wide zone like Charles Trombitás finds with human soleus.

Trinick: So I am just wondering if the antibodies determine the structure.

Greaser: Well, my counter argument to that point is that we used gluteraldehyde cross-linked myofibrils before we did the antibody labeling. And if you've ever looked at an SDS gel of a gluteraldehyde cross-linked myofibril, you don't find anything that comes out of the well. Everything is pretty well linked together. That doesn't guarantee that everything is positionally linked, but my inclination would be that it is pretty stable.

Linke: I don't see that there is a problem with the extensibility studies because as far as I understood, both you, Charles and Marion, and me, we first stretch the muscle and then label with the antibodies, so you don't really expect an effect on the extensibility. However, if we first label the myofibrils with antibodies and then stretch them, and compare with unlabeled myofibrils, we see a change in the stiffness, which probably reflects the cross-linking impact of the antibody.

Trinick: Well you're adding a cross linking agent. So my point is that I am not saying that it has to be a problem. But I just think that it would be quite good to keep an eye open because you are adding cross-linking agents, so if you are going to study extensibility with a cross-linking agent, then you might want to be a bit careful.

Labeit: Maybe I can just briefly comment on this argument of the antibodies. I agree it would make sense to do studies with monovalent antibodies; also, it would be of lower molecular weight and better penetrate into specimens.

Bullard: I just want to give an example of the effect of divalent antibodies on the structure of the thin filament. If you take divalent antibodies to troponin and label insect flight muscle, you get transverse bands. But if you take an Fab fragment, you get diagonal bands, which are expected from the symmetry of the actin filament.

Greaser: That is true, but coming back to the titin situation, I don't think it is going to perturb the epitope positions that much.

Bullard: That is true, as long as all the titins are in register.

Labeit: If I may summarize, we need some monovalent antibodies to at least check some conclusions. The other thing I was wondering about is whether the PEVK is forming folded structures or is permanently unfolded. How does the P-pack discovery factor in?

Greaser: One thing I forgot to mention was that, if we look at some of the splice patterns of the cardiac N2A, there is splice right in the middle of a domain. And the spliced sections complete a 28 amino acid repeat sequence. So there seems to be some importance to these P-pack repeats, and I don't know exactly what it is. My feeling also is that there is some sort of fold to these things, and it isn't entirely just a pile of random coils.

Baatsen: This question is not really directly related to titin.

Greaser: We won't take any hostile questions.

Baatsen: It amazes me that you had 4-fold symmetry in the Z-line and then 6-fold in the M-line. It seems to me that kind of hampers contraction at the shorter length. It seems to create kind of torsion in the thin filament or something like that.

Trombitás: Regarding the PEVK structure, we showed in the last Biophysical meeting two additional epitopes inside the soleus PEVK. And it turned out that the three PEVK subregions did not behave the same way.

Linke: I would like to comment also on the finding, Charles, that you just mentioned. Because this goes along with data we published last year (Linke *et al.*, *PNAS* 95, 8052). There seems to be some additional compliance in the PEVK region when you come to high stretches. It seems that at low stretch you may be able to explain the elastic properties of the PEVK region by an entropic spring mechanism, but at much higher stretches there seems to be additional compliance, which may be brought about by some opening of structures. So that would go along with what Charles just said.

Labeit: We need better structural info for PEVK.

Greaser: There is a large segment at the beginning of the PEVK that doesn't have the P-packs. Virtually all the rest of the molecule, except the latter parts of the N2B, have either P-pack or this polyglutamic acid sort of repeats. So there's, I'd say, nearly 80% of the soleus PEVK either in the P-pack or in the polyglutamic repeats; that's rather remarkable.

Vigoreaux:This is not a question or an answer; this is just an observation. I just want to go back to what you said earlier about the symmetry paradox. In insect flight muscle this is non-existent because the Z-band has the same symmetry compared to the M-line. And if you compare the sarcomere structure of insects and vertebrates, the I-band is much, much shorter in insect flight muscle, and as far as I know there is no nebulin in the insect flight muscle. So I think perhaps part of the answer to the paradox is in the presence or absence of nebulin.

Sugi: Extensibility and stiffness should be strictly defined, because extensibility could result from rupture.

Greaser: That is certainly a good point, and we have to realize when we stretch things a long ways that we can do damage, and some of the things I showed today were certainly damaged materials. However, one could still use such experiments to try to interpret what is happening, what is bound to what and so on. Most of the extensibility type experiments have been carried out at sarcomere lengths up to maybe 2.3 microns in cardiac muscle. You can stretch them out that far and let them come back and they behave the same way again. So that would indicate that we are working in a range where there is not extensive damage. It is a good point that we all need to be aware of when we are working in this area.

Bullard: Does the PEVK bind to actin?

Greaser: It is a dilemma. Certainly if PEVK binds very tightly to actin and remains anchored there, we are in rigor. It may sort of function as a damping system at high velocities.

MECHANICAL MANIPULATION OF SINGLE TITIN MOLECULES WITH LASER TWEEZERS

Miklós S.Z. Kellermayer[1], Steven Smith[2], Carlos Bustamante[2], and Henk L. Granzier[3]

[1]Dept. Biophysics, Pécs University Medical School, Pécs, Hungary;
[2]Dept. Physics, University of California, Berkeley, CA;
[3]Dept. VCAPP, Washington State University, Pullman, WA

Abstract: Titin (also known as connectin) is a giant filamentous polypeptide of multi-domain construction spanning between the Z- and M-lines of the vertebrate muscle sarcomere. The molecule is significant in maintaining sarcomeric structural integrity and generating passive muscle force via its elastic properties. Here we summarize our efforts to characterize titin's elastic properties by manipulating single molecules with force-measuring laser tweezers. The titin molecule can be described as an entropic spring in which domain unfolding occurs at high forces during stretch and refolding at low forces during release. Statistical analysis of a large number (>500) of stretch-release experiments and comparison of experimental data with the predictions of the wormlike chain theory permit the estimation of unfolded titin's mean persistence length as 16.86 Å (±0.11 SD). The slow rates of unfolding and refolding compared with the rates of stretch and release, respectively, result in a state of non-equilibrium and the display of force hysteresis. Folding kinetics as the source of non-equilibrium is directly demonstrated here by the abolishment of force hysteresis in the presence of chemical denaturant. Experimental observations were well simulated by superimposing a simple domain folding kinetics model on the wormlike chain behavior of titin and considering the characteristics of the compliant laser trap. The original video presentation of this paper may be viewed on the web at http://www.pote.hu/mm/prezentacio/mkpres/mkpres.htm.

Elastic Filaments of the Cell, edited by Granzier and Pollack
Kluwer Academic/Plenum Publishers, 2000

Introduction

Upon stretching a specimen of relaxed, non-activated striated muscle—be it a myofibril, myocyte, muscle fiber or fiber bundle—the specimen yields while force (passive force) is generated, signalling that muscle is elastic. The elastic properties of striated muscle have recently been found to be physiologically important for a variety of reasons: during muscle contraction, passive force limits sarcomere-length inhomogeneity along the muscle cell (Granzier *et al.*, 1991), and limits A-band asymmetry within the sarcomere (Horowits and Podolsky, 1987). Many observations have indicated that in the generation of passive force a significant role is played by a unique protein called titin. Titin is a ~3.5-million-dalton polypeptide that constitutes about 10% of the total muscle protein mass (Labeit and Kolmerer, 1995; Trinick, 1996; Wang, 1996; Maruyama, 1997; Gregorio *et al.*, 1999). This giant molecule spans the half sarcomere, from the Z-line to the M-line. Titin is anchored to the Z-line and to the thick filaments of the A-band (via strong myosin-binding property). The I-band segment of the molecule is constructed of serially-linked tandem immunoglobulin (Ig) domains interrupted by a proline (P)-, glutamate (E)-, valine (V)- and lysine (K)-rich segment called PEVK segment and other unique sequences (Labeit and Kolmerer, 1995). Upon stretch of the sarcomere, passive tension is generated by the extension of the I-band segment of titin (Horowits *et al.*, 1986; Higuchi *et al.*, 1992; Granzier and Irving, 1995). The A-band segment of titin on the other hand is composed of super-repeats of Ig and fibronectin (FN) domains (Labeit and Kolmerer, 1995). Titin's A-band segment does not participate in the generation of passive force under physiological conditions. Rather, this portion of the molecule seems to provide a structural scaffold for the thick filament (Whiting *et al.*, 1989; Trinick, 1991).

The elastic properties of titin have been demonstrated indirectly by immunoelectron microscopic (Fürst *et al.*, 1988; Itoh *et al.*, 1988; Trombitás and Pollack, 1993; Trombitás *et al.*, 1995) and mechanical experiments on muscle cells (Horowits *et al.*, 1986; Granzier and Irving, 1995). The molecular mechanisms of titin's extensibility, however, have long been elusive. Early theoretical works hypothesized that titin's elasticity may be attributed to the unfolding and refolding of its globular domains (Erickson, 1994). The discovery of titin's PEVK segment with prospective random coil properties indicated that a highly compliant spring might be present in the molecule that would allow extension even without domain unfolding (Labeit and Kolmerer, 1995). To sort out possible mechanisms one may ideally take a single titin molecule, hold it by the ends and stretch it while measuring the extension and the force generated. Indeed, such recent single-molecule mechanical studies (Kellermayer *et al.*, 1997; 1998; Rief *et al.*, 1997; 1998b; Tskhovrebova *et al.*, 1997) begin to shed light on the physical chemistry behind titin's elasticity.

In the present paper we summarize our recent efforts to characterize the elastic properties of the titin molecule with laser-tweezer technology. The molecule behaves as a wormlike chain in which the external force modulates the rates of domain unfolding and refolding.

Materials and Methods

Preparation of titin and the microscopic beads

Titin was prepared from various muscles (rabbit *longissimus dorsi*, rabbit heart, rabbit *soleus*, rat heart) by methods described in our earlier works (Kellermayer and Granzier, 1996; 1996b). To mechanically manipulate titin, each of its ends was attached to a different microscopic bead. The Z-line end of titin was bound to a latex bead coated with T12 anti-titin antibody. The M-line end of the molecule was attached to a silica bead coated with photo-reactive cross-linkers. The microscopic beads were prepared as described previously (Kellermayer *et al.*, 1997). Briefly, 3.18-μm carboxylated latex beads (SpheroTech) were coated with protein-A, then with T12. The antibody was then covalently cross-linked to protein-A with DMP (Dimethylpimelimidate-HCl, Pierce). Following blocking of the non-specific binding sites on the beads with 1% BSA and 0.2% Tween-20 in assay buffer (25 mM imidazole-HCl, 25 mM KCl, 4 mM $MgCl_2$, 1 mM EGTA, 1 mM DTT, pH 7.4), the beads were mixed and incubated with titin. The M-line-end bead was prepared by coating 2-μm amino-propyl silica beads (Bangs Laboratories) with the photoreactive cross-linker sulfo-SANPAH (Pierce). For attaching the M-line-end bead to titin, the two different beads were pushed and rubbed against each other in the laser tweezer setup (see below).

Mechanical manipulation of titin

Force measurement and the manipulation of one of the beads attached to titin were carried out with an optical trap. We used either a dual-beam, coaxial, counter-propagating laser tweezer system (Smith *et al.*, 1996) or a single-beam optical trap (Kellermayer *et al.*, 1998). In the dual-beam system two 200-mW diode lasers (830 nm, SDL) were used, giving a total optical power of ~100 mW at the specimen plane. Force in this system was obtained by measuring the change in the light momentum as the beam left the trap (Smith *et al.*, 1996). Using the dual-beam system allowed us to stretch titin with high forces (up to ~160 pN) with relatively low optical power. In the single-beam system a high-power laser beam (Nd:YAG, 1046 nm, 4W, Lee Laser) was directed in the sample chamber, where the final optical power was ~1.5 W. With the high optical power we were able to stretch titin with forces up to 400 pN. Force in the single-beam system was measured by following the bead displacement from its equilibrium position in the trap center (Kellermayer *et al.*, 1998). For force calibration we imposed preadjusted viscous drag forces on the trapped bead based on Stoke's equation (Svoboda and Block, 1994; Kellermayer *et al.*, 1998). For holding the microscopic bead attached at the other end of titin we used a glass micropipette (O.D. ~40 μm, tip diameter ~1 μm) mounted in the flow chamber (Smith *et al.*, 1996). The micropipette position was adjusted along the XYZ axes by moving the entire flow chamber with a piezoelectric stage (Thorlabs). With the M-line-end bead held by the moveable micropipette and the Z-line specific bead trapped in the force-measuring laser tweezers, titin was stretched by moving the pipette away from the trap. The titin tether was stretched at a constant rate to a maximum predetermined force, while continuously monitoring the force generated in the molecule. When the maximum force was reached, the process was reversed to obtain the release half-cycle. The length of the tethered molecule was calculated by measuring the distance between the beads. From the length and force data, force-extension curves were plotted.

Calculations and simulation

The raw force-extension data of titin were compared with the predictions of the freely-jointed (FJC) and wormlike chain (WLC) models. Theoretical FJC force-extension curves were generated using the equation

$$\frac{z}{L} = \coth\frac{Fb}{k_BT} - \frac{k_BT}{Fb} \tag{1}$$

where z is the extension (end-to-end distance) and L is the contour length of the chain, F is force, b is the Kuhn segment length, k_B is Boltzmann's constant, and T is absolute temperature (Smith *et al.*, 1992). Theoretical WLC force-extension curves were generated using the equation

$$\frac{FA}{k_BT} = \frac{z}{L} + \frac{1}{4(1-z/L)^2} - \frac{1}{4}, \tag{2}$$

where A is persistence length (measure of bending rigidity) of the chain (Bustamante *et al.*, 1994; Marko and Siggia, 1995). Contour length of the the titin tether was obtained using the WLC equation (2) as the length-axis intercept of the extrapolated $F^{-1/2}$-extension curve at high extension (release data) and the apparent persistence length (A_{app}) from the force-axis intercept $\left(2\sqrt{A_{app}/k_BT}\right)$ of the same curve (Bustamante *et al.*, 1994; Marko and Siggia, 1995; Kellermayer *et al.*, 1997). The persistence length of the single titin molecule (A_0) was obtained from the statistical distribution of A_{app} derived from many experiments. A_0 was identified as the length corresponding to the peak at the longest A_{app}. Considering that for multi-molecular tethers force at a given fractional extension is the integer multiple of the single-molecule force, using A_0 the number of molecules in any titin tether can be calculated from A_{app} derived from the experimental force-extension curve as

$$n = \frac{A_0}{A_{app}}.$$ (3)

For a Monte-Carlo simulation of titin's behavior during stretch and release the kinetics of globular domain unfolding/refolding were superimposed on the WLC behavior of the polypeptide chain and on the behavior of the bead in the optical trap of given stiffness (Kellermayer *et al.*, 1997)). The rates (k) of unfolding/refolding were calculated according to

$$k = \omega_0 e^{-\left(E_a - F\Delta x\right)/k_B T}$$ (4)

where ω_0 is the attempt frequency (10^8 s^{-1}), E_a is the energy barrier associated with the transition state (activation energy of the rate-limiting unfolding/refolding kinetic intermediate), F is force acting along the (unfolding or refolding) reaction coordinate, Δx is a characteristic distance (along the reaction coordinate) associated with the transition state (Bell, 1978; Evans and Ritchie, 1997) (or, width of the unfolding/refolding potential (Rief *et al.*, 1998a)), k_B is Boltzmann's constant, and T is absolute temperature. In the simulated experiment, for each step of micropipette displacement, extension and force were first calculated based on the WLC equation (2) and on iterative adjustments so that a balance of forces (chain opposing trap) on the trapped bead was achieved. Then, for the obtained force, the number of unfolded and folded domains were calculated using equation (4), with $\Delta x_{unfolding}$=0.3 nm and $\Delta x_{folding}$=8 nm ($E_{aunfold}$=28 pNnm; E_{afold}=5 pNnm). The simulated molecule contained 70 globular domains. The simulation results shown here were generated on a Power Macintosh G3 using object Pascal (CodeWarrior, Metrowerks). Simulation source codes are available for the PC (for TurboPascal) at http://alice.berkeley.edu/~cjblab and for the Mac at http://bfiz-molbiof.pote.hu.

Results and Discussion

Raw force-extension data of titin tethers

In the raw force-extension curves shown in Figure 1 for three different stretch-release experiments, several distinctive features of our data may be identified. First, force increases non-linearly upon stretch, indicating a departure from Hookean elasticity. Second, there is a considerable variation in the length at which a common maximum force is reached, probably due to a variation in the location of the bead attachment points within the titin tethers. Third, the tethers are extended to lengths greater than the ~1 μm length of the native, folded (but straightened) titin molecule, indicating that stretching may be accompanied by unfolding in the molecule. Fourth, normalizing the curves to the same length scale revealed that the force generated for a given fractional exten-

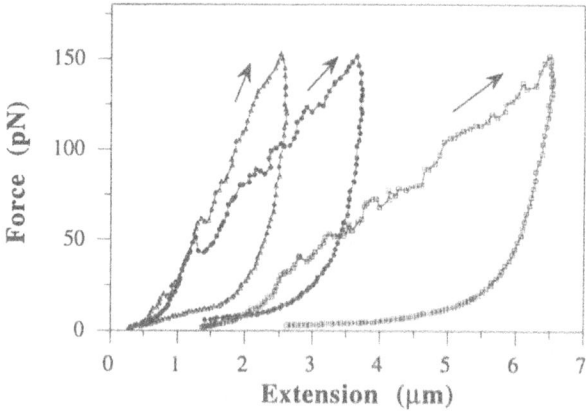

Figure 1. *Raw force-extension data obtained for rat cardiac titin tethers in three different stretch-release cycles. Stretch-release rate 50 nm/s. Arrows indicate direction of data acquisition (stretch).*

sion also varies from experiment to experiment, probably reflecting varying numbers of titin molecules within the tethers. And finally, all curves show hysteresis.

Comparison of data with predictions of entropic polymer models

The non-linear force response can be explained by the entropic polymer nature of titin. An entropic polymer shortens because of thermally driven bending movements that increase the chain's conformational entropy. To reduce entropy, the chain must be stretched with external force. We compared titin's force-extension curves with the predictions of two entropic elasticity models, the freely jointed chain (FJC) (Kuhn and Grün, 1942) and the wormlike chain (WLC) (Kratky and Porod, 1949) models. The FJC model describes the chain as a series of statistically independent, stiff segments (Kuhn segments). The WLC model on the other hand describes the chain as a deformable rod of a given persistence length, which is a measure of the chain's bending rigidity. As seen in Figure 2, the FJC model fails to describe the data while the wormlike chain model fits the stretch data (Fig. 2a) at low to moderate forces and the release data at moderate to high forces (Fig. 2b). Beyond these force regimes, the behavior of the molecule systematically deviates from that of the WLC, indicating the onset of structural transitions, unfolding and refolding, within the molecule.

By comparing the release data with the predictions of the WLC theory, the apparent persistence length (A_{app}) and the contour length (L) of the titin

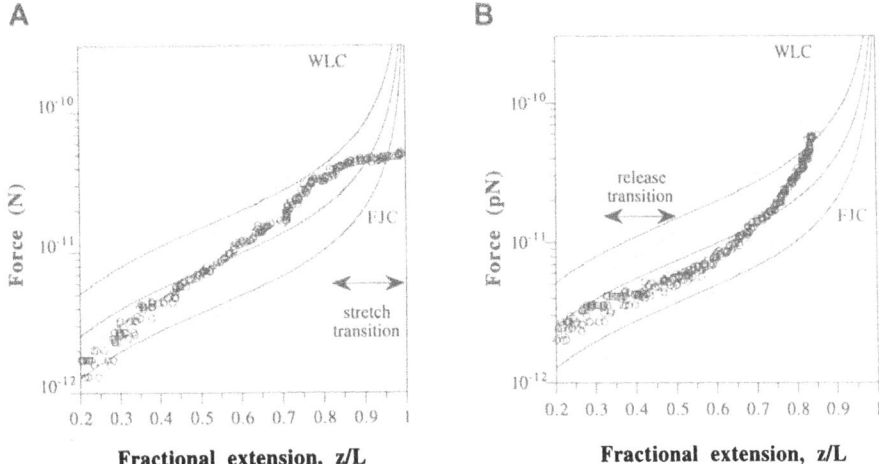

Figure 2. *Comparison of the stretch data (a) and the release data (b) with theoretical freely-jointed chain (FJC) and wormlike chain (WLC) curves. Dotted lines, WLC theoretical curves and continuous lines, FJC curves for chains of different stiffness, respectively. Stretch and release force transitions, where the data deviate from WLC behavior, are indicated with arrows.*

tethers were derived (Fig. 3). The A_{app} histogram (Fig. 3a) displays a multimodal distribution with peaks corresponding to the number of titin molecules within the tethers. Accordingly, the persistence length of the single (partially unfolded, see below) titin molecule (A_0), based on 517 stretch-release experiments is 16.86 Å (±0.11 SD). From the A_{app} histogram, the distribution of the number of molecules was obtained (Fig. 3b). The histogram of the number of molecules deviates from the expected Poisson distribution for several possible reasons: First, failures of attempted molecule attachments (which would correspond to 0 molecules) are not represented. Second, titin attachment to the M-line-end bead is not entirely a stochastic process, because pushing and rubbing the beads together places a bias toward encountering more molecules than would be encountered via diffusion. Third, A_{app} for a given number of molecules fluctuates depending on the ratio of the flexible unfolded (A_0=16.86 Å) and the stiff folded (A_0=150 Å (Higuchi *et al.*, 1993)) segments along the titin tether. Finally, multimolecular complexes of titin may be present.

The contour-length histogram obtained from the release data (Fig. 3c) has a peak at 4 μm. Considering that the release data correspond to unfolded titin (partial or complete unfolding, see below), and that the ~10-μm contour length of the completely unfolded titin (based on primary structure, (Labeit and Kolmerer, 1995)) was rarely observed, the factors influencing the contour-length distribution may be summarized in the following: (a) The M-line-end bead may bind somewhere along the length of the molecule either because of the bias caused by the titin-mounting procedure or because there is a high-affinity

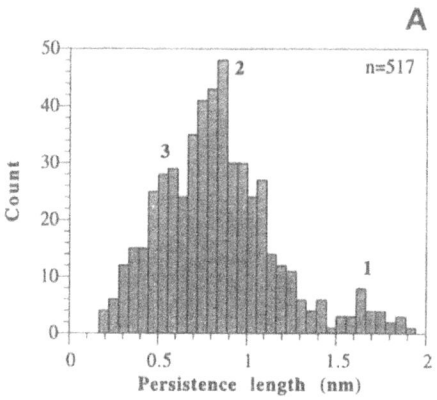

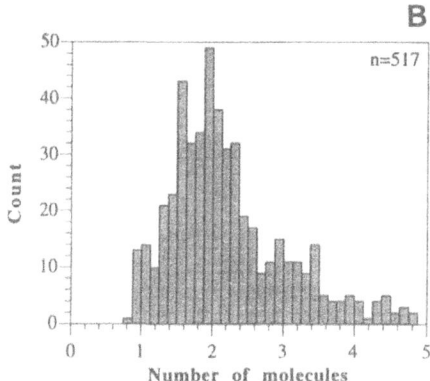

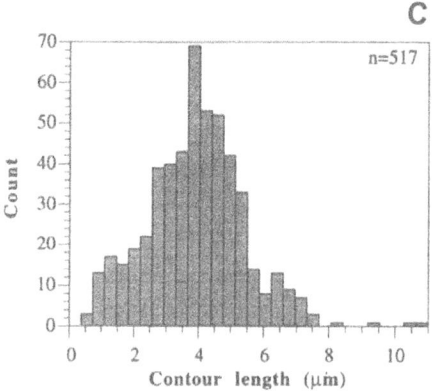

Figure 3. Histograms of the apparent pesistence length (a), the number of molecules within a tether (b) and the contour length (c) obtained for 517 different stretch-release cycles.

binding site (e.g., clusters of charged residues) along titin. Otherwise, if the statistical segments along the molecule had identical affinities for the M-line-end bead, and if the attachment of this bead was limited by diffusion only (as opposed to forceful pushing and rubbing), then the M-line tip of titin would have attached with the greatest probability, due to its highest degree of freedom. (b) Titin remains partially folded, and the contour length of the completely unfolded polypeptide is rarely seen. Incomplete unfolding may be explained by the limited force of the laser-tweezer method and by the presence of multi-molecular tethers that otherwise require proportionally greater forces for pulling.

Model of force-driven molecular changes in the titin molecule

The application of the WLC model allowed us to normalize our data to the single molecule. Accordingly, the observations can be reconciled with the following single-molecule model (Fig. 4): (a) At the beginning of stretch (between points A and B), part of the molecule is already unfolded. The molecule behaves as a wormlike chain whose properties are dominated by this pre-unfolded fraction. The pre-unfolded fraction may contain the PEVK domain (Labeit and Kolmerer, 1995), other unique sequences (Labeit and Kolmerer, 1995), or some unfolded globular domains as well. (b) As titin is stretched to high force (between 20 and 30 pN), transitions begin to occur in the molecule that convert part of the folded fraction into unfolded protein. The process con-

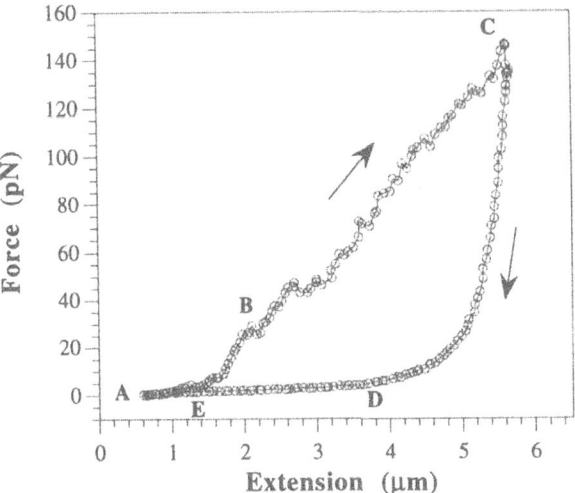

Figure 4. *Consensus force-extension curve for the single titin molecule.*

tinues until the maximum force allowed by the instrument is reached (at point C). (c) Upon releasing the unfolded titin (between points C and D), it does not refold initially, but behaves as a wormlike chain with the properties of an unfolded polypeptide. Thus, it has a persistence length between 15 and 20 Å. (d) Only as the molecule is allowed to shorten to about one half of its release contour length does refolding begin to take place, seen as a low-force transition between 2-3 pN (between points D and E).

In this model, the force-extension curve displays hysteresis because the rate at which we stretch or release the molecule exceeds the rate of unfolding and refolding at equilibrium at the given extension. Thus, application of the external force displaces the domain folding/unfolding reaction from the state of equilibrium. If the stretch and release protocols were carried out at considerably slower (*ad limitum* infinitely slow) rates, then the stretch and release curves would merge into a single, equilibrium unfolding/refolding force curve.

Test of the model I: complete mechanical denaturation

In the above model the force transitions seen during stretch are associated with domain unfolding. The transitions take place until the release begins or until there are no folded domains available. To denature the entire ensemble of domains present in the molecule, titin was stretched with high forces (Fig. 5) (Kellermayer *et al.*, 1998). At around 150 pN the force transitions stopped (point C), and the force began to rise sharply, following a WLC curve. Upon release

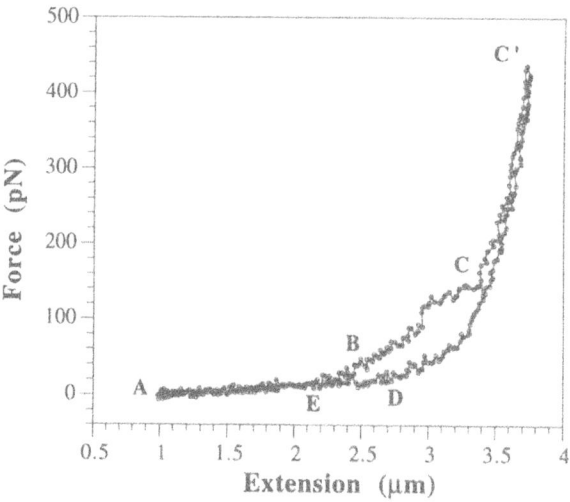

Figure 5. *Force-extension curve of a titin tether stretched with high forces.*

(between points C′ and C), force dropped rapidly, initially coinciding with the stretch curve. Between points C and C' the stretch and release curves are indistinguishable from each other, indicating that the conformational states of titin are in equilibrium throughout this region of the force-extension curve.

Test of the model II: partial release-restretch

According to the model, domain refolding does not begin until low forces are reached during the mechanical release of titin. Therefore, the initial phase of the release curve (between points C and D) is predicted to be reversible. Indeed, when restretching titin following partial release, but staying within the reversible force range (between points C and D), the restretch curve retraces the partial release curve (Fig. 6a). On the other hand, when partially entering the low-force transition (between points D and E), which is hypothesized to be associated with domain refolding, a partial recovery of the force hysteresis is expected to occur upon restretch. Such a partial release into the low-force range followed by restretch of titin indeed recovered part of the hysteresis (Fig. 6b).

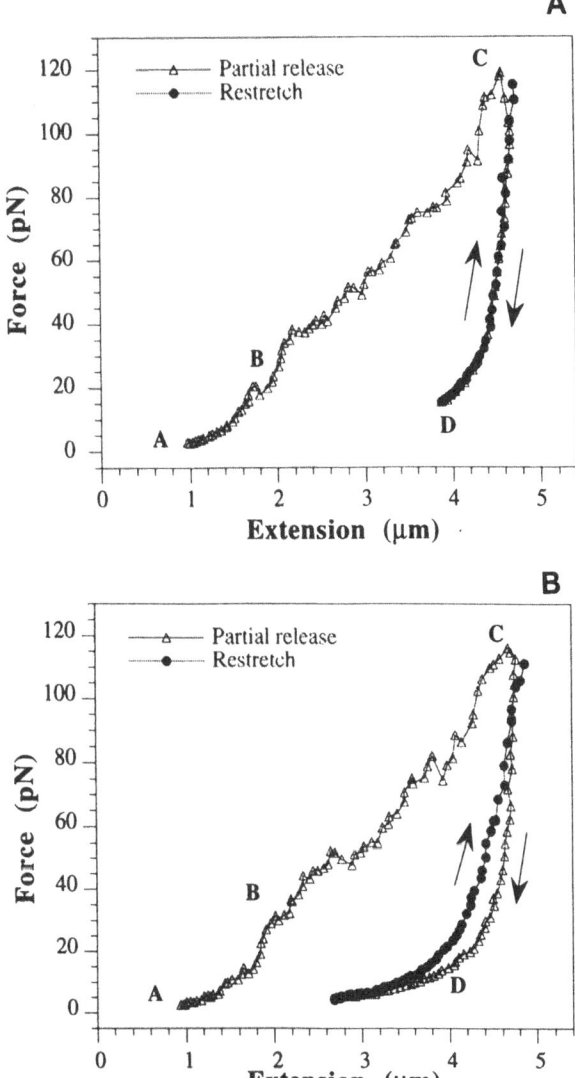

Figure 6. *Partial mechanical release followed by restretch. Titin was partially released to forces either in the reversible regime (between points C and D, see text) (a) or in the refolding transition (between points D and E) (b). Arrows indicate direction of data acquisition (stretch or release).*

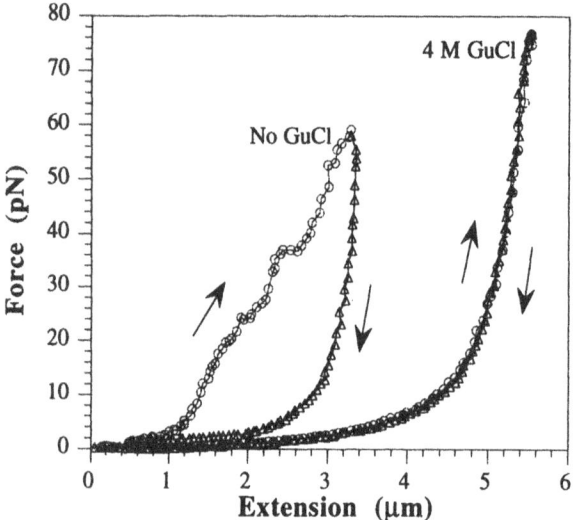

Figure 7. Effect of guanidine-HCl on force hysteresis. Shown are the force-extension curves prior to and following the addition of the denaturant. Arrows indicate direction of data acquisition (stretch or release).

Test of the model III: stretching in the presence of chemical denaturant

To abolish native domain structure with an independent method, titin was stretched in the presence of the chemical denaturant guanidine-HCl. Figure 7 shows the force-extension curve of a titin molecule before and after the addition of 4 M guanidine-HCl. The addition of the denaturant abolished hysteresis across the entire force spectrum. Since the release force curve retraces the stretch curve, the molecule is in a state of conformation equilibrium. Thus, the hysteresis in our experiments was indeed related to the domain folding/unfolding kinetics in titin.

Test of the model IV: computer simulation

Force hysteresis in the titin-stretching experiments appears because the rate at which the molecule is stretched or released exceeds the equilibrium rate of domain unfolding and refolding, respectively. Also, notably, the forces associated with unfolding of titin (20-300 pN (Kellermayer *et al.*, 1997; Rief *et al.*, 1997; Tskhovrebova *et al.*, 1997)) are far in excess of the equilibrium force predicted previously for the the unfolding of globular domains of titin (3.5 - 5 pN, (Erickson, 1994)). Apparently, the breakage and formation of the non-covalent bonds holding the globular domains of titin together are significantly affected by the presence of external force and the rate of change thereof, just as in the case of molecular and cellular adhesion bonds (Bell, 1978; Evans and Ritchie, 1997). Under external force, the rates of domain unfolding or refolding can be calculated according to equation (4) (see Materials and Methods). According to the relationship, the energy barrier associated with the rate-limiting folding (or unfolding) intermediate is discounted by the external force depending on the displacement (Δx) it induces along the molecular axis (Kellermayer *et al.*, 1997). Δx, however, is significantly different during stretch and release. While a relatively short, ~0.3 nm displacement is sufficient to result in the disruption of the bonds holding the domain together (Rief *et al.*, 1998b), for refolding the domain must shorten significantly (by ~ 8 nm) before the proper set of bonds can be re-established. According to equation (4) the unfolding rate should rise exponentially with force above a characteristic level (Evans and Ritchie, 1997), the "unfolding force" (F_u), because

$$k = k_0 e^{F/F_u}.$$ (5)

A Monte-Carlo simulation based on equation (4) reproduced all the essential features of our model, including the force hysteresis, the high-force stretch and low-force refolding transitions, and the exponential rate-dependence of the unfolding and refolding forces (Fig. 8). As predicted, these forces gradually

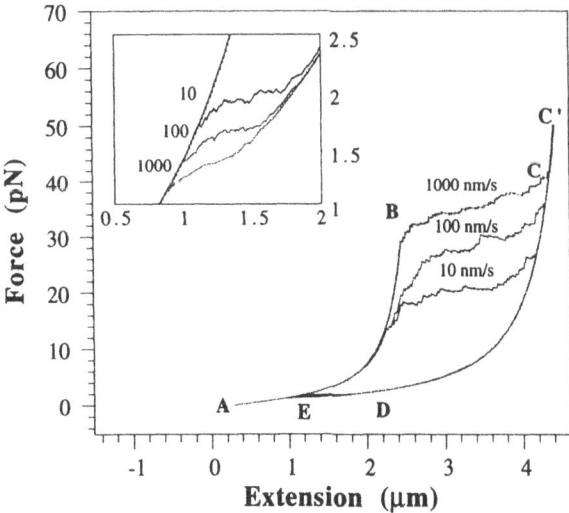

Figure 8. Monte-Carlo simulation of titin's behavior during mechanical manipulation. Effect of stretch rate on the force-extension curve. Inset, low-force data enlarged to reveal changes in the refolding force transition. Trap stiffness, 0.01 pN/nm.

approach a common intermediate value with lowering the rate of stretch/release. Conceptually, "unfolding force" arises solely within the context of a fixed-rate stretch experiment where the stretch rate matches the unfolding rate at a characteristic force. The deviation from such a characteristic force level is probably due to variation in the unfolding activation energies within the heterogenous domain population comprising titin (Kellermayer *et al.*, 1998). The apparent dissimilarities between the beautiful experimental and modeling work of Rief *et al.* (1997; 1998a; 1998b) and our results may derive from the different rates of stretch applied and differences in the stiffness of the AFM probe compared with the compliant optical trap.

In summary, single-molecule manipulation experiments were used to describe the elastic properties of the giant protein titin. The molecule behaves as an entropic polymer chain in which force-driven domain unfolding and refolding occurs. While the present work provides a good approximation for the overall mechanical behavior of titin, it is yet to be seen how the details of variation in stiffness along the molecule and cellular strategies to control domain unfolding/refolding determine the exact role titin plays in shaping the passive mechanical behavior of striated muscle.

Acknowledgments

This work was supported by grants from the National Institutes of Health, National Heart, Lung, and Blood Institute (HL61497 and HL62881) to H.L.G., by grants from the National Institutes of Health GM-32543 and the National Science Foundation MBC 9118482 to C.B., and grants from the Hungarian Science Foundation (OTKA F025353) and the Hungarian Ministry of Health (ETT T-06-021/97) to M.S.Z.K. H.L.G. is an Established Investigator of the American Heart Association.

References

Bell GI. Models for the specific adhesion of cells to cells. *Science* 1978;200:618-627.

Bustamante CJ, Marko JF, Siggia ED, Smith SB. Entropic elasticity of λ-phage DNA. *Science* 1994; 265:1599-1600.

Erickson, HP. Reversible unfolding of fibronectin type III and immunoglobulin domains provides the structural basis for stretch and elasticity of titin and fibronectin. *Proc Natl Acad Sci USA* 1994;91:10114-8.

Evans E, Ritchie K. Dynamic strength of molecular adhesion bonds. *Biophys J* 1997;72:1541-1555.

Fürst DO, Osborn M, Nave R, Weber K. The organization of titin filaments in the half-sarcomere revealed by monoclonal antibodies in immunoelectron microscopy: a map of ten nonrepetitive epitopes starting at the Z line extends close to the M line. *J Cell Biol* 1988;106:1563-72.

Granzier HLM, Akster HA, ter Keurs HED. Effect of thin filament length on the force-sarcomere length relation of skeletal muscle. *Am J Physiol* 1991;260:C1060-C1070.

Granzier HLM, Irving T. Passive tension in cardiac muscle: the contribution of collagen, titin, microtubules and intermediate filaments. *Biophys J* 1995;68:1027-1044.

Gregorio CC, Granzier H, Sorimachi H, Labeit S. Muscle assembly: a titanic achievement? *Curr Opin Cell Biol* 1999;11:18-25 (Review).

Higuchi H, Nakauchi Y, Maruyama K, Fujime S. Characterization of beta-connectin (titin 2) from striated muscle by dynamic light scattering. *Biophys J* 1993;65:1906-15.

Higuchi H, Suzuki T, Kimura S, Yoshioka T, Maruyama K, Umazume Y. Localization and elasticity of connectin (titin) filaments in skinned frog muscle fibres subjected to partial depolymerization of thick filaments. *J Muscle Res Cell Motil* 1992;13:285-94.

Horowits R, Kempner ES, Bisher ME, Podolsky RJ. A physiological role for titin and nebulin in skeletal muscle. *Nature* 1986;323:160-4.

Horowits R, Podolsky RJ. The positional stability of thick filaments in activated skeletal muscle depends on sarcomere length: evidence for the role of titin filaments. *J Cell Biol* 1987;105:2217-23.

Itoh Y, Suzuki T, Kimura S, Ohashi K, Higuchi H, Sawada H, *et al*. Extensible and less-extensible domains of connectin filaments in stretched vertebrate skeletal muscle sarcomeres as detected by immunofluorescence and immunoelectron microscopy using monoclonal antibodies. *J Biochem* 1988;104:504-508.

Kellermayer MSZ, Granzier HLM. Calcium-dependent inhibition of in vitro thin-filament motility by native titin. *FEBS Lett* 1996;380:281-286.

Kellermayer MSZ, Granzier HLM. Elastic properties of single titin molecules made visible through fluorescent F-actin binding. *Biochem Biophys Res Commun* 1996b;221:491-497.

Kellermayer MSZ, Smith SB, Granzier HL, Bustamante C. Folding-unfolding transitions in single titin molecules characterized with laser tweezers. *Science* 1997; 276:1112-1116.

Kellermayer MSZ, Smith SB, Bustamante C, Granzier HL. Complete unfolding of the titin molecule under external force. *J Struct Biol* 1998;122:197-205.

Kratky O, Porod G. *Rec Trav Chim* 1949;68:1106.

Kuhn W, Grün F. *Kolloid Z* 1942; 101:248.

Labeit S, Kolmerer B. Titins: giant proteins in charge of muscle ultrastructure and elasticity. *Science* 1995;270:293-296.

Marko JF, Siggia ED. Stretching DNA. *Macromolecules* 1995;28:8759-8770.

Maruyama K. Connectin/titin, giant elastic protein of muscle. *FASEB J* 1997;11:341-345.

Rief M, Fernandez JM, Gaub HE. Elastically coupled two-level systems as a model for biopolymer extensibility. *Phys Rev Lett* 1998a;81:4764-4767.

Rief M, Gautel M, Oesterhelt F, Fernandez JM, Gaub HE. Reversible unfolding of individual titin immunoglobulin domains by AFM. *Science* 1997;276·1109-1112.

Rief M, Gautel M, Schemmel A, Gaub H. The mechanical stability of immunoglobulin and fibronectin III domains in the muscle protein titin measured by atomic force microscopy. *Biophys J.* 1998b;75:3008-14.

Smith SB, Cui Y, Bustamante C. Overstretching B-DNA: The elastic response of individual double-stranded and single-stranded DNA molecules. *Science* 1996;271:795-799.

Smith SB, Finzi L, Bustamante C. Direct mechanical measurements of elasticity of single DNA molecules by using magnetic beads. *Science* 1992;258:1122-1126.

Svoboda K, Block S. Biological applications of optical forces. *Annu Rev Biophys Biomol Struct* 1994;23:247-285.

Trinick J. Elastic filaments and giant proteins in muscle. *Curr Opinion Cell Biol* 1991;3:112-118.

Trinick J. Cytoskeleton: Titin as a scaffold and spring. *Current Biology* 1996;6:258-260.

Trombitás K, Jin J-P, and Granzier HL. The mechanically active domain of titin in cardiac muscle. *Circ Res* 1995; 77:856-61.

Trombitás K, Pollack GH. Elastic properties of the titin filament in the Z-line region of vertebrate striated muscle. *J Muscle Res Cell Motil* 1993;14:416-22.

Tskhovrebova L, Trinick J, Sleep JA, Simmons RM. Elasticity and unfolding of single molecules of the giant muscle protein titin. *Nature* 1997;387:308-312.

Wang K. Titin/connectin and nebulin: giant protein rulers of muscle structure and function. *Adv Biophys* 1996;33:123-134.
Whiting A, Wardale J, Trinick J. Does titin regulate the length of muscle thick filaments? *J Mol Biol* 1989;205:263-8.

Discussion
(Presented by Henk Granzier)

Trinick: So the T12 antibody makes stable interaction with titin in guanidine-hydrochloride?

Granzier: Something is holding the molecule to the bead, and maybe T12 is involved.

Bullard: How do you know that the bead at the other.end (towards the M-line) is actually binding to the end of the molecule? It could bind anywhere couldn't it; it's not specific?

Granzier: Yes, it could. We have used, in various stages of this research, T51 antibody, myomesin and just bare silica beads. The data indicate that, typically, binding is in the A-band region of the molecule.

Greaser: How can you distinguish between a situation when you stretch two titin molecules held by their ends versus a single held by its end on one side and by its middle on the other? That is, could a double, full-length and a single, half-as-long tether generate similar forces if they are stretched to similar distances?

Granzier: We determine, for each tether, the apparent persistence length and contour length from the respective y(force)-axis and x(extension)-axis intercepts of the inverse-square-root force versus extension plot. The method works only if the tethers are stretched above 80% fractional extension, but that criterion is typically satisfied, and both parameters can be established. From a large number of experiments we determine the persistence length distribution. We believe it tells us that on occasion we have a single molecule with persistence lengths of about 20 Å. But more typically we have 2 molecules side by side.

Linke: You showed that even before it comes to the first transition you see a hysteresis in your stretch release curves; how do you explain it?

Granzier: At low force, in a well-rested molecule, there is indeed hysteresis. Since hysteresis occurs at low forces, this makes us think that unfolding of Ig or fibronectin domains is not involved. Instead, hysteresis may result from non-specific bonds within the molecule. In the contracted state at low force, different segments of the molecule have the opportunity to get near each other and electrostatic bonds may form between them. This could, for example, take place in the PEVK region.

Pollack: I have a question about the attachments at the ends to the bead. The bead is quite large relative to the diameter of the titin molecule. In order to attach the molecule the molecule needs to bend at right angles. What happens when you pull? How can you be sure that part of the titin molecule is not coming loose as you pull and increase the tension, and when you release the tension, the length of the segment that bonds with the microsphere is increasing. How could you control for the possibility that that is part of your response?

Granzier: The beads are extensively blocked with BSA and Tween-20. Nevertheless, it is difficult to exclude that some non-specific titin bead interaction occurs. It is worthwhile to note that tests revealed that the presence of guanidine-chloride does not affect non-specific binding of titin to the bead while the force-length traces are reversible and hysteresis is absent. Thus, when we see hysteresis in normal buffer, hysteresis is likely to arise from within the molecule.

TerKeurs: Are the extensibility properties of titin sensitive to second messengers, PK dependent phosphorylation for example?

Granzier: That is an interesting question. I speculate there may be some changes, but have no relevant data.

Vigoreaux:Is the guanidine effect reversible? Can you remove the guanidine?

Granzier: You can wash it out and the molecule recovers and the stretch-release curve displays again hysteresis.

Vigoreaux:And can temperature also duplicate that effect? Have you tried raising temperature?

Granzier: No, we haven't done that.

Qian: If you completely release the molecule, how can you still measure force?

Granzier: If you completely release it? Well, typically we don't fully release it. We keep the molecule under a small amount of tension.

Qian: So what are you controlling? You are still controlling a constant speed? Going back of the motor?

Granzier: Yes. We are controlling the stretch speed, the maximum force to which we stretch, the release speed, and the minimal force to which we release.

UNFOLDING FORCES OF TITIN AND FIBRONECTIN DOMAINS DIRECTLY MEASURED BY AFM

Matthias Rief[1], Mathias Gautel[2], and Hermann E. Gaub[3]

[1]*Stanford University School of Medicine, Stanford, CA;*
[2]*EMBL, Biological Structures Division, Heidelberg, Germany;*
[3]*Ludwig-Maximilians Universität München, Germany*

Abstract:　AFM-based Single Molecule Force Spectroscopy provides a new tool for probing the mechanical properties of single molecules. In this chapter we show that the unfolding forces of single protein domains can be directly measured. Unfolding forces give new insight into protein stability that cannot be deduced from thermodynamic measurements. A comparison between the unfolding forces measured in Ig domains of the muscle protein titin and those measured in fibronectin Type III domains reveals an extraordinarily high stability of titin domains.

Introduction

With the advent of piconewton instrumentation mechanical experiments with single molecules have become possible. Force induced conformational transitions in DNA (Cluzel *et al.*, 1996; Smith *et al.*, 1996) as well as in polysaccharides (Li, 1998; Rief *et al.*, 1997b) have been revealed and unbinding forces of receptor ligand complexes (Florin *et al.*, 1994; Hinterdorfer *et al.*, 1996) have been measured. Recently the unfolding forces of titin (Carrion-Vazquez *et al.*, 1999; Kellermayer *et al.*, 1997; Rief *et al.*, 1998; Rief *et al.*, 1997a; Tskhovrebova *et al.*, 1997), tenascin (Oberhauser *et al.*, 1998) and spectrin (Rief *et al.*, 1999) have been measured. These experiments have given insight into the forces stabilizing the complicated structure of biomolecules. In addition, recent molecu-

Elastic Filaments of the Cell, edited by Granzier and Pollack
Kluwer Academic/Plenum Publishers, 2000

lar dynamics calculations (Grubmüller *et al.*, 1995; Lu *et al.*, 1998) have contributed important information about the molecular details of these experiments.

From conventional calorimetric measurements information about the energies involved in processes like receptor-ligand binding or protein folding can be obtained. Conclusion about the stability of proteins under the influence of a force are not directly possible from such experiments. Force, however, is an important structural parameter especially for proteins that are designed to exert or resist forces, like for example, muscle or structural proteins. In this chapter we present a comparison of unfolding forces between immunoglobulin domains (Ig domains) of the muscle protein titin and type III domains of the protein fibronectin (Fn3 domains). The aim of this comparison is to elucidate whether titin which is constantly subject to stretching forces during muscle action exhibits a higher mechanical stability than the structurally similar cell adhesion protein fibronectin. Thermal denaturation studies showed similar melting temperatures for both the Fn3 domains of fibronectin and the Ig domains of titin (Plaxco *et al.*, 1996; Politou *et al.*, 1995).

Methods

Titin constructs

In this study measurements of the unfolding forces of a recombinant construct comprising 8 Ig domains (I27-I34) from the I band are presented (see Rief *et al.*, 1997a and Fig. 1a, black box). In order to improve anchoring to the gold surface, two carboxyterminal cysteines were appended to the construct. Each of the 8 Ig domains comprises 89 amino acids that fold into a compact β-barrel structure.

Fibronectin

Fibronectins are high molecular weight glycoproteins found in many extracellular matrices and in blood plasma (Hynes, 1990). They promote cell adhesion and affect cell morphology, migration and differentiation and cytoskeletal organization. Each fibronectin molecule is made up of a series of repeating units, the main part of which consists of 15-17 Fn3 repeats (see Fig. 1b). In contrast to fibronectin type III domains, the type I and type II domains are multiply disulfide bonded. On average each of the Fn3 repeats comprises 93.5 aminoacids folding into a β-barrel structure. In this study bovine plasma fibronectin (Calbiochem, Germany) was used.

The Force Spectrometer

An instrument capable of measuring the mechanical properties of single molecules must combine high force resolution (~1 pN) with high spatial reso-

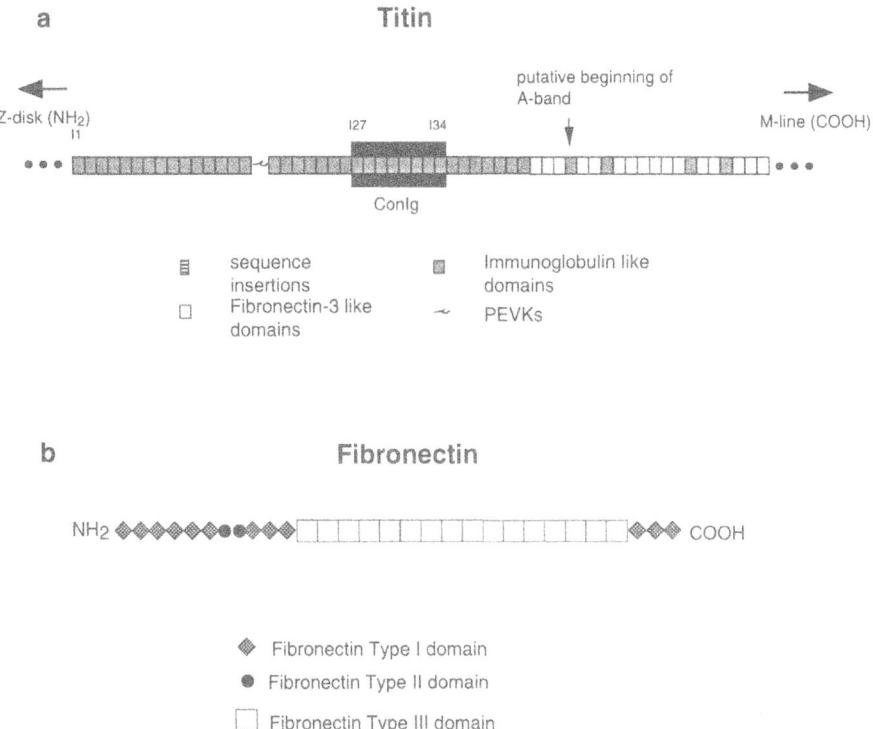

Figure 1. *(a) Schematics of the arrangement of domains in the I-band part of human cardiac titin. (b) Arrangement of domains in bovine Fibronectin.*

lution (~1Å). The Atomic Force Microscope (AFM) fulfills these two prerequisites in an ideal way: commercially available AFM force sensors have low spring constants (~10 mN/m) and very sharp tips (apex curvature ~20 nm) allowing to address single biological macromolecules.

A schematic of the custom built force spectrometer used in this study is given in Figure 2. The spectrometer is based on AFM technology; however, it is optimized for z-resolution. The x-y translation occurs manually. The instrument works like a clamp, consisting of two massive aluminum bars that are forced together with a strong cantilever spring and opened with a stack piezo with a built in strain gauge. Sample and cantilever are mounted on the opposing blocks of the clamp. The deflection sensing optics are mounted on the upper block and moved together with the cantilever. To avoid pointing instabilities creating force artefacts, the light is coupled into the optics via a monomode fiber. Tip and sample can be monitored through a microscope objective with a CCD camera. The cantilever holder is made of Plexiglas. The liquid drop is kept between holder and sample by surface tension without an additional O-

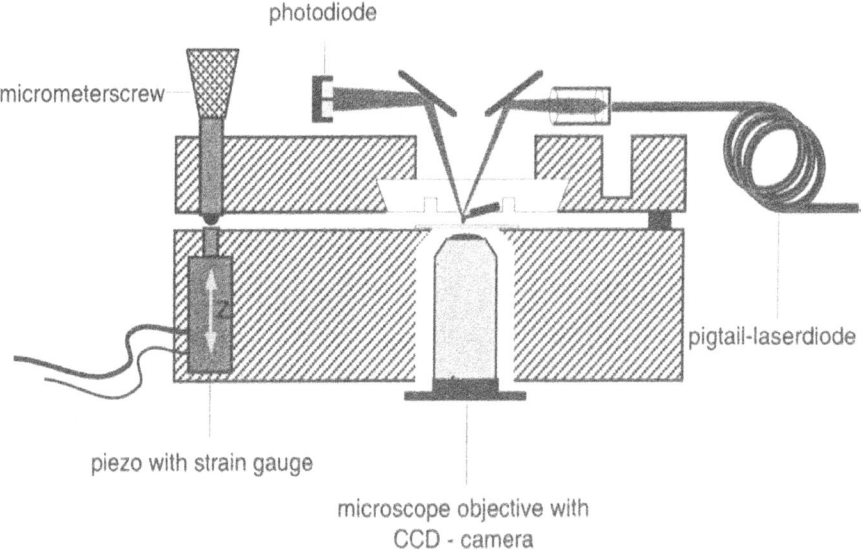

Figure 2. *Schematics of a force spectrometer.*

ring. Actuation and data acquisition are performed with 16-bit resolution. Typically, 4096 datapoints are taken per scan at a sampling rate of 60 kHz. The desired pulling rate was achieved by oversampling.

Sample preparation

Proteins were allowed to adsorb onto a freshly evaporated gold surface from a 50 μl drop of a 50 μg/ml solution in PBS (phosphate buffered saline, pH 7.4, 150 mM NaCl). After the incubation process (10 minutes) the sample was rinsed with PBS. Force measurements were carried out in PBS as well.

WLC fits

In order to model the force vs. extension characteristics of the unfolded polypeptide, the interpolation formula introduced in Bustamante *et al.* (1994) was used:

$$F(x) = \frac{k_B \cdot T}{p} \left(\frac{1}{4 \cdot (1 - x/L)^2} - \frac{1}{4} + \frac{x}{L} \right) \qquad (1)$$

The persistence length p describes the polymer stiffness, k_B is Boltzmann's constant, L the contour length and T the temperature. The black lines superimposed on the data in Figure 4 are fits using a persistence length of p=0.4 nm as in Rief *et al.* (1997a). This value describes the elasticity of the unfolded polypeptide strand best in a force range of 50-200 pN. Nevertheless, there are deviations, especially in the low force regime. This reflects the problems in describing by a single parameter p the complicated elasticity of a real polymer which is dominated by entropic as well as enthalpic contributions due to bond angle deformations.

Results and Discussion

Force measurements

Figure 3 shows a force spectroscopic experiment performed on a recombinant construct of a modular protein. For simplicity a construct comprising only 4 Ig domains is sketched. The construct is anchored to the gold surface via two carboxy terminal cysteines. The AFM tip is carefully brought into contact with the protein coated surface and the proteins are picked up by the tip via adsorption. It has been shown in various studies that adsorption can lead to very high connecting forces (> 700 pN) (Li *et al.*, 1998; Rief *et al.*, 1997a,b). The pickup can occur at any domain along the construct. Subsequently, the mechanical stability of the domains spanning the gap between tip and gold surface can be probed. Sequential unfolding of the domains results in a sawtooth

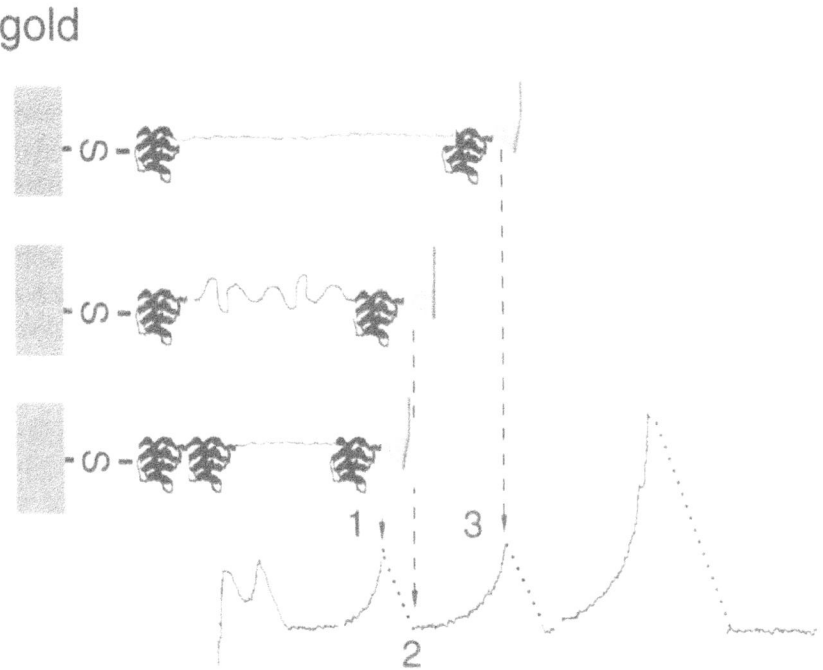

Figure 3. *Schematic of a force spectroscopic experiment performed on a recombinant construct comprising 4 Ig domains in series.*

pattern as is depicted in Figure 3. In a previous study (Rief *et al.*, 1997a) it could be shown that the slope leading up to each peak is mainly determined by the elasticity of the already unfolded polypeptide. At the peak force the weakest of the folded domains in the chain unfolds in an all-or-none event, adding an additional stretch of unraveled polypeptide to the chain. Hence, the absolute value of the peak force marks the unfolding force of a domain and the spacing to the following peak reflects the gain in length during a transition from the folded configuration to the fully unraveled polypeptide strand and can be correlated to the number of aminoacids folded in this domain.

Unfolding forces

The unfolding traces of the titin Ig construct exhibit a clear sawtooth pattern as can be seen from Figure 4a. Each of the sawteeth reflects the unfolding of an Ig domain. Peak forces rise from 190-250 pN. The forces are not identical because the individual domains in the construct are also not identical. In fact, measurements on the thermodynamic stability resulted in folding free energies that differed by up to a factor of three among the titin Ig domains (Politou *et al.*, 1995). The last peak in the unfolding pattern, which is distinctly higher than all the preceding peaks, does not reflect the unfolding of a domain but just shows the detachment of the protein from the AFM tip. Using Eq. (1) the left hand slopes of the unfolding pattern could be fitted (black solid lines in Fig. 4). The change in contour length $\Delta L=28.2$ nm from one peak to the next correlates well with the gain in length expected if the 89 amino acids folded in a domain are completely elongated after the unfolding process. The maximum force exerted on a single titin molecule at physiological conditions has been estimated to be 20-30 pN (Linke *et al.*, 1996). Consequently, domain unfolding should not take place during normal muscle action. It has been shown, however, that the unfolding force of domains also depends on the rate at which the force is applied (Rief *et al.*, 1997a). If a constant force of around 20 pN is applied to a titin molecule, domain unfolding will occur after several minutes. Hence, on the typical time scale of muscle cycles (~seconds), which is comparable to the time of the pulling cycle in Figure 4 unfolding is not expected to happen.

The unfolding of 10 fibronectin Fn3 domains is shown in Figure 4b. A similar sawtooth pattern as with the titin construct can be observed. Again, the last high peak reflects the detachment of the molecule from the AFM tip. Nevertheless, there are distinct differences from the titin trace. The most obvious differences are the much lower unfolding forces. Once again due to the variation in stability of the individual Fn3 domains peak, forces rising from 60-130 pN can be observed. This is 2 to 3 times lower than the forces in the titin construct. As already pointed out in the introductory paragraph, this difference in stability cannot be deduced from thermal denaturation experiments (Plaxco *et al.*, 1996; Politou *et al.*, 1995).

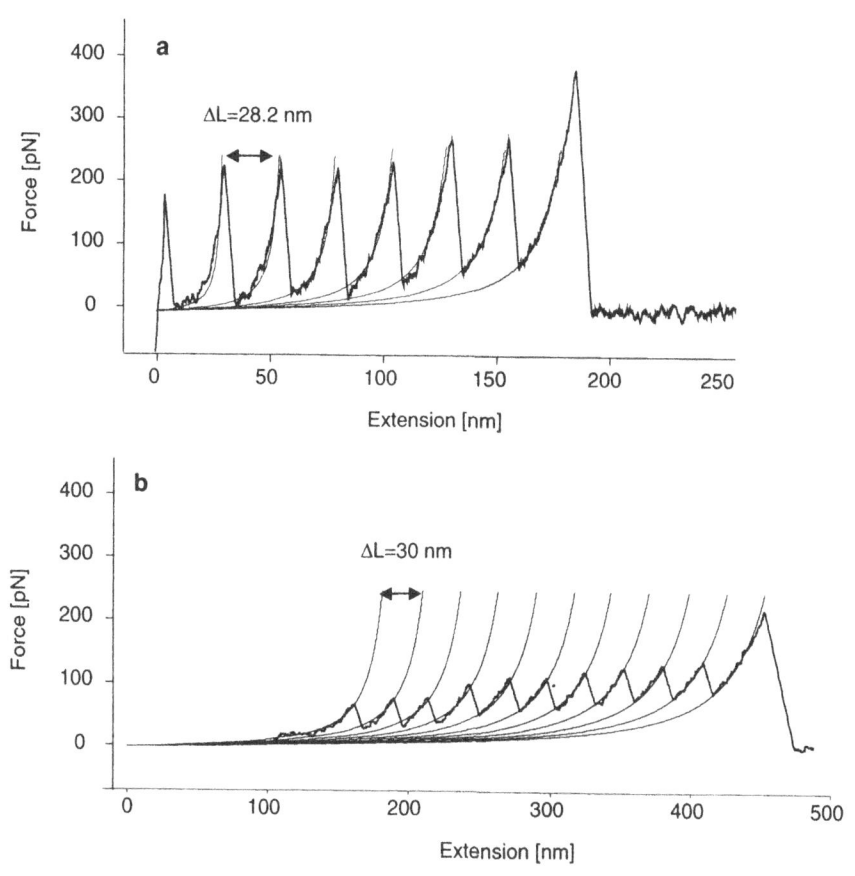

Figure 4. *(a) Force vs. extension curve reflecting the unfolding of 6 Ig domains of the muscle protein titin. The superimposed black lines are WLC fits according to Eq. (1). (b) Force vs. extension curves taken on native fibronectin.*

Another, more subtle, difference appears in the spacing of the peaks. The WLC-fits according to Eq. (1) yield an average peak spacing of ΔL=30 nm. This is due to the slightly higher number of amino acids (93.5 on average) folded in the Fn3 domains. This demonstrates the high spatial resolution of the technique.

The difference in unfolding forces between titin Ig and fibronectin Fn3 domains may be of physiological importance: the titin domains, which have to keep their structural integrity during muscle action, seem to be especially designed to resist mechanical unfolding. This conclusion is further supported by recent measurements of the unfolding forces of Fn3 domains in titin (Rief *et al.*, 1998) which, although structurally identical to fibronectin Fn3 domains, exhibit much higher unfolding forces (180 pN).

References

Bustamante C, Marko JF, Siggia ED, Smith S. Entropic Elasticity of l-Phage DNA. *Science* 1994;265:1599-1600.

Carrion-Vazquez M, Oberhauser AF, Fowler SB, Marszalek PE, Broedel SE, Clarke J, Fernandez JM. Mechanical and chemical unfolding of a single protein: a comparison. *Proc Natl Acad Sci USA* 1999;96:3694-9.

Cluzel P, Lebrun A, Heller C, Lavery R, Viovy J-L, Chatenay D, Caron F. DNA: an extensible molecule. *Science* 1996;271:792-794.

Florin E-L, Moy VT, Gaub HE. Adhesive forces between individual ligand-receptor pairs. *Science* 1994;264:415-417.

Grubmüller H, Heymann B, Tavan P. Ligand binding: molecular mechanics calculation of the streptavidin-biotin rupture force. *Science* 1995;271:997-999.

Hinterdorfer P, Baumgartner W, Gruber HJ, Schilcher K, Schindler H. Detection and localization of individual antibody-antigen recognition events by atomic force microscopy. *Proc Natl Acad Sci USA* 1996;93:3477-3481.

Hynes RO. *Fibronectins*. New York: Springer-Verlag, 1990.

Kellermayer MS, Smith SB, Granzier HL, Bustamante C. Folding-unfolding transitions in single titin molecules characterized with laser tweezers. *Science* 1997;276:1112-1116.

Li H, Rief M, Oesterhelt F, Gaub HE. Single-molecule force spectroscopy on xanthan by AFM. *Advanced Materials* 1998;10:316-319.

Linke WA, Ivemeyer M, Olivieri N, Kolmerer B, Rüegg JC, Labeit S. Towards a molecular understanding of the elasticity of titin. *J Mol Biol* 1996;261:62-71.

Lu H, Isralewitz B, Krammer A, Vogel V, Schulten K. Unfolding of titin immunoglobulin domains by steered molecular dynamics simulation. *Biophys J* 1998;75:662-71.

Oberhauser AF, Marszalek PE, Erickson HP, Fernandez JM. The molecular elasticity of the extracellular matrix protein tenascin. *Nature* 1998;393:181-185.

Plaxco KW, Spitzfaden C, Campbell ID, Dobson CM. Rapid refolding of a proline-rich all β-sheet fibronectin type III module. *Proc Natl Acad Sci* 1996;93:10703-10706.

Politou AS, Thomas DJ, Pastore A. The folding and stability of titin immunoglobulin-like modules, with implications for the mechanism of elasticity. *Biophys J* 1995;69:2601-2610.

Rief M, Gautel M, Schemmel A, Gaub HE. The mechanical stability of immunoglobulin and fibronectin III domains in the muscle protein titin measured by atomic force microscopy. *Biophys J* 1998;75:3008-14.

Rief M, Gautel M, Oesterhelt F, Fernandez JM, Gaub HE. Reversible unfolding of individual titin immunoglobulin domains by AFM. *Science* 1997a;276:1109-1112.

Rief M, Oesterhelt F, Heymann B, Gaub HE. Single molecule force spectroscopy on polysaccharides by AFM. *Science* 1997b;275:1295-1297.

Rief M, Pascual J, Saraste M, Gaub HE. Single molecule force spectroscopy of spectrin repeats: low unfolding forces in helix bundles. *J Mol Biol* 1999;286:553-61.

Smith SB, Cui Y, Bustamante C. Overstretching B-DNA: the elastic response of individual double-stranded and single-stranded DNA molecules. *Science* 1996;271:795-798.

Tskhovrebova L, Trinick J, Sleep JA, Simmons RM. Elasticity and unfolding of single molecules of the giant muscle protein titin. *Nature* 1997;387:308-312

Discussion

Pollack: I have a question regarding the force levels that you were describing. One of the issues in trying to measure the actual force level at which unfolding occurs is that the molecule you are examining is not in its natural environment surrounded by other proteins. Also one makes certain assumptions about, for example, the pH and the ionic environment that ordinarily surrounds the molecule. So my question is, given the expectation that the force level with which unfolding occurs is going to be sensitive to the environment, have you tried changing, for example, pH, or ionic strength or any of those, to see whether there is a substantial difference in force at which unfolding occurs?

Rief: Yes, in fact we have done that. I have a trace in my folder, where we changed pH. We did pH4 and pH7. What we learned is that the unfolding force itself, didn't change very much. What changed was the refolding kinetics. Refolding was slowed down considerably; we had to wait much longer in a relaxed state until the molecules could fold back again. The force itself only showed a slight dependence. The force was a little bit lower at pH4 than at pH7. We varied salt also. We haven't done that systematically, but we have tried 50 mM, as far as I remember, and there was no difference.

Vigoreaux: I think you partially answered my question, but I wasn't clear about the time resolution of your experiments whether the force changes are in phase with the stretch or whether there is a lag.

Rief: We can do one of these stretching cycles in 100 milliseconds. So each of the unfolding peaks may take 10 milliseconds; that is sort of the time resolution. And for a change in force we have a bandwidth that is determined by the resonance frequency of our cantilever, which is on the order of a kilohertz. Of course, every change that happens faster than that we will not be able to detect with the lever due to its inertia. That may be a reason why we can't really resolve very precise changes of first one strand unfolding, then the next one and so on. That it is definitely a reason for why these unfoldings seem to be so cooperative.

Vigoreaux: What about the kinetics of refolding?

Rief: We have done experiments for the refolding kinetics. It is not an ideal experiment, because we cannot, as was pointed out before, relax the molecule completely. So there is always a certain stretch because if we touch the surface again we may pick up more molecules, so we want to stay a little bit above the surface. We can model how it refolds under a force. All these things pointed toward refolding kinetics of maybe between 2 and 10 per second.

Qian: I noticed that you showed that there is discrepancy between the theoretical wormlike chain and your titin data. Have you tried to overlay all the different lines by normalizing them and determine the variance in the lower force range?

Rief: There is very good agreement between all these traces, and that is the criterion we use to determine whether we have single molecules. We normalize the traces and superimpose them, and there is very good agreement. I can't give you a number for how well they agree.

Qian: I mean the variance in the lower part of the data; does it agree with the theory?

Rief: The lower part does agree with the theory. It is only if we choose a persistence length of 4 nm that it doesn't agree. It just means that this parameter is wrong. I definitely think that the entropic elasticity models should predict the low force range.

Qian: You don't need any parameter to normalize the data and overlay them.

Rief: No they totally superimpose.

Qian: Second quick question, you studied a couple of different proteins by now; is their persistence more or less the same?

Rief: Yes, within 20% they are the same.

Qian: Third question, you mentioned you multiplied the well width with the force. Shouldn't that be the activation energy instead of comparing with melting temperature, which I assume is sort of unfolding free energy?

Rief: That is right. That is a very crude estimate. That was just to motivate. We don't draw conclusions from that. It is just interesting that that number comes out right. The problem is, anyway, that it should point toward some equilibrium free energy, as you say. But we show that our experimental conditions are non-equilibrium, so the right way to approach this is to measure speed dependence. I only want to draw conclusions from the width that we find out from speed dependence.

TerKeurs: You showed that when you stretch titin, force increases and then a molecular domain unravels. Can you give me a sense of where that happens during normal operation of muscle *in vivo*? Do you ever reach that unraveling force?

Rief: I don't think I am the right person to speculate about these physiological things. I think Wolfgang Linke has estimated the force range to be 25 picoNewtons that occurs under overstretch. And if you wait forever it will definitely start to unfold. Actually, we have an estimate, if you wait 100 seconds, keeping 100 Ig domains under load, one will unfold in these 100 seconds, on average. If you overstretch titin really for a long time unfolding will happen. But in a one-second cycle or so, I think unfolding shouldn't happen according to the data we have.

Trombitás: Regarding your last comment, we made an extensive study by just stretching the muscle for a long time before fixation. We used human soleus muscle which has a long PVEK, and we stretched the muscle far beyond the physiological range. The longest length was about 4.4 micron where the PEVK reached almost its estimated contour length. We keep the muscle in this state for up to 72 hours. At first we wanted to see some changes in the tandem Ig length. It was our biggest surprise that there is no change at all. The Ig domains in the I-band appear very stable.

Rief: I can totally agree with that. Just I think the estimate of the forces is a problem. So if you just have 5 picoNewtons; I think Miklos Kellermayer's data show that at 5 picoNewtons refolding could happen. The real problem is, what are the forces under these conditions?

Granzier: We estimated the forces at about 10 pN.

Rief: So another explanation could be that you have 1 or 2 domains unfolding. You would get a huge change in lengths because the unfolded length is 7-8 times higher than the folded. So as soon as one or two unfold maybe not much happens anymore.

Granzier: Whether one domain unfolds or not, we would be able to see that in Charles' experiments.

Rief: I think I should leave that I can't comment on that; that is just a result from our data.

Linke: You compared the unfolding forces between your data and Larissa's data, did you also compare your values to Miklos' data?

Rief: We have tried to model that. We come out at a little higher forces than they do, so if we model Miklos Kellermayer's experiment we get maybe 40 picoNewtons, instead the 20-30 pN they measure. It could be that we just haven't measured the really unstable domains, because we only see a small part of the molecule.

Linke: So the difference is twofold and not tenfold as it used to be.

I have a technical question. I was a little bit confused, maybe you can help me, with one of your pictures where you show results with a 4 domain fragment and then you model the results with a wormlike chain model. Do you always imply that you have already an unfolded protein fragment at the start of the stretch?

Rief: Are you talking about native titin?

Linke: No not a native titin, but constructs. If you begin to stretch, you always already see a peak, before you assign another peak to unfolding.

Rief: If we are close to the surface, and really indent into that surface, we will have non-specific interactions between the tip and the surface, and that's what we think happens here.

Linke: But my question remains when you have a 4-domain fragment of titin, and you model the initial part of the stretch before you come to the first unfolding, why do you use a wormlike chain model if you do not imply that there is pre-unfolded segment?

Rief: Well, I think we can use the worm-like chain model. There should also be contributions from the still-folded part, definitely yes. However, as the extension grows and grows, these traces are purely determined by worm-like chain behavior, so I don't see the problem.

Linke: We can talk about it later.

Granzier: How do you know that you have singlets and, for example, not doublets? This could be another reason why your data is different from ours at the same pulling speed.

Rief: We have a lot of data where we pull 2 or even 3 molecules. What happens in this case, is that we see double peaks, at higher forces. We see double peaks because it would be very unlikely that we always pick up these doublets exactly in register. Otherwise you would have to assume an interaction that keeps them in register. I think that such a scenario is very unlikely.

Granzier: How often do you have single molecules?

Rief: In many cases we have more than one. I would say maybe 5-10% or so are singlets.

Sanger: Matthias, I want to ask you a methological question. I am sort of thinking how the cell must deal with this molecule in assembling it. You mentioned that titin is very sticky. Have you tried substrates where there are lipids on there, then see how this molecule arranges itself on lipid surfaces?

Rief: We haven't tried this. It is maybe a good idea. We tried different surfaces, so we haven't just tried gold. For example we've tried mica as well.

COMPUTER MODELING OF FORCE-INDUCED TITIN DOMAIN UNFOLDING

Hui Lu[1], André Krammer[2], Barry Isralewitz[1], Viola Vogel[2], and Klaus Schulten[1]

[1]*Beckman Institute for Advanced Science and Technology*
University of Illinois at Urbana-Champaign, Urbana, IL
[2]*Departments of Physics and Bioengineering*
University of Washington, Seattle, WA

Abstract: Titin, a 1 μm long protein found in striated muscle myofibrils, possesses unique elastic and extensibility properties, and is largely composed of a PEVK region and β-sandwich immunoglobulin (Ig) and fibronectin type III (FnIII) domains. The extensibility behavior of titin has been shown in atomic force microscope and optical tweezer experiments to partially depend on the reversible unfolding of individual Ig and FnIII domains. We performed steered molecular dynamics simulations to stretch single titin Ig domains in solution with pulling speeds of 0.1 – 1.0 Å/ps, and FnIII domains with a pulling speed of 0.5 Å/ps. Resulting force-extension profiles exhibit a single dominant peak for each domain unfolding, consistent with the experimentally observed sequential, as opposed to concerted, unfolding of Ig and FnIII domains under external stretching forces. The force peaks can be attributed to an initial burst of a set of backbone hydrogen bonds connected to the domains' terminal β-strands. Constant force stretching simulations, applying 500 – 1000 pN of force, were performed on Ig domains. The resulting domain extensions are halted at an initial extension of 10 Å until the set of all six hydrogen bonds connecting terminal β-strands break simultaneously. This behavior is accounted for by a barrier separating folded and unfolded states, the shape of which is consistent with AFM and chemical denaturation data.

Elastic Filaments of the Cell, edited by Granzier and Pollack
Kluwer Academic/Plenum Publishers, 2000

143

Introduction

The giant muscle protein titin, also known as connectin, is a roughly 30,000 amino-acid-long filament that spans half of the sarcomere and plays a number of important roles in muscle contraction and elasticity (Labeit *et al.*, 1997; Maruyama, 1997; Kellermayer and Granzier, 1996; Wang *et al.*, 1993), as well as controlling chromosome shape in the cell nucleus (Machado *et al.*, 1998). During muscle contraction, titin, which is anchored at the Z-disk and at the M-line, exerts a passive force that keeps sarcomere components uniformly organized. The passive force developed in titin during muscle stretching restores sarcomere length when the muscle is relaxed. Titin is composed of about 300 repeats of two types of domains, immunoglobulin (Ig) domains and fibronectin type III (FnIII) domains, and the PEVK (70% proline, glutamic acid, valine and lysine residue) region (Labeit and Kolmerer, 1995). The FnIII domains are located only in the A-band of the molecule, the PEVK region is located in the I-band, while the Ig domains are distributed along the whole length of titin.

The region of titin located in the sarcomere I-band is believed to be responsible for titin extensibility and passive elasticity (Erickson, 1994; Linke *et al.*, 1994; Granzier *et al.*, 1996; Greaser *et al.*, 1996). The I-band region of titin consists mainly of two tandem regions of Ig domains, separated by the PEVK region. The Ig domains each form β-sandwich structures, but the PEVK region does not hold a stable conformation due to the charges on its glutamic acid and lysine residues. When muscle is stretched, the PEVK region unfolds and elongates. Under extreme conditions such as in overstretched muscle, the Ig domains in the titin I-band will unfold to provide the necessary extension. At further extension, the A-band of titin will unbind from myosin and contribute additional Ig and FnIII domains to the available pool of extendible domains. When forces are released, the unfolded Ig domains refold quickly (Rief *et al.*, 1997).

Titin domains have been observed, using single molecule techniques, atomic force microscopy (AFM) (Rief *et al.*, 1997) and optical tweezer experiments (Kellermayer *et al.*, 1997; Tskhovrebova *et al.*, 1997), to possess protection against strain-induced domain unfolding. The AFM experiments in particular have demonstrated that rather strong forces, on the order of 100 pN, must be exerted before Ig and FnIII domains rupture and unfold. We refer to proteins, which are designed to respond to force application under physiological conditions, as *mechanical proteins*. Other proteins, which do not encounter mechanical strain under physiological conditions, have been found to exhibit little resistance against strain-induced unfolding, as has been demonstrated through AFM experiments on the helical protein spectrin (Rief *et al.*, 1999).

The AFM force-extension profiles of multimers of Ig domains (from titin) and FnIII (from titin, and also from the protein tenascin) display a regularly repeating sawtooth pattern (Rief *et al.*, 1997; 1998; Oberhauser *et al.*, 1998). The spacing between any two force peaks matches the length of the completely extended polypeptide chain of one Ig or FnIII domain, proving that, when these multi-domain proteins are stretched, their domains unfold one by one. The high values of the force peaks (100–300 pN) imply that the Ig and FnIII domains are designed to withstand significant stretching forces. The peak values of the force depend on the type of domains being pulled and on the pulling speed adopted in an experiment. For a pulling speed of 1 μm/s, AFM unfolding of titin Ig domains requires about 200 pN, while unfolding of tenascin FnIII domains requires only about 140 pN (Rief *et al.*, 1997; Oberhauser *et al.*, 1998).

We would like to explain one-by-one domain unfolding in titin in terms of the structural properties of the proteins. Currently, only one experimental structure of titin I-band immunoglobulin (Ig) domains is available, the 27th Ig domain (I27) of cardiac titin (Improta *et al.*, 1996). Since no experimental structures are available for titin FnIII domains, we use a homologous FnIII domain, $FnIII_{10}$ of fibronectin (Leahy *et al.*, 1996). Figure 1 demonstrates that I27 and $FnIII_{10}$ both form β-sandwiches. AFM experiments studying Ig and FnIII stretching do not resolve atomic-level detail of the domains' conformational changes

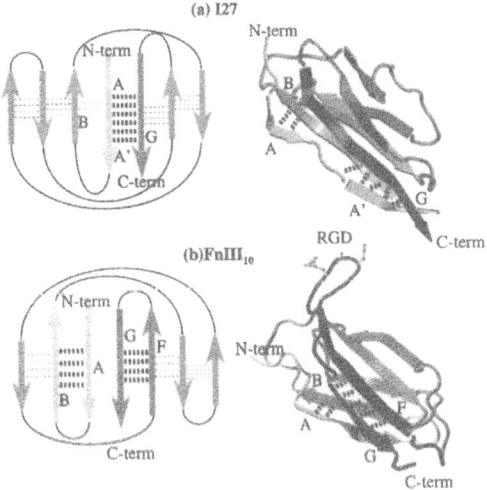

Figure 1. *Topology schematics and structures. The schematic representations on the left depict the topology of interstrand hydrogen bonding networks of (a) I27 and (b) $FnIII_{10}$. On the right are shown the corresponding proteins domains rendered with VMD (Humphrey et al., 1996) with the essential interstrand hydrogen bonds highlighted as in the schematic representations.*

during the unfolding. In this chapter we summarize the results of molecular dynamics simulations we have presented elsewhere (Lu *et al.*, 1998; Lu and Schulten, 1999a, 1999b; Krammer *et al.*, 1999) to provide an overall view of how SMD complements titin AFM observations, and to produce a detailed picture of titin domain stretching and unfolding.

Molecular dynamics (MD) simulations describe molecules as a set of atoms with known coordinates, masses, charges, and bond types (Allen and Tildesley, 1987). Motion is simulated by advancing through a series of time steps, producing atomic-level detail of molecular movements. At each time step, the forces on every atom from all other atoms due to Coulombic, van der Waals (vdW) and bonded interactions are calculated. The positions of each atom are updated for their resulting movement during one time step, then the forces are calculated again. Since the fastest oscillations in the system have a period of about 10^{-14} s, a time step of about 1/10 this size, 10^{-15} s, is often used. The CHARMM energy function (Brooks *et al.*, 1983) is used in this work to calculate forces on atoms; it has the form

$$V = \underbrace{\sum_{bonds} \frac{1}{2} k_b (r - r_0)^2}_{V_{bond}} + \underbrace{\sum_{angles} \frac{1}{2} k_\theta (\theta - \theta_0)^2}_{V_{angle}} + \underbrace{\sum_{dihedrals} k_\phi (1 + \cos(n_\phi \phi + \delta_\phi))}_{V_{dihedral}} +$$

$$\underbrace{\sum_{i \neq j} 4\epsilon_{ij} \left[\left(\frac{\sigma_{ij}}{r_{ij}} \right)^{12} - \left(\frac{\sigma_{ij}}{r_{ij}} \right)^6 \right]}_{V_{van-der-Waals}} + \underbrace{\sum_{i \neq j} \frac{q_i q_j}{\epsilon r_{ij}}}_{V_{Coulomb}}.$$

$$(1)$$

V_{bond} describes the oscillations about the equilibrium bond lengths. V_{angle} and $V_{dihedral}$ describe the other bonded interactions, oscillations of three atoms about an equilibrium angle and torsional rotations of four atoms about a central bond, respectively. The partial charges q_i and the energy parameters (e.g., k_b, σ_{ij}) are determined experimentally or by quantum chemical calculations. A set of these parameters is referred to as a *force field*.

Although many biological events take place on the millisecond-to-second time scale, the femtosecond (fs) time step taken in MD, combined with limits in available computational power, limit the time scales of simulations to the nanosecond (10^{-9} s = ns) range for systems with 10,000 to 100,000 atoms, such as the systems described in this chapter.

Steered Molecular Dynamics (SMD), reviewed in Izrailev *et al.* (1998), is a novel approach to the study of the dynamics of binding/unbinding events in bimolecular systems and of their elastic properties. The simulations reveal the details of molecular interactions in the course of ligand unbinding (Izrailev *et al.*, 1997; Isralewitz *et al.*, 1997; Wiggers and Schulten 1999; Stepaniants *et al.*, 1997; Kosztin *et al.*, 1999) or protein unfolding (Lu *et al.*, 1998; Krammer *et al.*, 1999; Lu and Schulten 1999a, 1999b), thereby providing important information about the molecular mechanisms underlying these processes. The advantage of SMD over conventional molecular dynamics is the ability to induce relatively large conformational changes in molecules on the ns time scales accessible to computation.

SMD simulations add external forces to conventional force fields, with the external forces imitating, for example, the effects of AFM cantilevers on protein domains. SMD simulations are ideally suited for the purpose of studying processes in which force is applied to molecules in actual systems, such as in force-induced protein stretching. We will show how SMD stretching trajectories, which are limited to the ns time scale, can provide insights into experimentally observed ms time scale stretching events. The problem of bridging the time scale gap has been discussed in detail in Gullingsrud *et al.* (1999).

Methods

In this work, we apply forces to systems with two SMD protocols: constant-velocity moving restraints and constant force restraints.

SMD using constant-velocity moving restraints simulates perturbing a protein with a moving AFM cantilever. At least one atom is held fixed during the simulation, which prevents the protein from simply translating in space when external forces are applied, and which corresponds to attaching the protein to a fixed substrate in the AFM experiments. At least one other atom from the protein is restrained to a point in space (restraint point) by an external, e.g., harmonic potential. The restraint point is then shifted in a chosen direction (Grubmüller *et al.*, 1996; Isralewitz *et al.*, 1997; Stepaniants *et al.*, 1997; Marrink *et al.*, 1998; Kosztin *et al*, 1999), forcing the restrained atom to move from its initial position in the protein. Assuming a single reaction coordinate x, and an external potential $U = K(x - x_0 - vt)^2/2$, where K is the stiffness of the restraint, and x_0 is the initial position of the restraint point moving with a constant velocity v, the external force exerted on the system can be expressed as

$$F = K(x_0 + vt - x). \tag{2}$$

This force corresponds to a molecule being pulled by a harmonic spring of stiffness K with one end attached to the restrained atom and the other end moving with velocity v. The force applied in constant force SMD, however, retains the same magnitude and direction regardless of the position of the restrained atom.

The energy-minimized average NMR structure of I27 (Improta *et al.*, 1996), which was obtained from the Brookhaven protein data bank entry 1TIT, was employed in this modeling study. This domain adopts the typical I-frame immunoglobulin superfamily fold (Politou *et al.*, 1995), consisting of two β-sheets packing against each other, as shown in Figure 1, with each sheet containing four strands. The first sheet comprises strands A, B, E, and D; the second sheet comprises strands A', G, F, and C. All adjacent β-strands in both sheets are anti-parallel to each other, except for the parallel pair A' and G. The β-strands A and A' belong to different sheets but are part of the N-terminal strand. The structure is stabilized by hydrophobic core interactions between the two β-sheets and by hydrogen bonds between β-strands.

To simulate a solvent environment, the I27 structure was surrounded by a sphere of water molecules of 31 Å radius, which covered the molecular surface of the domain by at least four shells of water molecules. The water molecules within 2.6 Å of the domain surface or within the volume occupied by the domain were deleted. About 3,300 water molecules were kept in the solvated I27 with the resulting protein-water system (Lu *et al.*, 1998) consisting of 11,400 atoms.

The SMD simulations were performed using the CHARMM22 force field (MacKerell *et al.*, 1998) with the programs NAMD (Nelson *et al.*, 1996) and XPLOR (Brünger, 1992), assuming an integration time step of 1 fs and a uniform dielectric constant of 1. The non-bonded Coulomb and vdW interactions were calculated with a cut-off using a switching function starting at a distance of 10 Å and reaching zero at 13 Å. The TIP3P water model was employed for the solvent (Jorgensen *et al.*, 1983).

SMD simulations with constant velocity stretching were carried out on I27 by fixing the C_α atoms of the first residue (Leu[1]), and by applying external forces to the C_α atom of the last residue (Leu[89]). The latter forces were implemented by restraining the C_α atom of the last residue harmonically to a restraint point and moving the restraint point with a constant velocity v in the desired direction. The procedure adopted is equivalent to attaching one end of a harmonic spring to the C_α atom of the last residue and pulling on the other end of the spring. This is similar to the procedure performed on titin and tenascin in AFM experiments (Rief *et al.*, 1997; Oberhauser *et al.*, 1998), except that the pulling speeds adopted in the simulations are about six orders of magnitude higher than those in the experiment. The value of K (see Eq. 2) was set to 10 $k_B T/\text{Å}^2$, corresponding to a spatial (thermal) fluctuation of the constrained C_α atom with variance $\delta x = \sqrt{k_B T/k} = 0.32$ Å at $T = 300$ K.

SMD simulations of constant force stretching were implemented by fixing the N-terminus of the domain I27 and by applying a constant force to the C-terminus along the direction connecting the initial positions of the N-terminus to the C-terminus.

The atomic coordinates of the entire system were recorded every picosecond. For constant velocity stretching, the elongation $d(t)$, defined as the increase of the end-to-end distance between the termini from that of the native fold, was monitored along with the force $F(t)$. For the analysis presented below, often the force is plotted as a function of extension d. The $(F(t),d(t))$ graphs will be referred to as the force-extension profile. In case of the constant force stretching simulation, the elongation $d(t)$ is recorded and plotted as $(d(t),t)$, which will be referred to as the extension curve.

Three pulling speeds have been adopted in our constant velocity stretching simulations, namely, 1.0 Å/ps, 0.5 Å/ps, and 0.1 Å/ps. The respective simulations will be referred to as I27-(1.0 Å/ps), I27-(0.5 Å/ps) and I27-(0.1 Å/ps). Four force values have been adopted for stretching I27 under constant force conditions: 500, 750, 900, and 1000 pN; the respective simulations will be denoted by I27-(500 pN), ..., I27-(1000 pN).

The system setup, solvation, equilibration, SMD simulations, and data recording performed for FnIII domains (Krammer *et al.*, 1999) are similar to those performed for I27. The X-ray crystallographic structure of $FnIII_{10}$ was obtained from the tetramer $FnIII_{7-10}$ (PDB code 1FNF) (Leahy *et al.*, 1996). All atoms, including hydrogens, were described explicitly. In the simulations, the N-terminal C_α atom (Leu[1]) of $FnIII_{10}$ was constrained to a fixed point, while the C-terminal C_α atom (Thr[94]) was attached to a constant velocity (0.5 Å/ps) moving SMD restraint.

Results

The force profiles for the I27 extensions are depicted in Figure 2a, which compare the simulations I27-(1.0 Å/ps), I27-(0.5 Å/ps) and I27-(0.1 Å/ps), as presented already in Lu and Schulten (1999a). The three profiles are qualitatively similar: at an extension of about 15 Å a dominant force peak arises that, in each individual profile, is about 2 to 3 times higher than forces at other extensions. At extensions larger than 20 Å the domain exhibits little resistance against stretching as is evident from the fact that only relatively weak forces are needed to increase extension. At an extension corresponding to a fully stretched polypeptide strand, i.e., at about 270 Å, the force required to stretch the now completely unfolded domain increases again as expected. The force-extension profiles in Figure 2a show also that the lower the pulling speed, the lower the forces needed to extend the domain. The decrease of the speed from 1.0 Å/ps to 0.5 Å/ps, to 0.1 Å/ps reduces the peak force from 2200 pN, to 2000 pN, to 1200 pN, respectively. All the simulated force-extension curves share the main qualitative feature with those observed in AFM stretching experiments: a single dominant force peak.

An analysis of the trajectories corresponding to the profiles in Figure 2a reveals that at the extension of the maximum force eight hydrogen bonds break concurrently, as pointed out in Lu *et al.*, (1998), as well as in Lu and Schulten (1999a). Figure 3 presents snapshots of the domain directly before and after the

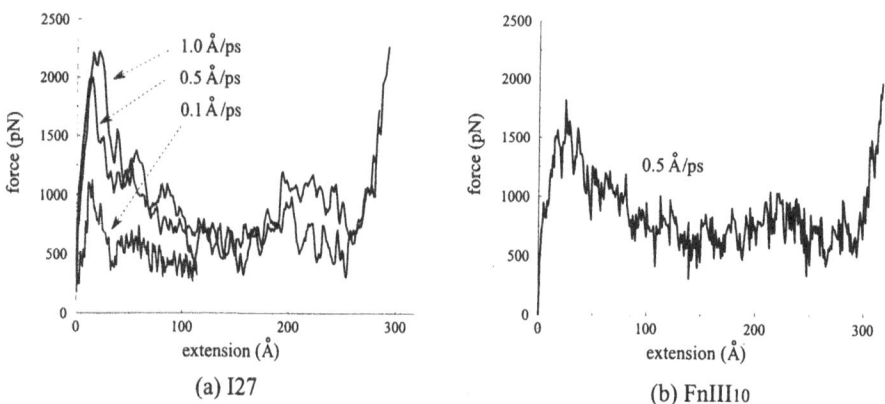

(a) I27   (b) FnIII10

Figure 2. *Force extension profiles of I27 and FnIII₁₀. (a) Simulation results of I27-(1.0 Å/ps), I27-(0.5 Å/ps) and I27-(0.1 Å/ps). The simulation times for pulling velocities 1.0 Å/ps, 0.5 Å/ps, and 0.1 Å/ps are 0.3 ns, 0.6 ns and 1.2 ns, respectively. The simulation of I27-(0.1 Å/ps) was stopped at an extension of 120 Å, whereas simulations I27-(1.0 Å/ps) and I27-(0.5 Å/ps) were stopped at about 300 Å. (b) Simulation results of FnIII₁₀ extension. The simulation time is 0.3 ns, with a velocity of 0.5 Å/ps.*

force peak in which the extensions measure 10 Å and 17 Å, during which 8 hydrogen bonds break concurrently: the 2 hydrogen bonds between strands A and B break during 10-14 Å of extension, and the 6 hydrogen bonds between strands A' and G break during 15-17 Å extension. During the following 190 Å of extension, the remaining 22 intra-β-strand hydrogen bonds break individually.

Simulations stretching I27 with constant force characterize the unfolding barrier responsible for the force peak described above. Simulations I27-(500 pN), I27-(750 pN), I27-(900 pN), and I27-(1000 pN) explore the influence of the stretching forces on the time between the start of force application and the beginning of domain unfolding. An analysis of the four simulations revealed that an initial 10 Å extension takes place, which is due to a straightening of the polypeptide chain near its terminal ends and due to a breaking of the two hydrogen bonds between β-strands A and B. This extension, shown in Figure 4, and presented originally in Lu and Schulten (1999a), continues until a plateau region (which is longer for weaker forces) is reached, at which point the six A'-G interstrand hydrogen bonds come under mechanical strain. The domain extensions fluctuate around a constant value during the plateau period, until the A'-G bonds are broken, after which extension rapidly increases. Snapshots of I27 structure at extensions smaller and greater than the plateau (dotted lines in Fig. 4) reveal the breaking event as described; the structural changes are roughly identical to those observed before and after the force peak seen in constant

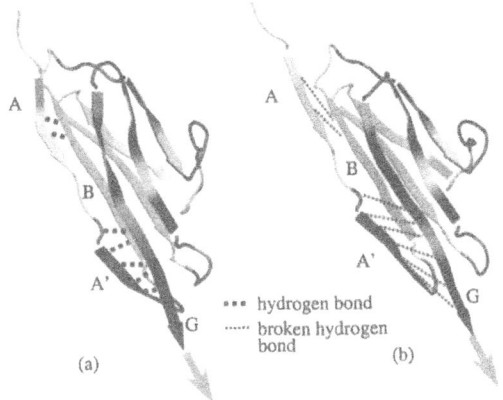

Figure 3. *Two snapshots of I27 unfolding. The domain is drawn in cartoon representation and key hydrogen bonds between strands A-B, and between strands A'-G are shown as dotted lines. (a) Snapshot of I27 before the force peak at 10 Å extension. The hydrogen bonds are all maintained. (b) Snapshot of I27 after the force peak at 17 Å extension. The hydrogen bonds between strands A-B, and between strands A'-G are broken, initiating unfolding.*

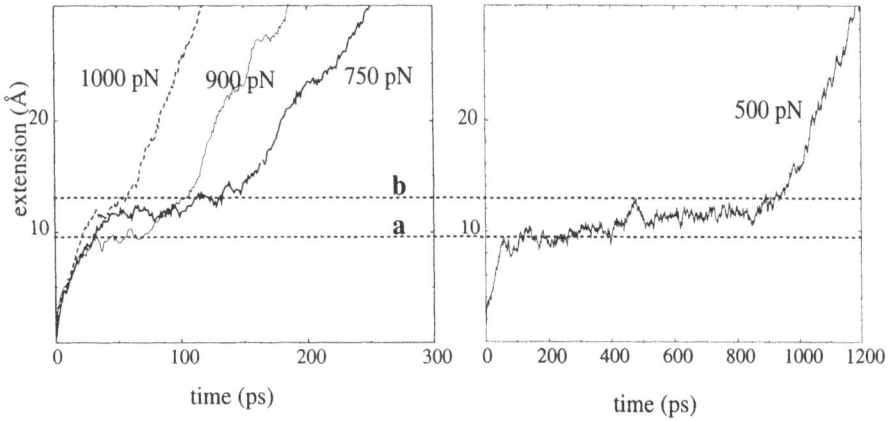

Figure 4. *Extension curves of I27 unfolding by means of constant forces of 500, 750, 900 and 1000 pN strength. In all cases the domain extended quickly by 10 Å, then remained approximately at constant length for long periods of time until the system unfolded (appearing as linear extension with time). Conformations at dotted lines a and b correspond to conformations shown in Figure 3a and 3b, respectively (see text).*

velocity stretching (Fig. 3). During the rapid post-plateau extension increase, the domain unravels, exhibiting relatively little resistance to extension.

We can model crossing the plateau as a barrier crossing process, with $\tau_{barrier}$ denoting the time spent at the plateau. The applied force effectively lowers the barrier such that stronger forces lead to faster barrier crossing ($\tau_{barrier}$ = 0.04 ns for 1000 pN) than weaker forces ($\tau_{barrier}$ = 0.9 ns for 500 pN). In all cases, the motion gets temporarily "stuck" in front of the barrier, which can then only be overcome by thermal fluctuations. This scenario can be described as Brownian motion governed by a potential, which is the sum of the indigenous barrier and a linear potential accounting for the applied force. The mean time to cross the barrier can be evaluated using the expressions for the mean first passage time developed in Schulten *et al.* (1980; 1981) and Szabo *et al.* (1980). By comparing the mean first passage times with the respective times $\tau_{barrier}$ for various forces, one can estimate the height of the indigenous potential barrier.

To estimate the shape of the indigenous potential barrier we assume a simple model for the energy $U(x)$ of the barrier: $U(x) = \Delta U (x - a)/(b - a)$ for $a \leq x \leq b$, with a reflective boundary condition at a, and an absorptive boundary condition at b. Here ΔU is the height of the barrier, and $b - a$ the barrier width. The choice of this potential function is dictated by the fact that for this barrier type the mean first passage time can be expressed analytically (Izrailev *et al.*, 1997):

$$\tau(F) = 2\ \tau_d \delta(F)^{-2}\ [e^{\delta(F)} - \delta(F) - 1] \tag{3}$$

We have introduced here $\tau_d = (b - a)^2/2D$ and $\delta(F) = \beta[\Delta U - F(b - a)]$. Assuming a width of 3 Å, estimated from the fluctuation of the extension curves in Figure 4, a least square fit procedure matching the data points in Figure 5 to the equation (3) results in the satisfactory match shown also in Figure 5. This curve corresponds to a barrier height of 1420 pNÅ, i.e., 20.3 kcal/mol.

In the constant velocity moving restraint SMD performed for $FnIII_{10}$, the domain was stretched from its initially compact and folded structure to its fully elongated configuration. The force-extension profile (Figure 2b) exhibits a single dominant force peak, following an unfolding process similar to that described above for Ig domain. This 1800 pN force peak, arising at 25 Å extension, implies again a barrier that must be overcome to initiate unfolding. Simulations presented in Krammer *et al.* (1999) reveal that, just as for the Ig domain, the respective barrier is caused by the need to simultaneously break several hydrogen bonds early in unfolding. In this case, the force peak is caused by 6 hydrogen bonds breaking during 8 Å of extension, a range longer than the 2 Å extension period required to break the same number of hydrogen bonds at the Ig force peak.

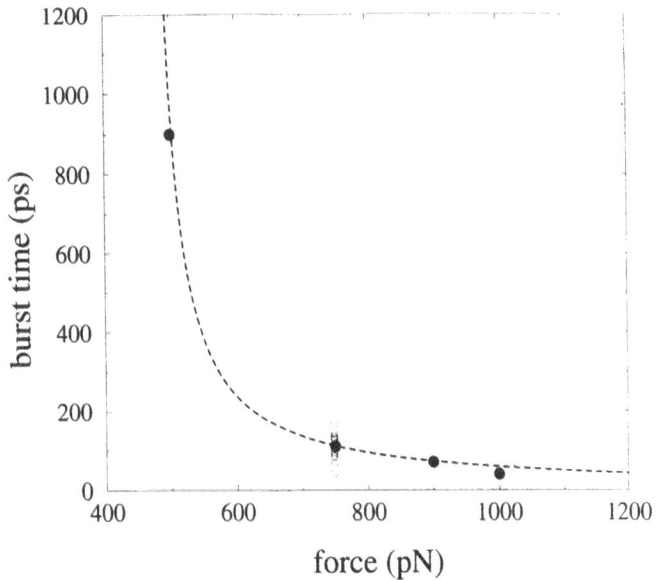

Figure 5. *Time needed for I27 to unfold vs. constant stretching forces. The solid dots represent simulation results and the dotted line is the least square fit of the mean first passage time described by Eq. (3). The small circles at the force of 750 pN demonstrate that the first passage time is widely distributed.*

In fibronectin, the $FnIII_{10}$ domain mediates cell adhesion to surfaces via its integrin binding motif, Arg^{78}, Gly^{79}, and Asp^{80} (RGD), which is placed at the apex of the β-hairpin turn connecting β-strands F and G. Fibronectin and other RGD-containing proteins serve as static anchoring sites against which cells can build up tensile strength.

In the above simulations, dramatic conformational changes of the RGD-loop are observed in the initial stages of the forced unfolding pathway of the module. In the unperturbed native state, the RGD-containing loop is located at ~12 Å away from the outer surface of the $FnIII_{10}$ module and is thus accessible to cell surface integrins. Optimal binding is obtained for distances between RGD and substrate in the range of 11 Å – 30 Å (Beer *et al.*, 1992). The RGD hairpin is bent at the C_{α} atoms of residues Thr^{76} and Ala^{83}, but upon domain extension past the force peak, this bend is straightened. A steep increase in the angle defined by the C_{α} atoms of residues Asp^{80}, Ala^{83}, and Ser^{89} is observed at 20 Å extension, followed by a movement of the apex of the loop closer to the surface of the domain from 12 Å to 8 Å while the module is extended from 25 Å to 35 Å. The RGD loop is distorted further as the extension continues. During the first 35 Å of extension of $FnIII_{10}$ we are likely seeing a force-regulated decrease in integrin accessibility of the RGD segment.

Discussion

Multi-domain proteins like titin involving immunoglobulin and fibronectin type III domains constitute a fascinating class of biopolymers with important cellular functions. Single molecule experiments based on AFM that probe the mechanical response of these protein systems provide a unique source of information that becomes more valuable in combination with steered molecular dynamics simulations. The latter provide atomic level pictures of the conformational processes governing the function of mechanical proteins which, however, need to be verified through comparison with AFM data.

Applied force experiments (Rief *et al.*, 1997; Kellermayer *et al.*, 1997; Tskhovrebova *et al.*, 1997) have elucidated the chief design requirements for titin I-band Ig domains under extreme stretch conditions: they must unravel one by one, and must increase the length of titin at each unraveling event by a set amount without affecting the stability of those domains that still remain folded. At small extensions of titin, when the link regions are pulled taut to form a straight chain, all Ig domains are at their resting contour length of about 40 Å. Upon SMD force application, every Ig domain exhibits a pre-burst increase in contour length of roughly 10 Å per Ig domain. With further extension, Ig domains continue to lengthen, but only after the force exceeds a given value. Our simulations provide an explanation for this bursting behavior. The links, which must be ruptured first to initiate the unfolding of the Ig domains, are the hydrogen bonds between β-strands A and B and between β-strands A' and G. Due to the topology of the Ig domain (Figure 1), until the A'G bonds break, force can not be transmitted along the backbone to unravel the rest of the protein. Only when the A'G and AB strands are separated, after breaking of all interstrand hydrogen bonds, can the unfolding of an Ig domain continue, involving rupture of the interstrand hydrogen bonds between the remaining β-strands.

The SMD stretching of FnIII proceeds similarly to the Ig stretching. The chief difference between Ig and FnIII domain SMD unfolding is the smaller rupture forces observed for FnIII as compared to Ig; the smaller forces are consistent with observations in three AFM experiments (Oberhauser *et al.*, 1998; Rief *et al.*, 1997; 1998). The AFM force peaks recorded for FnIII unfolding are 70% the size of those recorded for Ig domain unfolding. From analysis of structures and SMD trajectories (Krammer *et al.*, 1999) one can discern structural features, specifically differences in the topology of interstrand hydrogen bonding, likely responsible for the differences in force peak. In the case of Ig (Fig. 1a) there exist backbone hydrogen bonds directly between N-terminal strand A or A' and C-terminal strand G. In FnIII (Fig. 1b) backbone hydrogen bonds do not directly connect the termini, they connect the N-terminal strand A with its neighboring strand B, and the C-terminal strand G with its neighboring strand

F. In Ig, the breaking of interstrand hydrogen bonds occurs over a very small extension range (roughly the length of a hydrogen bond), whereas in FnIII, the breaking of the interstrand bonds occurs over a wider extension range. In FnIII the terminal segments where force is applied are not directly connected by hydrogen bonds; the FnIII topology of interstrand bonding renders the domain more flexible under stretching, allowing bond breaking over a longer extension range and hence with a reduced rupture force, as seen in Figure 2.

Both Ig and FnIII domains exhibit high (> 150 pN) dominant force peaks when stretched in AFM experiments; dominant force peaks are also seen for both domains when stretched with SMD. The domains are both members of a class of proteins, which we call Class I (Lu and Schulten, 1999), that exhibit resistance to forced unfolding; their fold topologies are such that interstrand hydrogen bonds must break in clusters in order to allow extension of the domain. Other domains (e.g., all-helical domains) have topologies that can be extended while breaking interstrand hydrogen bonds singly. These do not exhibit dominant force peaks when stretched in SMD simulations (Lu and Schulten, 1999b); we call these class II domains.

The current simulations suggest that AFM experiments on proteins that consist of several class II domains should show force-extension profiles different from those performed on proteins consisting of several class I domains. Either no dominant force peak (and thus no sawtooth pattern) or much lower force peak values than those observed for class I domains should be observed. The discernible-but-lower force peak values may arise from hydrophobic effects, which do not play an important role in force-induced unfolding of class I domains. Both possibilities, i.e., non-discernible (J. Fernandez, personal communication) and lower, discernible (Rief *et al.*, 1999) peaks have already been confirmed by recent AFM experiments.

The main obstacles to relating SMD simulations and AFM measurements, namely the time scale discrepancy and the related discrepancy in stretching forces required to induce unfolding, may be overcome through slower pulling speeds in the moving restraint SMD simulations. The slower I27-(0.1 Å/ps) and I27-(0.5 Å/ps) stretching simulations (Lu and Schulten, 1999a) produce the same scenario of clustered hydrogen bond breaking as the faster I27-(1.0 Å/ps) simulation. Figure 6 presents the measured AFM forces and their logarithmic extrapolation to faster pulling speeds together with the forces from simulations I27-(1.0 Å/ps), I27-(0.5 Å/ps) and I27-(0.1 Å/ps). One can discern that the extrapolated AFM forces correspond to a force of about 500 pN at a pulling speed of 0.1 Å/ps, whereas the simulated force is twice as large.

The results of constant force SMD simulations can be extrapolated to even more closely approach the forces observed in AFM. One may employ the crossing times $\tau_{barrier}$ for simulations I27-(500 pN), I27-(750pN), I27-(900 pN),

I27-(1000 pN) to estimate pulling velocities and extend the set of simulation data shown in Figure 6. We assume for this purpose that the forces applied are the peak forces and that the estimated barrier width of 3 Å provides a suitable estimate for the distance traveled. In case of simulation I27-(1000 pN), i.e., for a constant force of 1000 pN, one determines accordingly a velocity of 0.075 Å/ps. The corresponding data point in Figure 6 lies indeed close to the data point of simulation I27-(0.1 Å/ps), which exhibited a peak force of 1200 pN. The velocities corresponding to the constant forces 900 pN, 750 pN, and 500 pN are 0.05 Å/ps, 0.027 Å/ps, and 0.003 Å/ps, respectively, which contribute the corresponding data points in Figure 6. One can discern from these data points that according to our rough estimate the simulation data approach the measured AFM peak forces when scaled with *log*(velocity).

We also interpreted the results of the constant force simulations in terms of mean first passage time of barrier crossing (Eq. 3) and have found, in applying the theory of first passage times (Schulten *et al.*, 1980; 1981; Szabo *et al.*, 1980), that the estimated 3 Å barrier width, combined with the fitted 20 kcal/mol barrier height, match the simulation results; they correspond to the same barrier crossing time as observed in the simulations. These characteristics of the barrier separating the folded and unfolded forms of I27 agree well with AFM observations and chemical denaturation (Chevron plot) data (Carrion-Vazquez *et al.*, 1999) in which a potential barrier of 22 kcal/mol was measured.

Recently, several theoretical groups have also published studies on forced unfolding of protein domains, employing approaches and methods different from those applied in this chapter. Rohs *et al.* (1999) studied the stretching of α-helix and β-hairpin systems using molecular mechanics; they estimated the magnitude of forces involved in the unfolding of these secondary protein structures. Socci *et al.* (1999) studied the relation between force dependence and the reaction coordinate by stretching a lattice model. Evans and Ritchie (1999) modeled the Ig domain unfolding as a single bond-breaking event and approached the problem using the Kramers-Smoluchowski theory. Paci and Karplus (1999) studied FnIII unfolding by means of biased molecular dynamics using an implicit solvent model to reduce computational effort; the authors suggested that the dominant barrier against unfolding is due to vdW interactions, and not due to hydrogen bonds. Future SMD simulations will help clarify the differences among different theoretical approaches, including those described above, and will be aimed to provide a means to unite these approaches, while still relating them to experiment.

Our goal in this modeling of forced titin domain unfolding is to obtain an atomic-level view of the process. We approached this by incorporating into simulations two known properties of the system: the experimentally derived

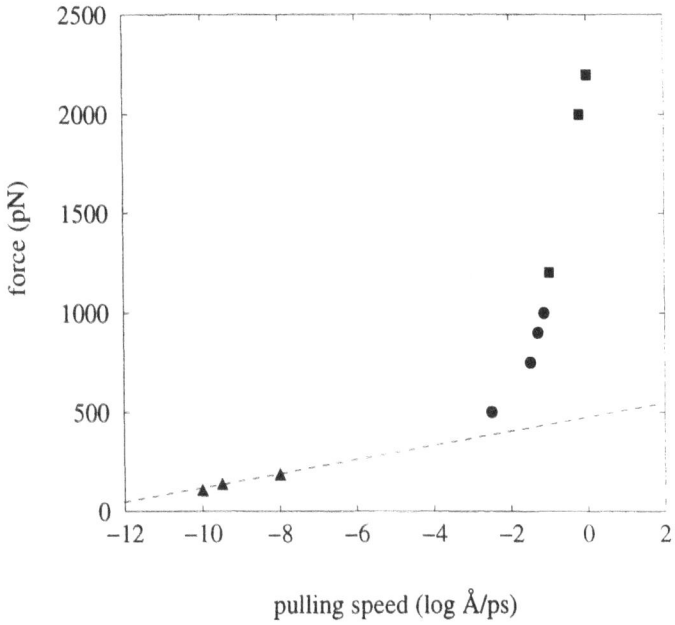

Figure 6. *Force peak value during unfolding vs. pulling speed. Triangles represent AFM data and the dotted line is an extrapolation of these data (Carrion-Vazquez et al., 1999); squares represent results from SMD simulations; circles represent forces estimated (see text) from simulations I27-(1000 pN), I27-(900 pN), I27-(750 pN, and I27-(500 pN).*

static structure and the experimentally known force-extension curve. Our resulting theoretical model should meet two criteria: first, the model should correspond closely to the experimental (AFM) situation, so as to provide a non-ambiguous check on the model's validity; second, the results should provide information on the process at atomic-level resolution that cannot be obtained from experiment. Our SMD method has satisfied both criteria listed above. Force-extension curves produced by simulations can be directly compared to AFM data to check validity. The SMD trajectories account for the unfolding process at atomic-level detail and, hence, reveal structure-function relationships for protein elasticity properties.

Future SMD simulations can be used to help design proteins with unfolding barriers of desired strength. Additionally, modeling force-sensitive proteins with SMD can provide insights into novel processes such as the RGD loop conformation changes observed for $FnIII_{10}$. SMD can contribute fundamentally to the understanding of structure-function relationships of mechanical proteins.

Acknowledgments

This work was supported by the National Institutes of Health (NIH PHS 5 P41 RR05969 and NIH GM49063) and the National Science Foundation (NSF BIR 94-23827EQ, NSF/GCAG BIR 93-18159, MCA93S028).

References

Allen MP, Tildesley DJ. *Computer Simulation of Liquids*. New York: Oxford University Press, 1987.

Beer JH, Springer KT, Coller BS. Immobilized Arg-Gly-Asp (RGD) peptides of varying lengths as structural probes of the platelet glycoprotein IIb/IIIa receptor. *Blood* 1992;79:117-128.

Brooks BR, Bruccoleri RE, Olafson BD, States DJ, Swaminathan S, Karplus M. CHARMM: A program for macromolecular energy, minimization, and dynamics calculations. *J Comp Chem* 1983;4:187-217.

Brünger AT. *X-PLOR, (Version 3.1): A System for X-ray Crystallography and NMR*. New Haven CT: Yale University Press, 1992.

Carrion-Vazquez M, Oberhauser AF, Fowler SB, Marszalek PE, Broedel SE, Clarke J, Fernandez JM. Mechanical and chemical unfolding of a single protein: A comparison. *Proc Natl Acad Sci USA* 1999;96:3694-3699.

Erickson HP. Reversible unfolding of fibronectin type III and immunoglobulin domains provides the structural basis for stretch and elasticity of titin and fibronectin. *Proc Natl Acad Sci USA* 1994;91:10114-10118.

Evans E, Ritchie K. Strength of a weak bond connecting flexible polymer chains. *Biophys J* 1999;76:2439-2447.

Granzier H, Helmes M, Trombitás K. Nonuniform elasticity of titin in cardiac myocytes: a study using immunoelectron microscopy and cellular mechanics. *Biophys J* 1996;70:430-442.

Greaser ML, Sebestyen MG, Fritz JD, Wolff JA. cDNA sequence of rabbit cardiac titin/connectin. *Adv Biophys* 1996;33:13-25.

Grubmüller H, Heymann B, Tavan P. Ligand binding: molecular mechanics calculation of the streptavidin-biotin rupture force. *Science* 1996;271:997-999.

Gullingsrud J, Braun R, Schulten K. Reconstructing potentials of mean force through time series analysis of steered molecular dynamics simulations. *J Comp Phys* 1999; 151:190-211.

Humphrey WF, Dalke A, Schulten K. VMD – Visual Molecular Dynamics. *J Mol Graphics* 1996;14:33-38.

Improta S, Politou AS, Pastore A. Immunoglobulin-like modules from titin I-band: extensible components of muscle elasticity. *Structure* 1996;4:323-337.

Isralewitz B, Izrailev S, Schulten K. Binding pathway of retinal to bacterio-opsin: a prediction by molecular dynamics simulations. *Biophys J* 1997;73:2972- 2979.

Izrailev S, Stepaniants S, Balsera M, Oono Y, Schulten K. Molecular dynamics study of unbinding of the avidin-biotin complex. *Biophys J* 1997;72:1568-1581.

Izrailev S, Stepaniants S, Isralewitz B, Kosztin D, Lu H, Molnar F, Wriggers W, Schulten K. "Steered molecular dynamics." In *Computational Molecular Dynamics: Challenges, Methods, Ideas,* volume 4 of *Lecture Notes in Computational Science and Engineering*, P Deuflhard, J Hermans, B Leimkuhler, AE Mark, S Reich, RD Skeel, eds. Berlin: Springer-Verlag 1998; 39-65.

Jorgensen WL, Chandrasekhar J, Madura JD, Impey RW, Klein ML. Comparison of simple potential functions for simulating liquid water. *J Chem Phys* 1983;79:926-935.

Kellermayer MS, Granzier HL. Elastic properties of single titin molecules made visible through fluorescent F-actin binding. *Biochem Biophys Res Commun* 1996;221:491-497.

Kellermayer MS, Smith SB, Granzier HL, Bustamante C. Folding-unfolding transition in single titin modules characterized with laser tweezers. *Science* 1997;276:1112-1116.

Kosztin D, Izrailev S, Schulten K. Unbinding of retinoic acid from its receptor studied by steered molecular dynamics. *Biophys J* 1999;76:188-197.

Krammer A, Lu H, Isralewitz B, Schulten K, Vogel V. Forced unfolding of the fibronectin type III module reveals a tensile molecular recognition switch. *Proc Natl Acad Sci USA* 1999;96:1351-1356.

Labeit S, Kolmerer B, Linke WA. The giant protein titin. Emerging roles in physiology and pathophysiology. *Circ Res* 1997;80:290-294.

Labeit S, Kolmerer B. Titins, giant proteins in charge of muscle ultrastructure and elasticity. *Science* 1995;270:293-296.

Leahy DJ, Aukhil I, Erickson HP. 2.0 Å crystal structure of a four-domain segment of human fibronectin encompassing the RGD loop and synergy region. *Cell* 1996;84:155-164.

Linke WA, Popov VI, Pollack GH. Passive and active tension in single cardiac myofibrils. *Biophys J* 1994;67:782-792.

Lu H, Isralewitz B, Krammer A, Vogel V, Schulten K. Unfolding of titin immunoglobulin domains by steered molecular dynamics simulation. *Biophys J* 1998;75:662-671.

Lu H, Schulten K. Steered molecular dynamics simulation of conformational changes of immunoglobulin domain I27 interpret atomic force microscopy observations. *Chem Phys* 1999a;247:141-153.

Lu H, Schulten K. Steered molecular dynamics simulations of force-induced protein domain unfolding. *Proteins Struct Funct Genet* 1999b;35:453-463.

Machado C, Sunkel CE, Andrew DJ. Human autoantibodies reveal titin as a chromosomal protein. *J Cell Biol* 1998;141:321-333.

MacKerell AD Jr, Bashford D, Bellott M, Dunbrack RL Jr, Evanseck J, Field MJ, Fischer S, Gao J, Guo H, Ha S, Joseph D, Kuchnir L, Kuczera K, Lau FTK, Mattos C, Michnick S, Ngo T, Nguyen DT, Prodhom B, Reiher IWE, Roux B, Schlenkrich M, Smith J, Stote R, Straub J, Watanabe M, Wiorkiewicz- Kuczera J, Yin D, Karplus M. All-hydrogen empirical potential for molecular modeling and dynamics studies of proteins using the CHARMM22 force field. *J Phys Chem B* 1998;102:3586-3616.

Marrink SJ, Berger O, Tieleman P, Jähnig F. Adhesion forces of lipids in a phospholipid membrane studied by molecular dynamics simulations. *Biophys J* 1998;74:931-943.

Maruyama K. Connectin/titin, giant elastic protein of muscle. *FASEB J* 1997;11:341-345.

Nelson MT, Humphrey W, Gursoy A, Dalke A, Kalé LV, Skeel RD, Schulten K. NAMD: a parallel, object-oriented molecular dynamics program. *Int J Supercomput Appl High Perform Comput* 1996;10:251-268.

Oberhauser AF, Marszalek PE, Erickson HP, Fernandez JM. The molecular elasticity of the extracellular matrix protein tenascin. *Nature* 1998;393:181-185.

Paci E, Karplus M. Forced unfolding of fibronectin type 3 modules: An analysis by biased molecular dynamics simulations. *J Mol Biol* 1999;288:441-459.

Politou AS, Thomas DJ, Pastore A. The folding and stability of titin immunoglobulin-like modules, with implications for mechanism of elasticity. *Biophys J* 1995;69:2601-2610.

Rief M, Gautel M, Oesterhelt F, Fernandez JM, Gaub HE. Reversible unfolding of individual titin immunoglobulin domains by AFM. *Science* 1997;276:1109-1112.

Rief M, Gautel M, Schemmel A, Gaub HE. The mechanical stability of immunoglobulin and fibronectin III domains in the muscle protein titin measured by atomic force microscopy. *Biophys J* 1998;75:3008-3014.

Rief M, Pascual J, Saraste M, Gaub HE. Single molecule force spectroscopy of spectrin repeats: Low unfolding forces in helix bundles. *J Mol Biol* 1999;286:553-561.

Rohs R, Etchebest C, Lavery R. Unraveling proteins: a molecular mechanics study. *Biophys J* 1999;76:2760-2768.

Schulten K, Schulten Z, Szabo A. Dynamics of reactions involving diffusive barrier crossing. *J Chem Phys* 1981;74:4426-4432.

Schulten K, Schulten Z, Szabo A. Reactions governed by a binomial redistribution process–the Ehrenfest urn problem. *Physica* 1980;100A:599-614.

Socci ND, Onuchic JN, Wolynes PG. Stretching lattice models of protein folding. *Proc Natl Acad Sci USA* 1999;96:2031-2035.

Stepaniants S, Izrailev S, Schulten K. Extraction of lipids from phospholipid membranes by steered molecular dynamics. *J Mol Model* 1997;3:473-475.

Szabo A, Schulten K, Schulten Z. First passage time approach to diffusion controlled reactions. *J Chem Phys* 1980;72:4350-4357.

Tskhovrebova L, Trinick J, Sleep JA, Simmons RM. Elasticity and unfolding of single molecules of the giant protein titin. *Nature* 1997;387:308-312.

Wang K, McCarter R, Wright J, Beverly J, Ramirez-Mitchell R. Viscoelasticity of the sarcomere matrix of skeletal muscles. *Biophys J* 1993;64:1161-1177.

Wriggers W, Schulten K. Investigating a back door mechanism of actin phosphate release by steered molecular dynamics. *Proteins Struct Funct Genet* 1999;35:262-273.

Discussion

Trinick: I can see that some of the parameters that you calculate are obviously in agreement, but how do we know that the detailed trajectory that you calculate is correct.

Krammer: I think you only have indirect evidence for that. I think what you have to keep in mind is that the energy landscape is very complex. What has been shown by Frauenfelder, is that there are multiple pathways that contribute to the unfolding of protein domains, and what we show here is one possible pathway. But what is being shown by Frauenfelder's experiments is that the unfolding pathways coincide fairly closely. So I think we can pick this as a representative. We don't say this is the only path.

Trinick: If you run the simulation again does it always go in exactly the same way, or can you get it going down different pathways?

Krammer: We did multiple simulations at various pulling speeds, and we always saw the same pathway. We also used different set-ups where we pulled on the N-terminal end and fixed the C-terminus, as well as pulled on both of the terminals, and again saw the same unfolding pathway. Correlating this with the AFM data, we only have indirect evidence that this could be a possible pathway. We cannot say this is the pathway that unfolding follows when you go to lower pulling speeds.

Rief: I am really happy to see the results and how good they agree and I'm really impressed with your work. Maybe a little bit nasty question. There was a recent *Journal of Molecular Biology* paper by the Karplus group that got a little bit different results. I think its an important question to answer. Is it the hydrogen bonds, or is it, as they say, rather van der Waals (interactions) that break off? Could you comment, why these results are so different?

Krammer: Actually, I think it was Emanual Paci (from Karplus' group) who did this work. We were talking to him quite extensively, and we actually exchanged our results. I think one of the problems they are facing is that they use an implicit water model and they don't use explicit water molecules in their system, so I don't know if this is really the reason or not why they see 2 plateaus. They determined that there are two intermediate unfolding stages where the domain stabilizes at. We were not able to replicate that result, so I think that is still an open question and I hope that we can answer it at some point.

Ter Keurs: You calculated the peak force as a function of the speed and most of your data were at the speed of sound. I saw three data points at much lower speeds. Were they calculated by you or did you take them from other data?

Krammer: I should maybe go back to that slide really quickly. Now, as I mentioned, we did two different set-ups. We used a harmonic potential where we applied a spring to the system and we also used a constant force where we waited for the domain to unfold. The black squares here are the data where we pulled with the harmonic potential, while the blue circles here represent the constant force data. What we did is we determined the pulling speed from the time it takes for the domain to unfold versus the extension at the point it unfolds, which is the barrier width. These red symbols are the AFM data.

Granzier: Could you explain why you use these high pulling speeds?

Krammer: Just one simulation takes approximately 2 months. I told you that we did 18. And you have to imagine that you wait in a queue until you finally are in and then it runs for an hour, maybe two hours, and then you are again kicked out of the loop; and then we wait again. So two months is just pure computation time. Until the job is finally finished, it takes about 4 or 5 months.

EXTENSIBILITY IN THE TITIN MOLECULE AND ITS RELATION TO MUSCLE ELASTICITY

Larissa Tskhovrebova[1] and John Trinick
School of Biomedical Sciences,
University of Leeds, Leeds, UK
[1]Permanent address: Institute of Theoretical and
Experimental Biophysics,
Pushchino, Moscow, Russia

Abstract: Studies of the origins of muscle passive tension have revealed a direct relationship between elasticity and the mechanical properties of the titin molecule. 'Molecular combing' has made it possible to visualize with high resolution changes in the configuration and structure of isolated titin caused by mechanical forces. The differential extensibility seen in individual molecules is consistent with the important role suggested for the PEVK-region in muscle elasticity. An additional factor emphasizing compliance of this part of the molecule in muscle may relate to the arrangement of the titin filament system in the sarcomere, in particular to titin interactions with thick and thin filaments. The branching of titin network near the PEVK-region suggests that, in addition to conferring extensibility, it may also be important in facilitating the transition of titin intermolecular interactions between the arrays of thick and thin filaments.

Introduction

The extensibility of the titin molecule and its role in muscle elasticity is a fascinating aspect in muscle function, with far-reaching consequences for the understanding of mechanical stability and elasticity in biopolymers generally. Titin is a giant protein (~3.5 MDa), whose molecules are ~1 μm long and span from M-line (C-terminus) to Z-line (N-terminus) (Furst *et al.*, 1988; Whiting *et al.*,

Elastic Filaments of the Cell, edited by Granzier and Pollack
Kluwer Academic/Plenum Publishers, 2000

1989). The A-band part of the molecule is integral with the thick filament, whereas the I-band part forms an elastic connection between the end of thick filament and the Z-line. These connections center thick filaments in the sarcomere and are the main mechanical connections through relaxed muscle fibers. They can be reversibly increased in length at least 3-4 fold, resulting in forces up to a few tens of picoNewtons (pN) per molecule (Linke *et al.*, 1996b; Wang *et al.*, 1993).

The titin molecule consists primarily of ~300 domains falling into two classes. These are similar to I-set immunoglobulin (Ig) and to fibronectin (Fn) type III domains (Labeit and Kolmerer, 1995), both of which are folded as β-sandwiches of 7 or 8 strands. The I-band part of the molecule also has several sequences of unknown structure of which the largest, composed mainly of proline, glutamate, valine and lysine residues, is known as the PEVK- region. This region is joined to Z-line and to the end of the thick filament by two groups of Ig domains. The A-band part of titin mainly consists of repetitive patterns of Fn and Ig domains, termed "super-repeats," whose periodicities reflect the complex structure of the thick filament (Labeit *et al.*, 1992; Labeit and Kolmerer, 1995).

Two types of mechanism have been suggested to explain the elasticity of I-band titin: one entropic involving straightening the molecule from an originally coiled state and the other enthalpic involving mechanical unfolding of the titin internal structure (Erickson, 1994; Helmes *et al.*, 1996; Labeit and Kolmerer, 1995; Linke *et al.*, 1996b; Soteriou *et al.*, 1993a). Mechanical measurements on fibers and myofibrils, correlated with antibody labeling, indicate that both entropic and enthalpic mechanisms occur (Helmes *et al.*, 1996; Linke *et al.*, 1996a). Immunolabeling also suggests non-uniform extensibility in the I-band region (Gautel and Goulding, 1996; Granzier *et al.*, 1996; Linke *et al.*, 1996b). The heterogeneity of titin structure and the fact that size of the PEVK region is inversely correlated with muscle stiffness led to the suggestion of differential extensibility along the molecule (Labeit and Kolmerer, 1995).

In addition to studies that monitored movement of titin epitopes *in situ*, extensibility has also been observed in single molecules. Direct measurements using atomic force microscopy (Rief *et al.*, 1997) and optical tweezers (Kellermayer *et al.*, 1997; Tskhovrebova *et al.*, 1997) demonstrated mechanical unfolding of the titin polypeptide and provided first quantitative data for the force-extension relationship of the molecule. Changes in configuration (Nave *et al.*, 1989; Trinick *et al.*, 1984) and structure (Tskhovrebova and Trinick, 1997) caused by mechanical force can also be visualized directly by electron microscopy after 'molecular combing.'

An important unresolved issue concerns the correlation of results obtained from the *in vitro* studies of individual titin molecules with the *in situ* observations (Linke *et al.*, 1996b; Yang *et al.*, 1998). In order for this comparison to be made, the arrangement of titin in the sarcomere and the influence of environment on its extensibility need to be clarified. These questions, together with visual observations of titin extensible properties, are discussed below.

Materials and Methods

Titin preparation and labeling procedure

Titin was prepared from rabbit skeletal muscle according to the procedure described previously (Soteriou *et al.*, 1993b). Chemical unfolding-refolding of the isolated protein was done by dialysis against a solution containing 6 M Gu-HCl for 24 hrs, and then for 24 hrs against 0.5 M KCl, containing 20 mM Tris, 1 mM DTT, 1 mM EGTA, 1 mM NaN_3, pH 7.9 (T-buffer).

Labeling with Cy3-dye (Amersham) was done as suggested by the company. 1 ml of titin solution (pH 9.0 -9.5) at 1.0 mg/ml was mixed with 1 vial of the dye. The protein-dye mixture was left at room temperature for 30 min and then passed through a Sepharose CL-2B gel filtration column (40cm x 0.5cm). The ratio of bound dye to protein was determined spectroscopically.

Light and electron microscopy

Fluorescence imaging of the labeled titin was done using a Dage 300 cooled CCD camera attached to a Leitz Laborlux optical microscope. Single fluorescent molecules were observed both in solution and after drying on a mica substrate. For observations in solution, 10-20 μl of labeled protein at a concentration 5-20 μg/ml in T-buffer was applied to a microscope slide.

Specimens of combed molecules for both light and electron microscopy were prepared according to Tskhovrebova and Trinick (1997). After drying on a mica substrate, the molecules were examined in the fluorescence microscope, or rotary shadowed with platinum and coated with a carbon film. The platinum/carbon replica was then transferred to electron microscope grids and examined using a Phillips 400T electron microscope operating at 80 kV.

Results

Titin straightening

Figure 1a illustrates Cy-dye-labeled titin on a glass substrate in high ionic strength buffer (0.5 M KCl). The molecules are seen as fluorescent spots with average diameter 0.3-0.5 μm. The round shape of the spots is consistent with the flexible nature and relatively small radius of gyration of the molecule in solution (Higuchi *et al.*, 1993; Trinick *et al.*, 1984). Variations in the size of the spots are probably due to titin oligomers; as well as monomers, oligomers containing two, three or four molecules joined at the head region were commonly seen in the titin preparations.

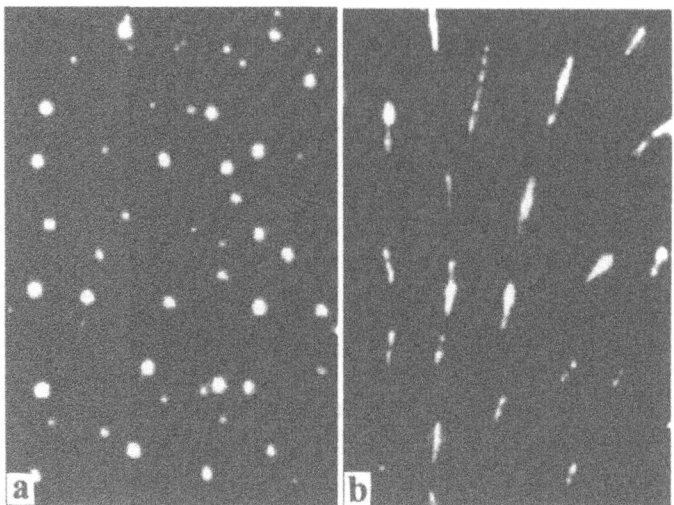

Figure 1. *Straightening of titin molecules due to combing by surface tension forces, visualized by fluorescence microscopy: (a) isolated unstressed molecules in solution on a glass substrate; (b) combed molecules on a mica substrate. Note the elongated contour of the spots after combing.*

Figure 1b shows labeled titin after molecular combing on a mica surface. The shape of the fluorescent spots is elongated, 2-3 micron in length, indicating that the molecules were straightened and extended.

Titin unfolding

We have previously described how 'molecular combing' results in straightening and extension of titin by surface tension forces (Tskhovrebova and Trinick, 1997). The force exerted on a molecule during this process may reach several hundred pN. This is high enough to break any type of bonds stabilizing tertiary structure and could therefore cause unfolding of the polypeptide.

A head-centered configuration of combed oligomers observed by electron microscopy illustrates most clearly titin flexibility and extensibility (Fig. 2). The head-centred configuration occurs when combing is initiated by attachment to the substrate of the tail end of one of the molecules in an oligomer. This results in the molecule being combed out upstream of the centrally placed head, with the other molecules in the oligomer combed downstream of the head. Inspection reveals consistent variations between the degree of extension of the up- and downstream molecules and there is often a segmented appearance, especially in the upstream molecules. The differences in lengths of the A-band parts of the molecules can be as large as 20-40% (Fig. 2a). This includes 10-

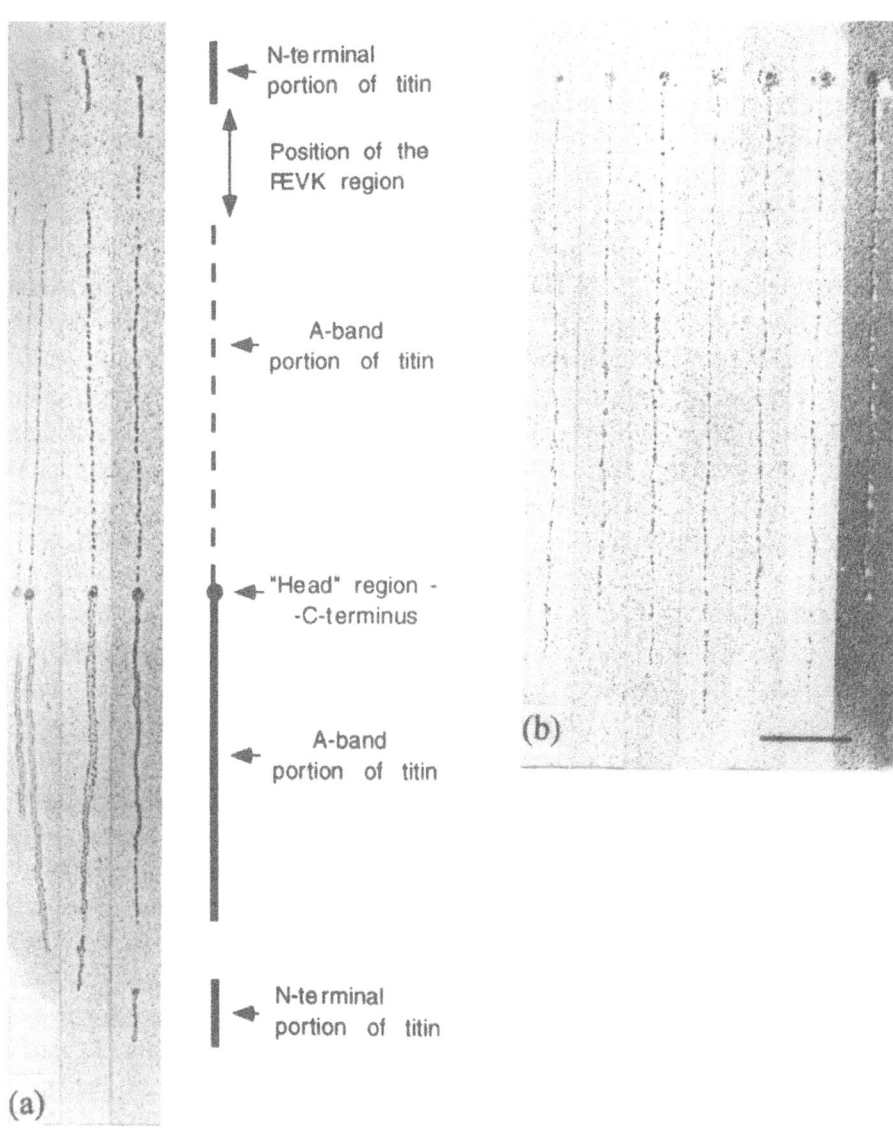

Figure 2. *Electron microscopy of combed titin molecules. (a) Variability in extension of molecules in the head-centred titin oligomers. The scheme on the right shows major functional and structural regions of the titin molecule. Note the longer length and punctuate appearance of the upstream (top) strands in comparison with the downstream strands. Note also the tendency to form a gap between the A- and N-terminal titin parts at the expected position of the PEVK region. (b) Micrographs showing A-band portions of molecules extended to more than 1.5 times, to ~1.5–1.7 μm (compared with the rest length of 0.9–1.0 μm). Note also the non-uniform diameter of the molecules. Bar – 0.25 μm.*

20% elongation for the upstream molecules and 10-20% shortening for the downstream molecules, compared to the unstressed contour length predicted from sequence. In some cases, the extension observed in the A-band region was as large as 150-200% (Fig. 2b). The most significant and consistent site of segmentation was at the expected position of the PEVK-region (Fig. 2a). Here an N-terminal section was often seen separated from the A-band portion of the molecule by a distance of up to 0.5 μm, which is comparable with the predicted length of the fully unfolded PEVK polypeptide. Segmentation was also observed throughout the A-band and N-terminal regions comprising mostly β-structure domains (Fig. 2a). The only difference was that these segments and the intervals between them were smaller and their positions were less regular.

Unfolded polypeptide is too small (diameter less than 10 Angstrom) to be resolved by the metal-shadow technique used in this work (probable resolution ~30-40 Angstrom). Therefore, although fine connections can sometimes be seen between the segments, there is little direct evidence of the presence of the unfolded polypeptide. However, the alignment of the segments and the restricted sizes of the gaps between them indicate that they remained bound to each other.

Titin refolding

That unfolding of titin domains is a reversible process was indicated by a number of studies, including the restoration of β-structure (Maruyama *et al.*, 1986) and the mechanical properties of the molecule (Carrion-Vazquez *et al.*, 1999; Kellermayer *et al.*, 1997; Rief *et al.*, 1997; Tskhovrebova *et al.*, 1997) after removal stress factors, such as chemical denaturant or mechanical force. Figure 3 shows electron micrographs of molecules before (a) and after (b) chemical denaturation by 6M Gu-HCl. The normal appearance was completely restored after removal of the denaturant, suggesting a high tendency to restore the original fold, which is in accord with the earlier observations. Refolded titin usually contained fewer oligomers, implying that associated M-line proteins, probably responsible for the titin-titin binding at the head region (Nave *et al.*, 1989), were separated during exposure to Gu-HCl. There was also a greater incidence of side-by-side association, which appeared to result from incorrect refolding joining two neighboring polypeptides. This may be similar to β-sheet interchange reported in crystals of Ig-like constructs (Murray *et al.*, 1995).

Discussion

The conclusions that can be made from the microscopy of molecules are in agreement with mechanical and biochemical *in vitro* and *in situ* studies of extensibility and, in general, are consistent with titin being the major determinant of muscle elasticity. The visual observations show that titin can be easily extended by mechanical force to at least twice its estimated rest length. The

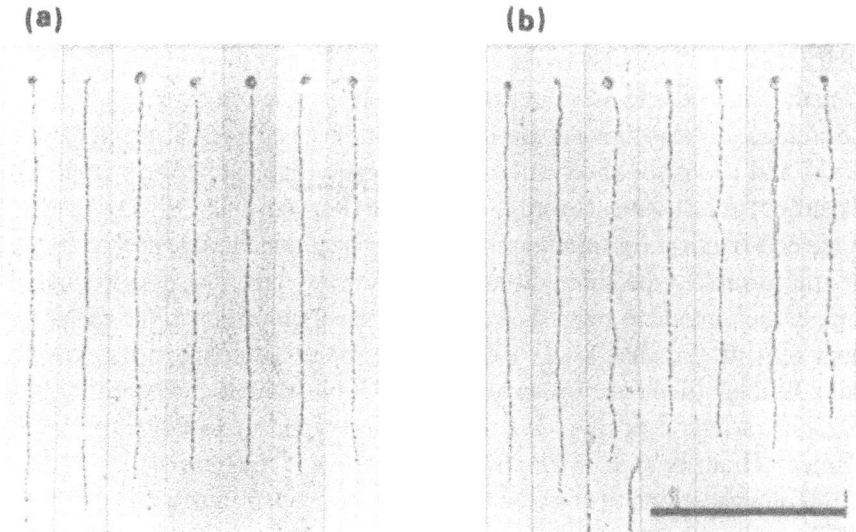

Figure 3. *Electron micrographs illustrating restoration of titin appearance characteristic for the native isolated molecules (a), and after renaturation following Gu-HCl treatment (b). Bar – 0.5 μm.*

extension is not uniform and maximum compliance is seen at or near the position of the PEVK-region, which appeared to be easily extended to at least five times the estimated unstressed length. The resolution of metal-shadowing electron microscopy does not usually allow visualization of the unfolded polypeptide; however, the amount of the observed extension and the overall appearance of the molecules suggest that extension was due to mechanical unfolding of the polypeptide.

Although there is qualitative agreement between the *in vitro* and *in situ* observations, several factors preclude direct quantitative comparison. One factor concerns the effect of environment on the molecule properties. Electron microscopy of titin is usually done at relatively high ionic strength (~0.5 M) and high pH (~7.9), where the tendency of the isolated molecules to aggregate is inhibited. Decreasing the ionic strength to physiological levels causes both self-aggregation and increased binding to the substrate, as can be judged by electron microscopy.

Moreover, under physiological ionic conditions, titin binds to several other myofibrillar proteins (Houmeida *et al.*, 1995; Labeit *et al.*, 1992; Soteriou *et al.*, 1993b) and, reflecting these interactions, different parts of the molecule are bound to different parts of the sarcomere *in situ*. Attachment of the A-band region of the molecule to the thick filament probably prevents extension of this region *in situ* (Wang *et al.*, 1993; Whiting *et al.*, 1989). Structural and mechanical studies (Granzier *et al.*, 1996; Linke *et al.*, 1997; Trinick, 1981) suggest that situation in the I-band region is likely to be even more complicated. Each thick filament has several titin molecules bound to it, and these probably emerge from the filament end as a single bundle. Self-association here is indicated by the presence of "end-filaments" at the tips of the isolated thick filaments (Trinick, 1981), and by *in situ* observations of the elastic filament network connecting the ends of thick filaments with Z-line region (Funatsu *et al.*, 1993). The length of the end-filament (or titin bundle near A-/I-band junction) is around 0.1 μm. At approximately the same distance from the A/I-junction, electron micros-copy of the sarcomere suggests a change in the organization of the elastic fila-ment network (Funatsu *et al.*, 1993). At this point the connections extending from the thick filaments tips appear to branch into narrower filaments. Closer to Z-line, a region of titin ~0.1 μm long is thought to be bound to the thin filament (Granzier *et al.*, 1996; Linke *et al.*, 1997; Trombitás *et al.*, 1997).

   Thus, according to this scheme, three zones with different organization of titin are present in the I-band (Fig. 4): (1) outside of the tip of thick filament 0.1 μm long segments of titin molecules are associated side-by-side in register in bundles consisting of several molecules; (2) between approximately 0.1 μm from the end of thick filament and 0.1 μm from the center of Z-line the mol-ecules appear separate; and (3) at approximately 0.1 μm from the Z-line center titin binds to thin filaments.

   One can speculate what happens when this network is stretched (Fig. 4c). It is evident that, even if the mechanical properties of the separate titin mol-ecule were uniform along its length, the mechanics of the I-band region will not be uniform *in situ*. The middle of the I-band part, between 0.1 μm from the A/I-junction and 0.1 μm from Z-line center, is therefore likely to be most vul-nerable to extension. Interaction with thick and thin filaments probably makes the corresponding titin regions essentially inextensible. Similarly, the self-as-sociation to form end-filaments is likely to be the cause of the increased stiff-ness of the connecting filaments near the A/I-junction.

   The rearrangement of the titin network in the middle of the I-band region is likely to be related to symmetry considerations. The thick and thin filament lattices in vertebrate striated muscles, as well as the thick and thin filaments themselves, have different symmetries and the intermolecular interactions of titin in these regions must in some way reflect this. Branching of the titin bundles in the middle of the I-band region may well facilitate or initiate the transition from the hexagonal/threefold symmetry of the A-band to the tetragonal/two-fold symmetry near and in the Z-line.

   The branching and rearrangement of titin appears to occur near the PEVK region. This is an appropriate site since there is a sharp structural change here and rise of net charge. A bundle comprising several molecules could be ex-

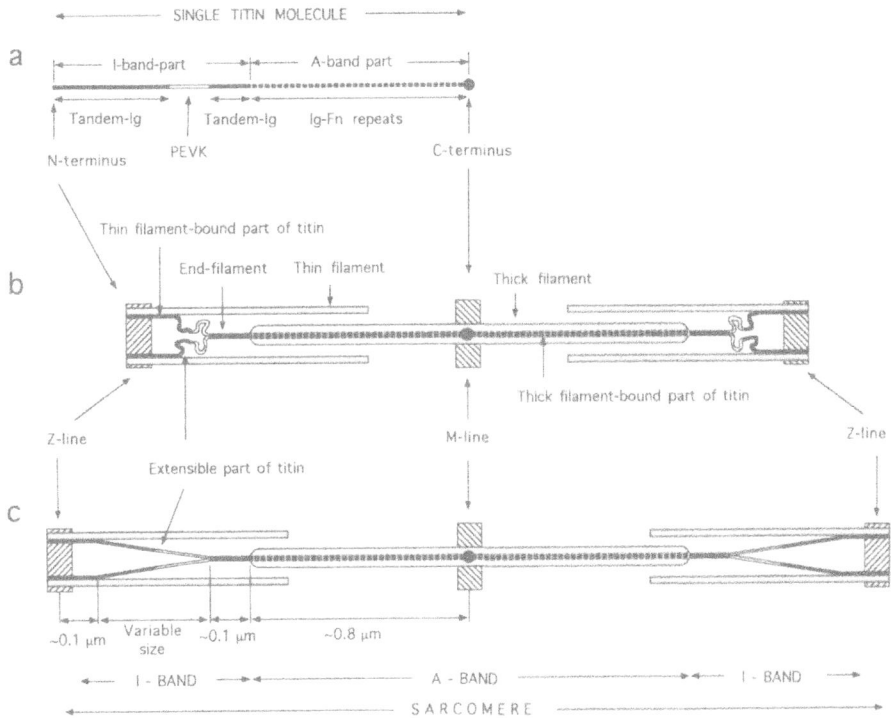

Figure 4. *Diagram showing titin structure and organization in muscle sarcomere. (a) Major functional and structural regions of the titin molecule. (b) Probable titin arrangement in muscle at slack sarcomere length, and (c) during sarcomere extension and developing passive tension. For details see text.*

pected to be very unstable in the PEVK region (immediately N-terminal to the end-filament), because the molecules will experience strong electrostatic repulsion from one another. The molecules will therefore tend to separate beyond the end-filaments. Thus, in addition to the role it plays in elasticity, the PEVK-region may also facilitate the rearrangement of the titin network in order to unite the thick and thin filament lattices.

Acknowledgments

Supported by The Wellcome Trust and BBSRC (UK).

References

Carrion-Vazquez M, Oberhauser AF, Fowler SB, Marszalek PE, Broedel SE, Clarke J, Fernandez JM. Mechanical and chemical unfolding of a single protein: a comparison. *Proc Nat Acad Sci USA* 1999;96:3694-3699.

Erickson HP. Reversible unfolding of fibronectin type-III and immunoglobulin domains provides the structural basis for stretch and elasticity of titin and fibronectin. *Proc Nat Acad Sci USA* 1994;91:10114-10118.

Funatsu T, Kono E, Higuchi H, Kimura S, Ishiwata S, Yoshioko T, Maruyama K, Tsukita S. Elastic filaments in situ in cardiac-muscle - deep-etch replica analysis in combination with selective removal of actin and myosin-filaments. *J Cell Biol* 1993;120:711-724.

Furst DO, Osborn M, Nave R, Weber K. The organization of titin filaments in the half-sarcomere revealed by monoclonal antibodies in immunoelectron microscopy: a map of ten nonrepetitive epitopes starting at the Z line extends close to the M line. *J Cell Biol* 1988;106:1563-1572.

Gautel M, Goulding D. A molecular map of titin/connectin elasticity reveals 2 different mechanisms acting in series. *FEBS Lett* 1996;385:11-14.

Granzier H, Helmes M, Trombitás K. Nonuniform elasticity of titin in cardiac myocytes - a study using immunoelectron microscopy and cellular mechanics. *Biophys J* 1996;70:430-442.

Helmes M, Trombitás K, Granzier H. Titin develops restoring force in rat cardiac myocytes. *Circ Res* 1996;79:619-626.

Houmeida A, Holt J, Tskhovrebova L, Trinick J. Studies of the interaction between titin and myosin. *J Cell Biol* 1995;131:1471-1481.

Kellermayer MSZ, Smith SB, Granzier HL, Bustamante C. Folding-unfolding transitions in single titin molecules characterized with laser tweezers. *Science* 1997;276:1112-1116.

Labeit S, Gautel M, Lakey A, Trinick J. Towards a molecular understanding of titin. *EMBO J* 1992;11:1711-1716.

Labeit S, Kolmerer B. Titins: giant proteins in charge of muscle ultrastructure and elasticity. *Science* 1995;270:293-296.

Linke WA, Bartoo ML, Ivemeyer M, Pollack GH. Limits of titin extension in single cardiac myofibrils. *J Musc Res Cell Motil* 1996a;17:425-438.

Linke WA, Ivemeyer M, Labeit S, Hinssen H, Ruegg JC, Gautel M. Actin-titin interaction in cardiac myofibrils: Probing a physiological role. *Biophys J* 1997;73:905-919.

Linke WA, Ivemeyer M, Olivieri N, Kolmerer B, Ruegg JC, Labeit S. Towards a molecular understanding of the elasticity of titin. *J Mol Biol* 1996b;261:62-71.

Maruyama K, Itoh Y, Arisaka F. Circular dichroism spectra show abundance of β-structure in connectin, a muscle elastic protein. *FEBS Lett* 1986;202:353-355.

Murray AJ, Lewis SJ, Barclay AN, Brady RL. One sequence, 2 folds - a metastable structure of Cd2. *Proc Nat Acad Sci USA* 1995;92:7337-7341.

Nave R, Furst DO, Weber K. Visualization of the polarity of isolated titin molecules: a single globular head on a long thin rod as the M-band anchoring domain. *J Cell Biol* 1989;109:2177-2187.

Rief M, Gautel M, Oesterhelt F, Fernandez JM, Gaub HE. Reversible unfolding of individual titin immunoglobulin domains by AFM. *Science* 1997;276:1109-1112.

Soteriou A, Clarke A, Martin S, Trinick J. Titin elasticity and folding energy. *Proc Roy Soc Lond B* 1993a;254:83-86.

Soteriou A, Gamage M, Trinick J. A survey of the interactions made by titin. *J Cell Sci* 1993b;14:119-123.

Trinick J. End-filaments: a new structural element of vertebrate skeletal muscle thick filaments. *J Mol Biol* 1981;151:309-314.

Trinick J, Knight P, Whiting A. Purification and properties of native titin. *J Mol Biol* 1984;180:331-356.

Trombitás K, Greaser ML, Pollack GH. Interaction between titin and thin filaments in intact cardiac muscle. *J Musc Res Cell Motil* 1997;18:345-351.

Tskhovrebova L, Trinick J. Direct visualization of extensibility in isolated titin molecules. *J Mol Biol* 1997;265:100-106.

Tskhovrebova L, Trinick J, Sleep JA, Simmons RM. Elasticity and unfolding of single molecules of the giant muscle protein titin. *Nature* 1997;387:308-312.

Wang K, McCarter R, Wright J, Beverly J, Ramirez-Mitchell R. Viscoelasticity of the sarcomere matrix of skeletal muscles. The titin-myosin composite filament is a dual-stage molecular spring. *Biophys J* 1993;64:1161-1177.

Whiting A, Wardale J, Trinick J. Does titin regulate the length of muscle thick filaments? *J Mol Biol* 1989;205:263-268.

Yang P, Tameyasu T, Pollack GH. Stepwise dynamics of connecting filaments measured in single myofibrillar sarcomeres. *Biophys J* 1998;74:1473-1483.

Discussion
(Presented by John Trinick)

Andrew: I need to ask a naïve question. Could you just have these bundles together and have 12 in the Z-line region.

Trinick: You have got 4 lattice points on the Z-disk, and you could imagine that only some of them are occupied. But I think a take-home message from biology is that it is much easier to imagine you have the same rule in all cases. So if you were going to have some attachment points occupied and not others, then that is harder to imagine.

Andrew: Yes, if your simple model does not work.

Trinick: You can imagine more complicated hypotheses. You could imagine adaptor molecules which have say 3 sites on one end and 4 sites on the other end, but that is quite a complicated hypothesis.

Linke: What muscle did you use?

Trinick: This was rabbit psoas, and we did frog thigh muscle as well. The two muscle types gave similar results. A-band titin sequences are very highly conserved and A-band structure generally in vertebrate-striated muscle is rather highly conserved, so I don't think the numbers of titin molecules is going to vary.

Linke: The only point would be that perhaps the Funatsu 1993 paper would show some sort of interactions in the I-band titin.

Trinick: Well, those are freeze-fracture results. My personal prejudice is that I think the evidence for an interaction between titin and actin near the Z-line is quite strong. But when you see titin bound to thin filaments in the I-band by freeze-fracture, you've got to worry that everything is waving around and maybe they just get stuck together.

Linke: So you'll exclude that titin filaments merge from say 6 filaments into 4 at the N2 line?

Trinick: The numerology still doesn't work out, I think. You have an essential 2-ness about the Z-line, so if you've got an essential 3-ness coming out of the thick filament, you have still got a problem.

Pollack: Well my question is an extension of what Wolfgang just said. Isn't it true that your data, while extremely interesting, doesn't really address the issue of what happens at the Z-line?

Trinick: Well, it is going to be the same number of molecules, so I think the problem doesn't really go away. You've got to dispose of that number of molecules at the Z-disk unless half of the molecules turn around and go in the opposite direction, which seems unlikely.

Vigoreaux: If I understand it correctly, then the end filaments result from titin molecules that have been broken.

Trinick: The easiest way to make separated thick filaments is to leave the muscle post mortem for a while, and I think this allows it to be proteolytically cleaved. You make separated thick and thin filaments by homogenizing relaxed muscle. If you do that with completely fresh muscle, it doesn't work. So I think there needs to be a bit of proteolysis first.

Vigoreaux: The assumption of your measurements would be that all titin molecules are broken at the same place.

Trinick: The negative stain electron microscopy, of particularly Roger Craig, shows very, very regular stalk-like structures at the end of the A-band. There are also quite a lot of sections in the literature where you can see what looks like a stalk-like projection beyond the end of the thick filament. Further into the I-band it's more difficult to see what happens then, but the guess is that the molecules separate. There is a huge amount of charge at the PEVK region at that point, so the molecules may well separate then on the way to the Z-line.

Granzier: I think that distal tandem Ig segment in the sarcomere, without external force, collapses and contracts. So it is puzzling that the end filaments always look so straight.

Trinick: If the titin molecules are packed in register there, then I imagine that it would be reasonable to expect that the titin assembly would be stiffer than a single molecule.

Granzier: I was referring to observations in the sarcomere.

Trinick: So you mean that there are antibodies to both ends of the tandem Ig region and they superimpose.

Granzier: Yes.

Trinick: I don't know, but there are quite a lot of end filament reports. End filaments have been reported in scallops and...

Greaser: Are these all negatively stained?

Trinick: There is also section data. There are structures that you could well think were end filaments. You can see stalk-like structures of about the right length, but the trouble is the sectioning doesn't have resolution to pick of the band patterning, but they look quite plausibly like that, so the data is not all negative stain.

Krammar: I want to know what is the number ratio in the unit cell of thick and thin filaments?

Trinick: It's two to one, so in the unit cell there would be two actin filaments to every thick filament in the sarcomere. That is in the vertebrate striated muscle.

Krammar: So then what's the mentioning of 12?

Trinick: There are 12 titins per whole thick filament, six in each half filament. You've got multiples of 3 titins coming off each thick filament, and you've got to divide those between the structure of the Z-disk, which has essentially multiples of 2, so I think there is still a problem.

Sugi: I want to make a suggestion. I am very interested in your beautiful electron micrographs. It appears that the titin filament is bound tightly to the thick filament. One experiment I would like to suggest is to make thick filaments and study their periodicities, using X-ray diffraction, before and after myosin extraction. This may make many issues clear.

Trinick: I think the trouble is you want to solve the structure of the thick filament which is a hard problem.

Trombitás: The end filament is long enough to be labeled with T11. Have you ever tried to label the end filaments with T11? This would prove that the end filaments are titin.

Trinick: No I haven't, but we have made an antibody to I27 recently so we were hoping to pursue that question.

Discussion
(Presented by Larissa Tskhovrebova)

Greaser: Do you have evidence of PEVK length in any of your micrographs?

Tskhovrebova: Extended?

Greaser: Extended; in other words, have you compared different titin sources and looked at the apparent lengths.

Tskhovrebova: In the PEVK region, a structure can't be resolved in electron micrographs. In molecules 1.3 micron long, you can't resolve this structure as something separate. The presence of this structure reveals itself only in segmented molecules, so you can see at this point that there is something different. I think that the resolution is not enough.

Andrew: How do you purify the titin?

Tskhovrebova: We dissolve muscle in high ionic strength buffer followed by precipitation at low ionic strength, and then purification by gel filtration.

Andrew: Have you tried protein made in Ecoli, different fragments and doing the same kind of measurements with different domains of the protein?

Tskhovrebova: To study flexibility of the fragments?

Andrew: Exactly.

Tskhovrebova: Well, I suppose it is possible. Actually you need long molecules in order to get enough force—forces directly relate to the length of the molecule.

Andrew: What is the smallest size of expressed titin fragment that you could use for your work?

Tskhovrebova: Well, I suppose you could calculate this, but we did not do this.

Bullard: Does this message rely on one end of the titin molecule, that is the M-line end, sticking to the substrate so that the rest can be stretched out? If that is the case, then using shorter constructs wouldn't necessarily work because they will just all wash off the grid.

Tskhovrebova: This is a good question. It depends on the charge distribution along this constructs and if you will find something different. For example, the charge distribution along the titin molecule is such that the ends of the polypeptide have a higher positive charge, and because you have negatively charged mica they will tend to stick stronger to the substrate than the rest of the molecule. So it depends on the charge distribution.

Bullard: Only the M-line end is sticking? These intermediate charges are not sticking?

Tskhovrebova: No, actually I show head-centered oligomers where the N-terminal end bound first. The extension happened because the N-terminal end portion, not head region, bound first. After this, force straightens the molecule and the second sticky point, the head region stabilizes the stretched molecule.

Sanger: What is the salt concentration of the solution that your rhodamine titin's are in?

Tskhovrebova: It is the same high ionic strength. It is 0.5 M.

Greaser: If you look at your extended molecules you usually see a head. But the ones that were directly placed on the mica, the heads are not as visible. How do you explain that?

Tskhovrebova: No, actually you can find heads everywhere. But I suppose when they freely lay on the substrate it is more difficult to find where the head is because the molecule is folded.

Greaser: The heads should be at the ends.

Tskhovrebova: Yes, but the ends are folded inside the molecule. Actually you could see heads in some occasions. They are quite clear in the negative-stained preparations.

Greaser: The ones where they were not extended, the slide at least that you showed, there were rarely heads on the ends of those molecules. So is the head due to part of the titin or is it due to excessory proteins?

Tskhovrebova: I suppose what can happen, even if it is one single string it still shows a small head. This just might be M-line proteins. It was shown by D. Fürst and his colleagues that the head contains these proteins. What else gives rise to the head is not clear.

TITIN ELASTICITY IN THE CONTEXT OF THE SARCOMERE: FORCE AND EXTENSIBILITY MEASUREMENTS ON SINGLE MYOFIBRILS

Wolfgang A. Linke

Institute of Physiology II, University of Heidelberg, Heidelberg, Germany

Abstract: Skeletal-muscle titin contains in its I-band section two main elastic elements, stretches of Ig-like domains and the PEVK segment. Both elements contribute to the extensibility and passive force development of relaxed skeletal muscle fibers during stretch. To explore the nature of elasticity of the segments, their force-extension relation was determined with immunofluorescence and immunoelectron microscopy, combined with isolated myofibril mechanics. The results were then fitted with recent models of biopolymer elasticity. Whereas an entropic-spring mechanism may account for the elasticity of the Ig-domain segments, PEVK-titin elasticity appears to have both entropic and enthalpic origins. The modeling explains why the two elements extend sequentially upon stretch: elongation of the Ig-domain regions (with folded modules) is followed by unraveling of the PEVK domain.

   I-band titin in cardiac muscle is expressed in two main isoforms, N2-A and N2-B. The N2-A isoform is similar to that found in skeletal muscle, whereas the N2-B titin is distinguished by cardiac-specific Ig-motifs and nonmodular sequences within the central I-band section. By examining the extensibility of N2-B titin, it was found that this isoform extends by recruiting three distinct elastic elements: poly-Ig regions and the PEVK domain at low to modest stretch, and in addition, a unique 572-residue sequence insertion at higher physiological stretch. Extension of all three elements allows cardiac titin to stretch fully reversibly at physiological sarcomere lengths, without the need to unfold individual Ig domains.

Elastic Filaments of the Cell, edited by Granzier and Pollack
Kluwer Academic/Plenum Publishers, 2000
179

Introduction

Over the past two decades, a wealth of evidence has been amassed indicating that the sarcomeres of vertebrate striated muscles contain a third filament system—besides thin and thick filaments—which is made up of giant proteins, the titin/connectin molecules (Maruyama, 1997; Wang, 1996; Trinick, 1996; Gregorio *et al.*, 1999). Arguably one of the best-known properties of titin filaments is their ability to act as molecular springs that provide non-activated ("relaxed") myofibrils with elasticity. The >3 MDa-large titin polypeptides span, as continuous strands, the half of a sarcomere from the Z-disk (N-terminus) to the A-band center (C-terminus), but only a relatively short segment of titin confined to the I-band is functionally elastic. The nature of titin elasticity has long remained unknown, but the past 3-4 years have seen remarkable progress in the field that began with the determination of the full-length human titin sequence (Labeit and Kolmerer, 1995). For the first time, the elastic I-band titin region was shown to consist of two main structural elements, stretches of tandemly arranged immunoglobulin-like (Ig) domains that flank a unique sequence rich in proline (P), glutamate (E), valine (V) and lysine (K) residues, the PEVK segment (Fig. 1). Both elements are expressed in specific length variants, depending on muscle type. Before the discovery of the PEVK segment, early concepts on the nature of titin elasticity had assumed that the molecule's elastic behavior could result from reversible unfolding of globular modules, such as Ig domains (Erickson, 1994). On the other hand, biophysical

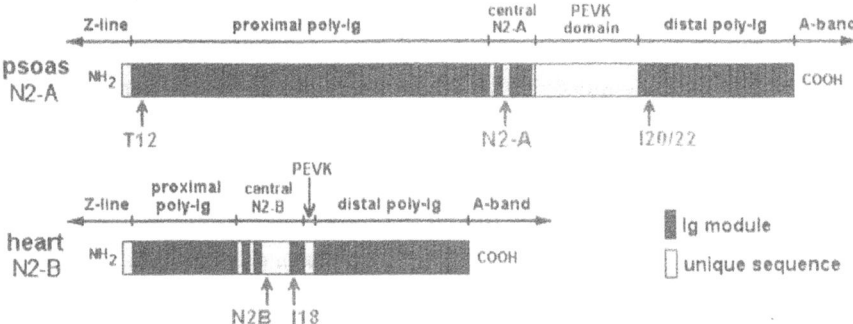

Figure 1. Domain architecture of the elastic I-band titin region (after Labeit and Kolmerer, 1995). Schematically shown is the structure of the two main titin isoforms, N2-A and N2-B, expressed in human striated muscle. The N2-A isoform shown is found in psoas muscle; the N2-B isoform is specific to cardiac muscle. In both isoforms a PEVK domain is flanked by stretches of tandemly arranged Ig-like modules. Unique to each isoform is the central I-band region, which is made up of isoform-specific Ig-domains and nonmodular sequences, notably a 572-residue insertion in N2-B. The epitope locations of the titin antibodies used in this study are indicated by the arrows.

studies clearly demonstrated that titin's Ig modules fold into thermodynamically stable domains (Politou *et al.*, 1995). It was therefore proposed that the Ig-domain regions might represent quite stiff components. With the primary structure of I-band titin in hand, the concepts were now testable.

Two main approaches have since been taken to study the nature of titin elasticity. One approach consists in the use of single isolated molecules to determine the mechanical properties of titin *in vitro*. A variety of elegant methods has been employed, including a "molecular combing" technique where titin is stretched with meniscus force (Tskhovrebova and Trinick, 1997) and micromechanical manipulation of molecules by optical tweezers (Kellermayer *et al.*, 1997; Tskhovrebova *et al.*, 1997) or the atomic force microscope (Rief *et al.*, 1997) (please see the authors' contributions to this volume). In another approach, the mechanism of titin elasticity can be probed in preparations in which the orderly structure of the sarcomere is preserved. Soon after the full titin sequences became available, Gautel and Goulding (1996) and Linke *et al.* (1996) immunostained stretched sarcomeres with sequence-assigned I-band titin antibodies to investigate the extensibility of individual titin regions. They found that both the Ig-domain regions and the PEVK segment contribute to titin extensibility. Moreover, the two elements extended sequentially. The molecular basis of such sequential extension behavior was subsequently investigated in the context of the sarcomere by Granzier *et al.* (1997), Trombitas *et al.* (1998) and Linke *et al.* (1998a,b). The following overview aims to summarize the principal results of the work done in the Linke laboratory.

Results and Discussion

Rationale

The results obtained by Linke *et al.* (1996) demonstrated that in skeletal myofibrils, low stretch forces extended titin's poly-Ig chains, whereas much higher forces were needed to extend the PEVK domain. It was suggested that upon stretching myofibrils from slack sarcomere length (SL), the "contracted" state of I-band titin changes first by straightening of poly-Ig regions, followed by unraveling of the PEVK domain. Unfolding of individual Ig domains was considered unlikely, because the Ig-domain regions almost ceased to extend at modest stretch. A first clue as to why the structurally distinct titin segments exhibit such differential extensibility then came from the results of single-molecule studies on titin.

The micromechanical studies suggested that at stretch forces similar to those occurring during normal muscle function, the elasticity of titin may be explained by an "entropic spring" mechanism (Kellermayer *et al.*, 1997; Rief *et al.*, 1997; Tskhovrebova *et al.*, 1997). The entropic-chain characteristics of the molecule would result in a nonlinear force response to stretch, similar to the well-known nonlinear rise in passive force seen during stretch of myofibrils. At low forces, titin's force-extension relation could be well fitted with a worm-like chain (WLC) model of entropic elasticity. An entropic chain undergoes thermally induced bending movements that tend to shorten its end-to-end length. Stretching such a chain reduces its conformational entropy and thus, requires an external force. Taken simply, the stiffness of a WLC may be parameterized by its contour length, L, and a persistence length, A, which is a measure of the chain's bending rigidity (also see the section "Application of polymer elasticity theory"). The larger the persistence length, the more rigid the polymer and the smaller the external forces required for stretching out (straightening) the molecule. In the single-molecule studies on titin it was proposed that the persistence length of the Ig-domain region might be relatively large, whereas that of the PEVK segment might be much smaller, resulting in titin's differential extension behavior.

The *in-vitro* studies on titin also indicated that the molecule's Ig domains are capable of unfolding, at least at high external forces. Whether Ig-domain unfolding was a mechanism relevant to the physiological function of titin, remained a matter of dispute. To answer this issue and to probe the entropic-spring concept, it was necessary to examine whether the elastic behavior of titin is similar in the single isolated molecule and in the sarcomere, where titin filaments are arranged in a three-dimensional network. In addition, *in vivo*, titin filaments are functionally elastic only along part of their length, and their extensible I-band section has a heterogeneous structure (Fig. 1) consisting of poly-Ig regions and the PEVK domain. Therefore, it was desirable to "dissect" I-band titin into the structurally distinct regions and measure the elastic properties of each segment in the environment of the sarcomere.

Experimental approach

The concept of sequential titin extension suggested by Linke *et al.* (1996) was based on observations made on mechanically manipulated single myo-fibrils. The myofibrils can be isolated from both skeletal and cardiac muscle tissue according to standard protocols and viewed under an inverted micro-scope (Bartoo *et al.*, 1993; Linke *et al.*, 1994). With the help of micromanipu-lators, a preparation is picked up at its ends by the glue-coated tips of fine glass needles. By connecting the glass needles to a sensitive force transducer and a micromotor, respectively, it is possible to measure the forces that develop, for example, during stretch of a relaxed myofibril (Linke *et al.*, 1994; 1996). More-over, the specimen can also be immunolabeled with antibodies specific to sar-comeric proteins, such as titin, and visualized in the fluorescence mode of the microscope by using a fluorophore-conjugated secondary antibody (Fig. 2a).

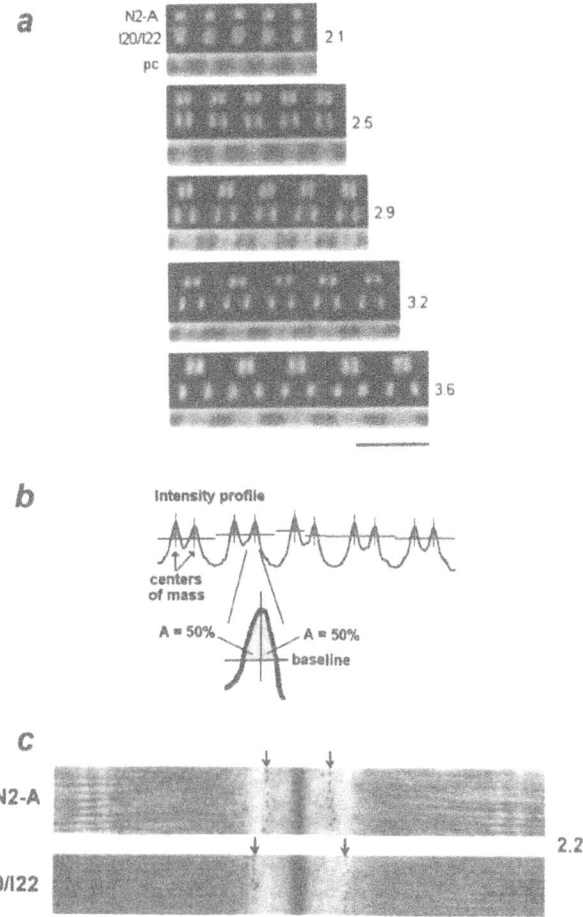

Figure 2. *Examples of immunostainings of rat psoas myofibrils with titin antibodies flanking the PEVK region. (a) Immunofluorescence images of single psoas myofibrils stretched to different SLs (the actual degree of stretch in μm units is indicated on the right) and labeled with either the N2-A or the I20/I22 titin antibody. As secondary antibody, Cy-3-conjugated IgG was used. pc, phasecontrast images. Scale bar, 5 μm. The panel in (b) shows the intensity profile of a fluorescence image to demonstrate how the spacing between intensity peaks was determined by performing a center-of-mass analysis (for detailed explanation, see the section "Measurement of epitope spacing"). (c) Immunoelectron micrographs of psoas-muscle sarcomeres at 2.2 μm SL, stained with N2-A or I20/I22 titin antibodies. The nanogold particles indicate the respective epitope positions (arrows). Scale bar, 0.5 μm.*

Because the position of titin epitopes in the I-band varies with myofibril stretch, the method can be applied to analyze the extension behavior of titin segments under defined experimental conditions. These techniques were used to identify the relative contribution of the two structurally distinct I-band titin segments—poly-Ig regions and the PEVK domain—to myofibrillar elasticity.

Antibodies

A prerequisite for the study of titin-segment elasticity is the availability of sequence-assigned antibodies. Figure 1 indicates the epitope position of antibodies to I-band titin used in our analysis; the titin structure is shown schematically according to Labeit and Kolmerer (1995) for two length isoforms of I-band titin expressed in psoas skeletal muscle (N2-A isoform) and in the heart (N2-B isoform). (For details, please see the contribution of S. Labeit to this volume.) For the analysis of psoas-muscle titin, three different antibodies were used which flank the proximal Ig-domain region (T12, N2-A) and the PEVK segment (N2-A, I20/22) at their NH_2-termini and COOH-termini, respectively. T12 is a monoclonal antibody (Fürst *et al.*, 1988), N2-A and I20/22 are affinity-purified polyclonal antibodies (Linke *et al.*, 1996; 1998a;b). The immunostaining of the cardiac-specific N2-B isoform will be discussed below (see "Cardiac titin elasticity" section).

Immunofluorescence microscopy

A myofibril is stretched in relaxing buffer to a desired SL and is labeled with primary titin antibody and Cy3-conjugated secondary IgG (for details, see Linke *et al.*, 1996; 1998a;b). The technique works on both unfixed and fixed myofibrils; in case of fixation, the stretched specimen is treated for 20 minutes with a 3.5% paraformaldehyde solution added before the staining procedure. Identical results were obtained using both fixed and unfixed myofibrils, although fixation tended to decrease fluorescence intensity. The dilution of antibodies (in relaxing buffer) depends on the reactivity of the antisera, and was between 1:5 and 1:50; for secondary antibodies, 1:50. In most experiments the incubation time was 20 min for both primary and secondary antibodies. Images were recorded with a sensitive black and white CCD camera in the epifluorescence mode of the microscope (100x, 1.4NA oil immersion objective) and digitized using a frame grabber and image acquisition software (Global Lab Image, Data Translation, Marlboro, MA). Typical images of rat psoas myofibrils extended to different SLs and stained with antibodies specific to N2-A or I20/I22 are shown in Figure 2a.

Measurement of epitope spacing

To map the epitope positions in a half-sarcomere, we plotted the intensity profiles along the myofibril axis and calculated the center-of-mass pixel position for each intensity peak with custom-written software (Fig. 2b). The program calculates the area under an intensity maximum, above a "baseline" drawn parallel to the abscissa. This baseline was either set to pass through the relative minimum between peaks reflecting the unlabeled Z-line region or was positioned by eye to above that minimum. The latter case is shown in Figure 2b. Next, the program calculates those points on the abscissa at which a vertical straight line can be constructed that divides the area under each maximum into two equal parts (vertical lines in Fig. 2b). These points were then used to measure the epitope spacings across the Z-line. The calculated peak distances were divided by two to obtain the Z-disk-center to epitope distance. Because the method could be applied successfully only when peaks did not overlap, reliable detection required epitope spacings of at least ~200 nm. Excellent results with this analysis method were obtained at modest-to-long SLs, whereas at short SLs, the two epitopes around the Z-disk frequently were seen to merge into one broader strip. Therefore, epitope spacings particularly at short SLs were analyzed by immunoelectron microscopy (IEM; example in Fig. 2c). A standard nanogold-particle technique was used and is explained in detail elsewhere (Tokuyasu, 1989; Mundel *et al.*, 1991).

Extensibility of titin segments in skeletal muscle

A summary of results of immunostaining experiments on rat psoas myofibrils is shown in Figure 3a. For the antibodies N2-A and I20/22, data from both IEM and immunofluorescence microscopical measurements are included. Results obtained with the two methods were similar within experimental error. The data sets for these antibody types were fitted by third-order regressions. The T12 data points (IEM results only) were fitted with a linear regression curve that has a very flat slope, indicating that the contribution of the Z-disk-T12 segment to the extensibility of I-band titin is very small.

From the results shown in Figure 3a, the extensibility of the structurally distinct segments within psoas I-band titin can be deduced: the T12 to N2-A distance reveals extension of the proximal Ig-domain region, the N2-A to I20/22 spacing that of the PEVK segment, and the distance between I20/22 and the A-band edge that of the distal Ig-domain region. Lengthening of the proximal Ig-domain segment took place mainly at smaller extensions: at 2.1 μm SL, the region's end-to-end length was ~70 nm, whereas at 2.8 μm it was ~200 nm; stretch to 3.2 μm SL increased the end-to-end length to 215 nm. The PEVK-segment length at the average slack SL of ~2.1 μm was approximately 45 nm, and the segment began to elongate substantially at ~2.4 μm SL. PEVK extension then continued over most of the SL range investigated. The distal Ig-domain segment contributed up to ~100 nm to the length of I-band titin at modest to high sarcomere stretch.

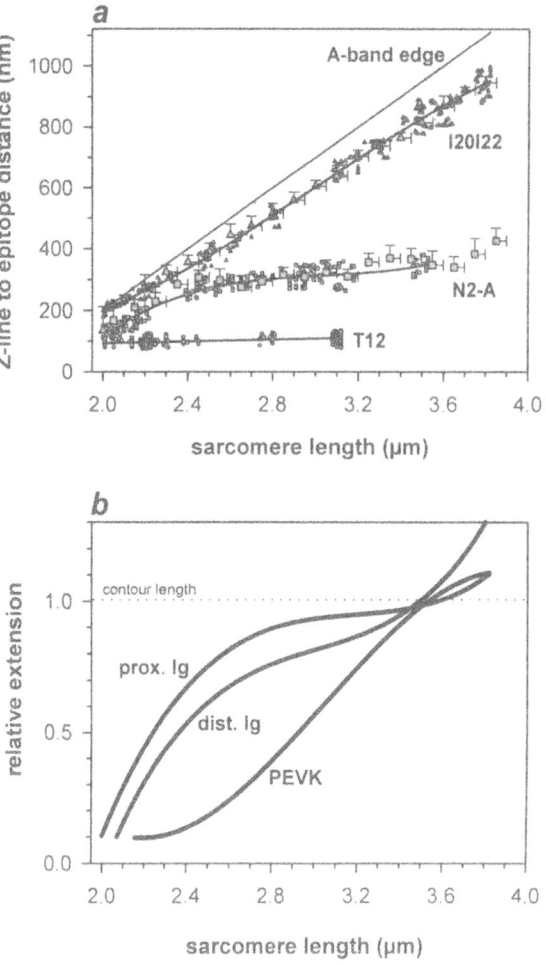

Figure 3. *Extension behavior of I-band titin segments in rat psoas muscle. (a) Summary of results of immunolabeling experiments. Data points for the Z-line center to epitope spacing at different SLs, as determined by immunoelectron microscopy, are shown for the antibodies T12 (small circles), N2-A (small squares) and I20/ I22 (small triangles). The larger shaded circles and error bars indicate the mean distances from the Z-disk center (in 0.1 μm SL bins) and standard deviations for both N2-A and I20/I22, measured by immunofluorescence microscopy. Data sets were fitted as follows: T12 by linear regression, N2-A and I20/I22 by third-order regressions. (b) Extension of a given titin segment versus SL, relative to each segment's predicted contour length (225 nm for the proximal Ig-domain region, 475 nm for the PEVK domain, 115 nm for the distal Ig-domain segment). The curves were calculated from the fit data in (a).*

To obtain a better estimate of the extension capacity of individual titin segments, we plotted the relative (or fractional) extension of each segment *versus* SL (Fig. 3b). This can be done by relating the end-to-end length of a titin segment at a given SL to the contour length predicted from the segment's sequence structure. Contour-length values were predicted as follows (Linke *et al.*, 1998a;b): proximal Ig-domain region, 225 nm (50 folded Ig domains times 4.5 nm for the assumed maximal domain spacing); PEVK domain: ~475 nm (1,400 residues per psoas PEVK segment times 0.34 nm for the maximal residue spacing); and distal Ig-domain region, ~115 nm (22 Ig domains plus a few extra modules at the A/I junction, times 4.5 nm). From the plot shown in Figure 3b it can be concluded that (i) extension of the Ig-domain regions precedes that of the PEVK segment; only when the Ig-segment extension reaches 70-80% does the PEVK domain begin to stretch substantially; and (ii) the end-to-end length of all titin segments stays below the predicted contour-length value, up to ~3.5 μm SL, *i.e.*, in a physiologically relevant range of SLs.

Force-Extension curves of titin segments

Force per titin

To measure titin elasticity, two main parameters are needed: extensibility of titin segments and the force needed to stretch the molecule. The first parameter was obtained as described above. The second can be deduced from the passive length-tension curve of single myofibrils or small bundles of myofibrils (Fig. 4a; see Linke *et al.*, 1998a;b). The method of measuring the nanonewton forces developed by single myofibrils has been applied successfully for some time in this laboratory, and details have been described (*e.g.*, Linke *et al.*, 1994; 1997). Somewhat less straightforward is the inferral of force per titin strand from passive tension measurements on myofibrils, because the exact number of titin filaments per cross-sectional area is not known. However, it is reasonable to assume that approximately 1000-2000 parallel filaments exist in a single myofibril (Linke *et al.*, 1994; 1998a;b). Other assumptions must be made to obtain force per titin: (i) all titin molecules behave independently of one another and have comparable elastic properties; (ii) all parallel titin strands or titin segments within a molecule bear the same force at a given extension; the sum of these parallel forces is the force measured at the ends of the myofibril. The force per titin then calculated is in the range of 5-10 pN at 3 μm SL and 20-40 pN at 3.5 μm SL. The shape of the force per titin *versus* SL curve is the same as that of the passive length-tension curve shown in Figure 4a.

Force-extension relations

To define the elastic properties of the titin segments, force per titin strand was plotted *versus* relative extension, for both the proximal Ig-domain region

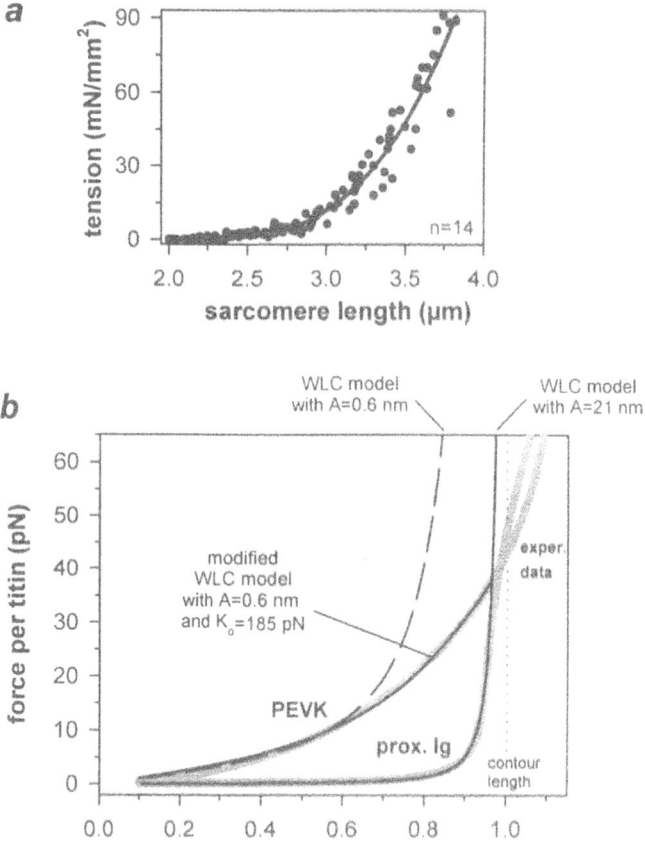

Figure 4. *Force-extension relations and fits according to polymer-elasticity theory. (a) Passive tension versus SL relation of isolated rat psoas myofibrils. Quasi steady-state tension (following stress relaxation) was measured. Data points shown include only those force values obtained during the first stretch of the specimens; some preparations were not extended to above 3.0 μm SL. The data were fitted by a third-order regression. (b) Force per titin deduced from the tension data in (a) versus relative extension of the proximal Ig-domain region and the PEVK segment. Thick shaded lines indicate fits to the experimentally determined data. These data were then simulated according to a standard WLC model of entropic elasticity (thin black line for prox. Ig curve; dashed line for PEVK curve). The PEVK data were also fitted with a modified WLC model of entropic-enthalpic elasticity, which revealed a much better fit above 60% relative extension. For further details, see text.*

and the PEVK segment (Fig. 4b, thick gray curves). It appears that the PEVK curve increases much more steeply than the Ig-segment curve, at least at relative Ig-segment extensions below 80-85%, *i.e.*, up to 2.6-2.7 μm SL. Thus, low stretch forces will straighten out the Ig-domain region but barely the PEVK segment, whose extension requires much higher forces. On the other hand, above 90% relative extension, the Ig-segment curve begins to rise dramatically, indicating that the segment-length increase at stretch forces acting on the filament above 2.8 μm SL is reduced greatly. Titin now elongates mainly by extension of the PEVK domain. Finally, it is important to note that the slope of the force-extension curve of the Ig-domain segment decreased again when force reached values of several tens of piconewtons (see below).

Application of polymer elasticity theory

WLC model

Because the single-molecule experiments (Kellermayer *et al.*, 1997; Rief *et al.*, 1997; Tskhovrebova *et al.*, 1997) had suggested that titin may be an entropic spring showing WLC behavior, we tried to fit the experimental data with a standard WLC model of entropic elasticity (Bustamante *et al.*, 1994; Marko and Siggia, 1995). According to this model, the external force (*f*) is related to the fractional extension (*z/L*) of the chain by

$$f = \left(\frac{k_B T}{A}\right)\left[\frac{1}{4(1 - z/L)^2} - \frac{1}{4} + \frac{z}{L}\right] \tag{1}$$

where *A* is the persistence length, k_B is the Boltzmann constant, *T* is absolute temperature (300 K in our experiments), *z* is the end-to-end length, and *L* is the chain's contour length. Entropic compliance results when $L \gg A$, due to the numerous configurations the polymer may adopt. The force necessary to extend one titin strand was deduced from the myofibrillar passive tension-SL curve by using a value of $1.2 * 10^9$ titins per mm^2 cross-sectional area (Higuchi *et al.*, 1993; Maruyama, 1994). The respective values for (*z/L*) were obtained from the data shown in Figure 3b. Force was then calculated as a function of persistence length; curve fitting was done by using a nonlinear least-squares method (Levenberg-Marquardt algorithm; Origin 5.0 software by Microcal).

Entropic elasticity of the Ig-domain segment

For the proximal Ig-domain region, best fits were returned for a persistence length of ~21 nm (Fig. 4b, curve named "WLC model with A=21 nm"). Importantly, to obtain good fits ($\chi^2 < 1$), we had to exclude the data for a fractional extension >0.96; otherwise, even the best fit did not correctly describe the measured force-extension curve at forces higher than ~35 pN. Such systematic deviations between fits and data at large extensions indicate the onset of structural alterations in the molecule, in our case presumably unfolding of Ig modules. However, the fit results confirmed that unfolding is unlikely to take place at physiologically relevant stretch forces, or at SLs below ~3.4 µm in rat psoas muscle. It can be concluded that the proximal poly-Ig region may behave as a WLC over the entire physiological SL range. Preliminary calculations (not shown) suggest that WLC behavior can also be ascribed to the distal Ig-domain region.

Because the contour length of the WLC affects the above calculations (in which L was 225 nm), we performed the calculations also with different L-values ranging from 200 to 230 nm. The range of persistence lengths thus obtained was 15 to 27 nm, for $L = 230$ to 215 nm. For $L < 215$ nm, there were large systematic deviations between data and fits, particularly above 85% relative extension. Still, the best fit faithfully reproducing the experimentally determined curve, was that shown in Figure 4b, with $A = 21$ nm. This persistence-length value is close to that reported by Higuchi *et al.* (1993) in a dynamic light scattering study on isolated titin molecules: the authors inferred a number of 15 nm. On the other hand, a mechanical study on single titin molecules had suggested a considerably smaller persistence length of ~5 nm for the Ig-domain region (Tskhovrebova *et al.*, 1997). These differences could well be due to the very different experimental approaches taken.

The approach taken by us has advantages, because titin elasticity can be studied in the context of an intact sarcomere structure. On the other hand, a number of uncertainties in the data have to be considered. First, a critical step in the force analysis is the estimate of the true cross-sectional area of the specimens. In the past, this issue has been extensively dealt with (Bartoo *et al.*, 1993; Linke *et al.*, 1994): it was found that the variability in myofibrillar diameter, measured by light microscopy, may well reach ±20% of the average value (~1.0 µm), implying that the error in cross-sectional area measurements is even higher. Thus, a significant part of the nearly twofold variation in passive tension at a given SL (Fig. 4a) might be due to systematic error in the diameter estimates. However, with a large enough number of specimens used for the force measurements (>10), the approach should reveal valid estimates of passive tension. A second limitation of the analysis is due to the fact that titin degradation cannot be inhibited completely during an experiment, even with protease inhibitors (particularly leupeptin) present in the buffer (Linke *et al.*, 1998b). Although the damage is limited, it will affect force to some degree. Such effect cannot be adequately corrected for, because the magnitude of force reduction resulting from titin degradation may vary between experiments. Assuming that the relative force reduction is similar at all SLs, then the fit to WLC theory will still be valid, with the true persistence length being slightly overestimated. Finally, a third restriction to the calculation of force per titin also comes from the uncertainty as to how many titin strands exist per half thick filament. This uncertainty does affect the calculation of a persistence length perhaps within a factor of two, but has no consequences for the fit quality (Linke *et al.*, 1998b). Therefore, the major conclusions of this study are valid regardless of the limitations to the force analysis: (1) a WLC model can correctly describe the elastic properties of the poly-Ig region of psoas titin; (2) WLC behavior may be found over the entire range of physiological SLs; and (3) individual Ig domains are unlikely to unfold at physiologically relevant stretch forces.

Modeling elasticity of the PEVK domain

We also tried to fit the PEVK force-extension data (Fig. 4b) with the standard WLC model of entropic elasticity (Eq. 1). The resulting curve is shown in Figure 4b (dashed curve). This model described the experimental data curve reasonably well for forces below ~12 pN or relative extensions below 60%; a best fit returned a persistence length of ~0.6 nm. This value is approximately 35 times smaller than the persistence length calculated for the proximal Ig-domain region. Hence, due to its relatively low bending rigidity (persistence length), the PEVK domain is a much stiffer spring than the Ig-domain region. This conclusion is consistent with that of Trombitás *et al.* (1998), obtained by fitting the PEVK-extension data of human soleus muscle to a standard WLC model.

At higher forces, there were large systematic deviations between experimental data and WLC fit. In search of a possible theoretical explanation for this observation, we tried to fit the data with a number of more recent polymer elasticity models. Much better results for the moderate and high force regimes were obtained with three entropic-enthalpic models, a freely-jointed chain (FJC) model (Smith *et al.*, 1996), a model according to Odijk (1995), and a modified WLC model (Wang *et al.*, 1997). Of these, the latter revealed by far the best fits. This model uses similar parameters as the standard WLC model to describe entropic elasticity, but also defines a stretch modulus, K_0, to describe enthalpic contributions to elasticity. The stretch modulus measures a polymer's intrinsic resistance to longitudinal strain. According to this modified WLC model, the force is related to the fractional extension through

$$f = \left(\frac{k_\mathrm{B}T}{A}\right)\left[\frac{1}{4(1 - z/L + f/K_0)^2} - \frac{1}{4} + \frac{z}{L} - \frac{f}{K_0}\right] \qquad (2)$$

Curve fitting was done again with a nonlinear least-squares method, with both A and K_0 calculated as dependent variables.

Using Eq. 2, the PEVK fit at lower forces was comparable to that obtained with Eq. 1, but the data curve was particularly well reproducible for forces >12 pN (Fig. 4b, curve named "modified WLC model..."). For $L = 475$ nm, the best fit returned a persistence length of ~0.6 nm and a stretch modulus of 185 pN. The analysis was performed also with different L-values ranging from 450 to 500 nm. These variations affected the stretch modulus, but had little impact on the persistence length, while the quality of the fits was similar to that shown in Figure 4b. Importantly, in no case was it possible to describe the high-extension data with the purely entropic model (Eq. 1). Thus, even in light of an uncertain contour length, the above analysis is qualitatively valid. In conclusion, entropic theory alone does not appear able to describe the elastic properties of the PEVK region. Rather, PEVK stiffness may also be based on enthalpic contributions to elasticity. Enthalpic elasticity at high sarcomere stretch may be important in that it provides an additional extensible "buffer" element that is recruited before the onset of potentially more destructive events, such as Ig domain unfolding.

Entropic and enthalpic forces

The results of the entropic-enthalpic PEVK fit, *i.e.*, the values, $A = 0.6$ nm and $K_0 = 185$ pN, respectively, were combined with the results of the entropic poly-Ig-segment fit (for both the proximal and distal Ig regions; $A = 21$ nm) and used to predict the forces at a given titin extension according to Eq. 1 and Eq. 2. From the calculated force per titin, a "theoretical" passive length-tension curve was then extrapolated, by using the value of $1.2* 10^9$ titins per mm^2 cross-sectional area. The predicted curve correctly reproduced the experimentally determined passive length-tension curve of rat psoas myofibrils over the whole range of physiological SLs (Fig. 5a).

How could enthalpic contributions to PEVK elasticity be explained? Linke *et al.* (1998a) provided evidence for electrostatic stiffening of the PEVK region by comparing the change in titin-based myofibril stiffness of skeletal and cardiac specimens upon reduction in electrostatic screening of charges, achieved by lowering ionic strength. Importantly, the stiffness increase was much more pronounced in skeletal myofibrils than in cardiac specimens, which was interpreted to be due to the intrinsic properties of the skeletal PEVK segment. On the other hand, Linke *et al.* (1998a) also suggested that electrostatic stiffening might not be the sole source of enthalpic elasticity; elastic anisotropy, and/or hydrophobic interactions might also play a role. In this regard, an observation is mentioned which can be made when relaxed, isolated, psoas myofibrils are stretched from a modest SL by ~0.1 µm SL units, and held at the new length (Fig. 5b). Essentially all of the passive force response of these specimens should result from the elasticity of titin filaments (Bartoo *et al.*, 1997). Following a stretch completed within 8 ms, the preparations exhibit substantial stress relaxation (Fig. 5b). The force decay may not be due to Ig domain unfolding, which is very unlikely to occur at the relatively short SLs of 2.4 to 2.9 μm. The time course of force decay is also not explainable by a purely entropic mechanism of elasticity. Rather, we argue that the observed viscoelastic behavior may arise, at least in part, from the entropic-enthalpic elasticity of some titin region(s), presumably the PEVK domain.

Fundamental questions about the nature of titin elasticity can now be answered. A summary is given by the model shown in Figure 6. At low stretch (stages 1-2 in Fig. 6), the poly-Ig regions of titin elongate by straightening out, thus conferring much of the extensibility to titin while developing a very small entropic net force. At modest stretch (stages 2-3), the Ig-domain regions are almost maximally straightened, whereas the PEVK domain unravels and now contributes most of the extensibility to titin; the sequential extension is ex-

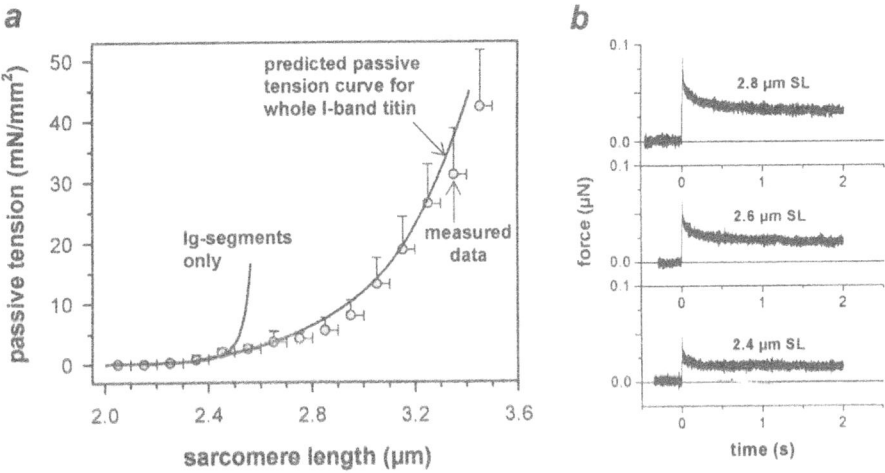

Figure 5. *Entropic-enthalpic forces. (a) Calculation of a "theoretical" passive length-tension curve from the predictions of entropic elasticity theory (Ig-domain regions) and entropic-enthalpic theory (PEVK domain), respectively. The predicted curve correctly follows the experimentally determined passive tension data during stretch of rat psoas myofibrils up to 3.4 μm SLs. Also shown for comparison is the curve obtainable by considering entropic elasticity of Ig-domain regions only (assuming a stiff PEVK segment). (b) Force response of relaxed isolated psoas myofibrils to a stretch of ~0.1 μm SL units, completed within 8 ms. SL before the stretch was set to 2.4, 2.6, and 2.8 μm. After the stop of the stretch, the preparations exhibit stress relaxation, which becomes more pronounced with increasing SLs. In all experiments, the relaxing buffer contained 20 mM 2,3-butanedione monoxime to suppress force-generating actin-myosin interactions possibly remaining even under relaxing conditions.*

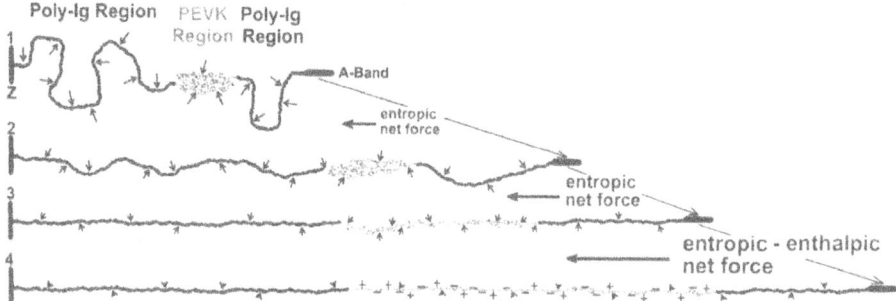

Figure 6. *Model of sequential extension of skeletal-muscle titin. Schematically shown is the elastic I-band section of psoas titin, at different degrees of stretch. Arrows indicate thermally induced fluctuations of the polymer chain, giving rise to entropic elasticity; the + and – symbols refer to the finding that electrostatic interactions may contribute to the enthalpic component of PEVK-segment elasticity (Linke* et al., *1998a).*

plainable by the entropic-spring behavior of the segments and the relative difference in bending rigidity between Ig-domain regions and the PEVK domain. At high physiological stretch (stages 3-4), enthalpic stiffening becomes increasingly important; the developed net force is of entropic-enthalpic nature. Ig domains may not unfold under physiological conditions.

Validity of the approach

Some potential limitations resulting from the application of polymer elasticity theory to the current analysis should be discussed. First, for valid application of WLC theory, one must assume independent mechanical behavior of individual titin strands—a situation not obvious in the sarcomere, which contains many parallel filaments. For example, titin may interact weakly with other filaments, which could affect passive force transiently. The presence of titin-actin interactions has been demonstrated *in vitro* (*e.g.*, Kellermayer and Granzier, 1996), but not in the elastic I-band region of the intact sarcomere. Preliminary biochemical evidence has been published indicating interaction between the PEVK segment and actin (Gutierrez *et al.*, 1998). As far as our analysis is concerned, transient effects of protein/protein interactions on force may not play a role, since in all calculations we used the force values after stress relaxation, *i.e.*, the quasi steady-state force. On the other hand, filament-packing constraints in the I-band could hinder the random motion of titin molecules, thereby modifying entropic force. Currently it is unknown whether steric hindrance must indeed be considered. Finally, it should be pointed out that the entropic-spring concept for titin has not yet been proven experimentally. It will be interesting to see whether the hypothesis will be confirmed, for example, by measuring the temperature-dependency of force development upon titin stretch.

Cardiac titin elasticity

In cardiac muscle, titin exists in different length isoforms expressed in the molecule's I-band region (Labeit and Kolmerer, 1995; Labeit *et al.*, 1997). Both isoforms, termed N2-A and N2-B, comprise stretches of Ig-like modules separated by the PEVK domain. It is not unlikely that the structurally distinct segments of the N2-A isoform may have comparable elastic properties in cardiac and skeletal muscles. Therefore, the conclusions made from the analysis described in the previous sections may also apply to the cardiac N2-A isoform. What appears to be clear is that in cardiac titin, both the poly-Ig segments and the PEVK region represent extensible elements contributing to myofibrillar elasticity (Trombitás *et al.*, 1995; Linke *et al.*, 1996; Granzier *et al.*, 1997; also see H. Granzier's and K. Trombitás' contributions to this volume). On the other hand, significant structural differences are found between the cardiac N2-A and N2-B isoforms. The proximal Ig-domain region of N2-B titin is much shorter than that of the N2-A isoform, and also the N2-B-PEVK segment is very short (only 163 residues in humans). In addition, the central part of N2-B titin contains isoform-specific Ig motifs and nonmodular sequences, notably a 572-residue insertion (see Fig. 1). It may be an intriguing task to find out whether these structural differences give rise to certain functional differences between the isoforms.

Extensibility of the N2-B unique sequence insertion

A first approach to be described here is the investigation of a possible functional significance of the long sequence insertion within the central N2-B region (refer to Fig. 1). It was studied whether the insertion might contribute to the extensibility of N2-B titin (Linke *et al.*, 1999), especially in light of the isoform's short PEVK segment. Sequence-assigned antibodies against the NH_2-terminus and the COOH-terminus of the insertion were kindly provided by T. Centner from S. Labeit's group and M. Gautel from the EMBL Heidelberg (please see their contributions to this volume). The epitope position of the antibodies is indicated in Figure 1. The affinity-purified polyclonal antibody, N2B, is directed against an epitope within a 162-residue-long sequence close to the NH_2-terminus of the insertion (Linke *et al.*, 1999). The monoclonal antibody, I18, stains the first Ig domain COOH-terminal to the insertion (Gautel *et al.*, 1996). Using these antibodies, the extensibility of the insertion was probed both by immunofluorescence microscopy on single myofibrils and immunoelectron microscopy.

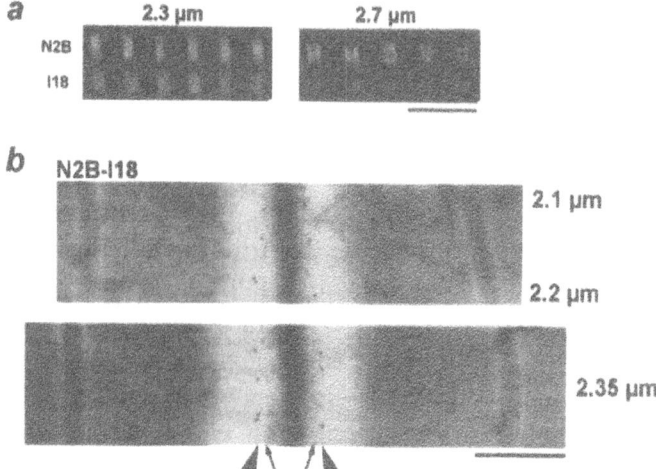

Figure 7. *Examples of immunostained sarcomeres from rabbit left ventricle, using titin antibodies flanking the unique N2-B sequence insertion. (a) Immunofluorescence images of stretched single myofibrils, labeled at 2.3 and 2.7 μm SL. Scale bar, 5 μm. (b) Immunoelectron micrographs of stretched sarcomeres double-stained with both N2B and I18 antibodies. The secondary antibody for N2B was conjugated to 10 nm gold particles (arrows), that for I18 to 15 nm particles (arrowheads). Epitopes begin to separate at SLs between 2.1 and 2.2 μm. Scale bar, 0.5 μm.*

Figure 7a shows typical images of single rabbit cardiac myofibrils extended to two different SLs, 2.3 and 2.7 μm, and stained with the respective primary antibody and Cy-3-conjugated secondary antibody. At least at high stretch it was obvious that N2B and I18 stained different positions on N2-B titin. Both antibodies were subsequently used also for immunoelectron microscopy, to determine the distance of a given epitope from the center of the Z-disk at shorter SLs, and to determine the SL at which the unique sequence insertion begins to extend. Rabbit cardiac sarcomeres were double-stained with both N2B and I18, and secondary antibodies conjugated to different sized nanogold particles were used. As shown in Figure 7b, the epitopes could not yet be separated at 2.1 μm SL. However, separation began shortly above 2.1 μm SL and was clearly detectable in sarcomeres stretched to 2.35 μm.

The results of the immunoelectron microscopical analysis are summarized in Figure 8a. For each antibody type, individual data points were pooled in 50 nm wide SL bins, plotted as mean±SD, and fitted by a third-order regression. At a given SL bin, the epitope positions measured by immunoelectron microscopy were similar within experimental error to those determined by im-munofluorescence microscopy (data not shown). The N2B antibody stained a position clearly different from that labeled by I18, at least at modest to high sarcomere stretch. The unique sequence flanked by I18 and N2B was con-firmed by statistical analysis (unpaired Student's t-test) to begin to lengthen significantly at ~2.15 µm SL (p<0.001; n=17 for the I18 data set and n=19 for the N2B data set at SL bin 2.15-2.19 µm). Thus, the N2-B titin isoform recruits a third spring-like element besides the PEVK segment and the Ig-domain re-gions: the long unique sequence insertion of the central I-band section.

The distance between the fit curves at a given SL reveals extension of the N2-B unique sequence *versus* SL. The insertion began to stretch substantially above ~2.15 µm SL, reached a length of 60-70 nm at 2.3-2.4 µm SL, before the extension curve approached a plateau at ~2.6 µm SL. Similar to the analysis of extension of the N2-A titin segments (see section "Extensibility of titin seg-ments in skeletal muscle"), we used the available sequence data for cardiac titin (Labeit and Kolmerer, 1995) to predict the contour length of the sequence flanked by the N2B and I18 titin antibodies and to plot the relative extension of the sequence against SL (Fig. 8b). When we assumed that the unique inser-tion, which contains between 310 and 436 residues (depending on where the epitope is located within the 126 residue-long fragment used to raise the anti-body), is capable of complete unfolding with a maximum residue spacing of 0.38 nm, the extension curve never reached the predicted contour-length value (dashed lines in Fig. 8b; for more details, see caption). On the other hand, because the segment's extension curve did approach a plateau, it is reasonably to assume that the region may not unfold entirely. Rather, it may adopt some permanent structural fold, which would lead to a shorter contour length; the resulting curve is shown in Figure 8b (solid line).

Model of sequential N2-B titin extension

From these results (see Linke *et al.*, 1999), a model for titin N2-B exten-sion in cardiac muscle can be proposed (Fig. 9). Low sarcomere extension may be brought about by elongation of tandem-Ig segments and the PEVK domain (Fig. 9, stages 1-2). Above 2.15 µm SL, also the 572-residue N2-B region in the middle of I-band titin begins to elongate substantially (Fig. 9, stages 2-3) and may compensate for the relatively short length of the PEVK segment, whose extensibility would soon be exhausted. Elasticity of the unique N2-B sequence may also help the shorter N2-B titin isoform to adjust its range of extension to that of the longer N2-A isoform, since both isoforms are co-expressed at the sarcomere level (Linke *et al.*, 1996). During all three stages of extension (Fig. 9; up to 2.35 µm SL in stage 3), elongation of poly-Ig segments will be brought about by straightening, rather than by domain unfolding. With maximum SLs in the working cardiac muscle presumably no longer than 2.3-

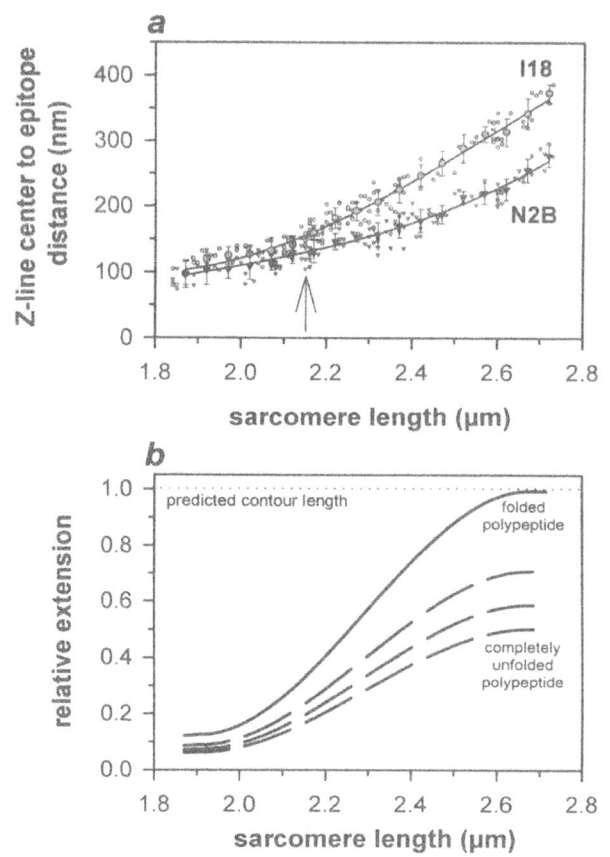

Figure 8. Antibody epitope mobility and extension of cardiac titin segments. (a) Summary of results of immunoelectron microscopical measurements at SLs ranging from 1.84 to 2.74 μm. Small symbols indicate individual data points, larger shaded symbols show mean values calculated in 50 nm wide SL bins; standard deviations are also shown. Data sets for each antibody type were fitted by third-order regressions. The arrow indicates the onset of statistically significant separation between curves. (b) Extension of the unique sequence insertion between N2B and I18 antibodies, relative to the segment's predicted contour length. The polypeptide was either assumed to adopt a (partially) folded state even at the highest stretch applied (solid curve) or to be completely unfolded (dashed curves). Contour-length predictions were as follows: solid curve: 84 nm; dashed curves: 118, 142, and 166 nm, from upper to lower curve, respectively.

2.4 μm (*e.g.*, Allen and Kentish, 1985), cardiac titin will be able to stretch reversibly even to long physiological SLs without requiring Ig domains to unfold.

It will be useful to model the elastic properties of cardiac titin according to polymer-elasticity theory, as has been done previously by Granzier *et al.* (1997), who assumed two different WLC elements, the Ig-domain segments and the PEVK domain, in series. With the knowledge that a third type of elastic element exists in the cardiac N2-B isoform, it is obvious that the WLC modeling for cardiac titin needs modification. A complication is that recent results from the laboratory of C.C. Gregorio (please see her contribution to this volume) raise the possibility that part of the central N2-B-titin region is involved in protein/protein interactions (Linke *et al.*, 1999). The gray oval in Figure 9 schematically indicates the position on N2-B titin implicated in these interactions. Thus, cardiac titin may not be allowed to move freely in the sarcomere, which could affect the elastic properties of the protein chain. Because it is unsettled how this possible association will change the (hypothesized) WLC behavior of cardiac N2-B titin, a WLC analysis was not performed at this stage.

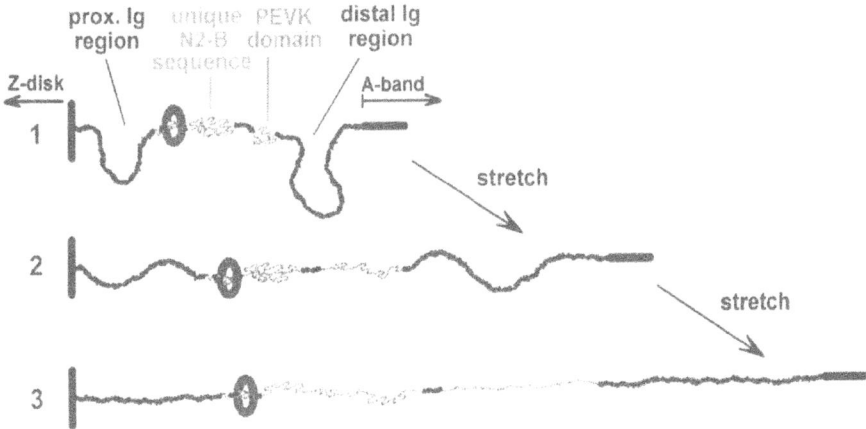

Figure 9. *Extension model for the N2-B titin isoform of rabbit cardiac muscle. Shown is the elastic I-band region at three different stages of extension within a physiologically relevant SL range. Upon low stretch (first stage, 1.95 μm SL, to second stage, 2.15 μm SL), proximal and distal Ig-segments, as well as the PEVK domain, begin to extend. On further stretch, the unique sequence insertion also elongates (second to third stage, 2.15 to 2.35 μm SL). This may help prevent potentially catastrophic events such as irreversible unfolding of Ig-domains at high physiological stretch. The elastic behavior of cardiac titin might be constrained by protein/protein interactions involving specific Ig domains of the central N2-B region (indicated by the oval; cf., Linke et al., 1999).*

Conclusions

In skeletal muscle, I-band titin represents a nonlinear spring composed of two elastic elements in series, the poly-Ig chains and the PEVK domain. The elastic properties of the Ig-domain regions can be correctly simulated by a worm-like chain model of entropic elasticity. The poly-Ig regions appear to behave as entropic springs over the entire range of physiological sarcomere lengths; individual Ig domains may not unfold in normally functioning skeletal muscle. PEVK-titin elasticity appears to have both entropic and enthalpic origins. Whereas entropic elasticity is most relevant at low to moderate sarcomere stretch, enthalpic elasticity may be the dominant factor at larger extensions. The polymer-elasticity modeling explains why during stretch of titin, the Ig-domain regions extend before the PEVK domain.

Cardiac titin contains, within its N2-B isoform, a unique sequence insertion, which extends toward the high end of the physiological sarcomere-length range. Thus, cardiac titin is a molecular spring consisting of three distinct elements: the unique N2-B sequence, the PEVK domain, and stretches of (folded) Ig modules.

Acknowledgments

The author would like to thank S. Labeit, B. Kolmerer, T. Centner, and M. Gautel for providing antibodies to titin, M. Ivemeyer and M. Stockmeier for performing WLC calculations, H. Hosser for help with immunoelectron microscopy, and T. Welsch for writing a computer program. The financial support of the Deutsche Forschungsgemeinschaft (Li 690/2-2, SFB 320) is gratefully acknowledged.

References

Allen DG, Kentish JC. The cellular basis of the length-tension relation in cardiac muscle. *J Mol Cell Cardiol* 1985;17:821-40.

Bartoo ML, Popov VI, Fearn LA, Pollack GH. Active tension generation in isolated skeletal myofibrils. *J Muscle Res Cell Motil* 1993;14:498-510.

Bartoo ML, Linke WA, Pollack GH. Basis of passive tension and stiffness in isolated rabbit myofibrils. *Am J Physiol* 1997;273:C266-76.

Bustamante C, Marko JF, Siggia ED, Smith S. Entropic elasticity of l-phage DNA. *Science* 1994;265:1599-1600.

Erickson HP. Reversible unfolding of fibronectin type III and immunoglobulin domains provides the structural basis for stretch and elasticity of titin and fibronectin. *Proc Natl Acad Sci USA* 1994;91:10114-8.

Fürst DO, Osborn M, Nave R, Weber K. The organization of titin filaments in the half-sarcomere revealed by monoclonal antibodies in immunoelectron microscopy: a map of ten nonrepetitive epitopes starting at the Z-line extends close to the M-line. *J Cell Biol* 1988;106:1563-72.

Gautel M, Goulding D. A molecular map of titin/connectin elasticity reveals two different mechanisms acting in series. *FEBS Lett* 1996;385:11-4.

Gautel M, Lehtonen E, Pietruschka F. Assembly of the cardiac I-band region of titin/connectin: expression of the cardiac-specific regions and their relation to the elastic segments. *J Muscle Res Cell Motil* 1996;17:449-61.

Granzier H, Kellermayer M, Helmes M, Trombitas K. Titin elasticity and mechanism of passive force development in rat cardiac myocytes probed by thin-filament extraction. *Biophys J* 1997;73:2043-53.

Gregorio CC, Granzier H, Sorimachi H, Labeit S. Muscle assembly: a titanic achievement? *Curr Opin Cell Biol* 1999;11:18-25.

Gutierrez G, van Heerden A, Wang K. Interactions and conformational studies of PEVK segment of human fetal skeletal muscle titin. *Biophys J* 1998;2(2):A349.

Higuchi H, Nakauchi Y, Maruyama K, Fujime S. Characterization of β-connectin (titin 2) from striated muscle by dynamic light scattering. *Biophys J* 1993;65:1906-15.

Kellermayer MS, Granzier HL. Calcium-dependent inhibition of in vitro thin-filament motility by native titin. *FEBS Lett* 1996;380:281-6.

Kellermayer MSZ, Smith SB, Granzier HL, Bustamante C. Folding-unfolding transitions in single titin molecules characterized with laser tweezers. *Science* 1997;276:1112-6.

Labeit S, Kolmerer B. Titins, giant proteins in charge of muscle ultrastructure and elasticity. *Science* 1995;270:293-6.

Labeit S, Kolmerer B, Linke WA. The giant protein titin. Emerging roles in physiology and pathophysiology. *Circ Res* 1997;80:290-4.

Linke WA, Ivemeyer M, Mundel P, Stockmeier MR, Kolmerer B. Nature of PEVK-titin elasticity in skeletal muscle. *Proc Natl Acad Sci USA* 1998a;95:8052-7.

Linke WA, Ivemeyer M, Olivieri N, Kolmerer B, Rüegg JC, Labeit S. Towards a molecular understanding of the elasticity of titin. *J Mol Biol* 1996;261:62-71.

Linke WA, Popov VI, Pollack GH. Passive and active tension in single cardiac myofibrils. *Biophys J* 1994;67:782-92.

Linke WA, Rudy DE, Centner T, Gautel M, Witt C, Labeit S, Gregorio CC. I-band titin in cardiac muscle is a three-element molecular spring and is critical for maintaining thin filament structure. *J. Cell Biol* 1999;146:631-44.

Linke WA, Stockmeier MR, Ivemeyer M, Hosser H, Mundel P. Characterizing titin's I-band Ig domain region as an entropic spring. *J Cell Sci* 1998b;111:1567-74.

Marko JF, Siggia ED. Stretching DNA. Macromolecules 1995;28:8759-70.

Maruyama K. Connectin, an elastic protein of striated muscle. *Biophys Chem* 1994;50:73-85.

Maruyama K. Connectin/titin, giant elastic protein of muscle. FASEB J 1997;11:341-5.

Mundel P, Gilbert P, Kriz W. Podocytes in glomerulus of rat kidney express a characteristic 44 kD protein. *J Histochem Cytochem* 1991;39:1047-56.

Odijk T. Stiff chains and filaments under tension. Macromolecules 1995;28:7016-8.

Politou AS, Thomas DJ, Pastore A. The folding and stability of titin immunoglobulin-like modules, with implications for the mechanism of elasticity. *Biophys J* 1995;69:2601-10.

Rief M, Gautel M, Oesterhelt F, Fernandez JM, Gaub HE. Reversible unfolding of individual titin immunoglobulin domains by AFM. *Science* 1997;276:1109-12.

Smith SB, Cui Y, Bustamante C. Overstretching B-DNA: the elastic response of individual double-stranded and single-stranded DNA molecules. *Science* 1996;271:795-9.

Tokuyasu KT. Use of poly(vinylpyrrolidone) and poly(vinyl alcohol) for cryoultramicrotomy. *Histochem J* 1989;21:163-71.

Trinick J. Cytoskeleton: Titin as a scaffold and spring. *Curr Biol* 1996;6:258-60.

Trombitás K, Greaser M, Labeit S, Jin J-P, Kellermayer M, Helmes M, Granzier H. Titin extensibility in situ: Entropic elasticity of permanently folded and permanently unfolded molecular segments. *J Cell Biol* 1998;140:853-9.

Trombitás K, Jin J-P, Granzier H. The mechanically active domain of titin in cardiac muscle. *Circ Res* 1995;77:856-61.

Tskhovrebova L, Trinick J. Direct visualization of extensibility in isolated titin molecules. *J Mol Biol* 1997;265:100-6.

Tskhovrebova L, Trinick J, Sleep JA, Simmons RM. Elasticity and unfolding of single molecules of the giant muscle protein titin. *Nature* 1997;387:308-12.

Wang K. Titin/connectin and nebulin: giant protein rulers of muscle structure and function. *Adv Biophys* 1996;33:123-34.

Wang MD, Yin H, Landick R, Gelles J, Block SM. Stretching DNA with optical tweezers. *Biophys J* 1997;72:1335-46.

Discussion

Pollack: Is it possible with cardiac muscle, as you stretch to fairly long sarcomere lengths, that you are pulling the titin filament off the thick filament? Could the tensions be high enough for that to happen?

Linke: That is definitely happening. There is a study from Henk's lab ('95) in the *Biophysical Journal*, saying that this may happen at a sarcomere length of approximately 2.9 to 3 µm. At a shorter length it could happen perhaps as a result of using degraded titin, as may be the case when one uses glycerinated muscle. We do see a difference when we compare glycerinated and freshly prepared muscle. We now use freshly prepared muscle for our studies.

Trombitás: It was a nice presentation. Tomorrow I will show something about the last question. We labeled N2-B titin and used the MIR antibody which labels the A/I junction of the sarcomere. The MIR antibody stays in the right place at least until a sarcomere length of 2.6 micron is reached. So titin does not come off the thick filament.

Linke: So 2.6 µm would be the length at which detachment begins.

Granzier: How did you determine the contour lengths in your skeletal muscle experiments?

Linke: For the Ig domain region we used the prediction from the sequence. And we also plotted $\frac{1}{\sqrt{force}}$, at the high extension range. As you know, you can from the crossing of the extrapolated curve with the x axis, determine the contour length. We concluded that it is about 225 nm for the Ig domain region. And for the PEVK we predicted the contour length from the sequence.

Granzier: Which sequence is that then? Human psoas?

Linke: Well the rat sequence is not known. I think that it is pretty reasonable to assume similar sequences for different species. When we compare the elasticity, for example, of rat and rabbit myofibrils, we do see a very similar passive tension curve of these two specimens.

What I didn't show in my talk is that we varied the contour lengths in the mathematical analysis. And very importantly, even if we tried out shorter or longer contour lengths, we could never fit our data with a pure wormlike chain equation. It never worked out at the high extensions. We went down to 420 nm (from about 475 nm initially) and up to 520 nm—we considered more variation unlikely. But you have to keep in mind that we used 0.34 nm for the maximum residue spacing. If we went to 0.38 nm, the contour lengths would be longer than the one we predicted. So I think this contour length issue will affect the calculation of persistence length but it will not change the main conclusion that there is a big difference in the persistence length between PEVK region and Ig domain region. Furthermore, the ionic strength dependency data and the fact that we do see stress relaxation at relatively low forces where you would not yet expect Ig domain unfolding, provides evidence that a wormlike chain model may not be entirely correct for the modeling of titin.

Labeit: You have this nice data on ionic strength influence on stiffness. Did you also look at calcium? Because there are all the speculations that calcium might alter passive stiffness.

Linke: We looked at that briefly; that was some years ago, maybe four years ago. We extracted actin from the sarcomere and then added calcium, so that we didn't get contraction. Sometimes we also added some BDM. And we didn't see any change in elastic properties, although this was just a first shot and it was never really published except in the form of an abstract. But we didn't see any indication that calcium affects stiffness. It is a difficult protocol to do in intact muscle. It has been proposed from *in vitro* data that there is a calcium dependency of actin-titin binding; there is data by Miklos Kellermayer and Henk Granzier ('96). But we didn't see that in intact sarcomeres.

TerKeurs: I would just like to make a comment about his very beautiful data on the cardiac myofibrils. You show that at sarcomere lengths of say, 2.2 micrometer, stress is on the order of 20 mN/mm². In intact muscle this would amount to approximately 10 mN/mm², because 50% of the cross-sectional area of muscle is mitochondria. So one can compare your number to the measurement in intact muscle itself. And it is interesting to note that the elasticity of an intact muscle has exactly the same properties as you would calculate on the basis of your data. It is also interesting to note that the muscle contains a collagen meshwork around the myocytes, collagen in between myocyte struts and large collagen fibers along the myocytes. Now I can see what large collagen fibers are doing along the myocytes or along the muscle, and that is they limit stretch of the muscle to about 2.4 micron. Apparently, given the comparison between your data and these intact muscle data, the collagen meshwork and the struts are doing nothing elastically. So what's your explanation for that?

Linke: I wouldn't conclude that. Of course the collagen network does a lot for the stiffness of the intact cardiac muscles, the intact heart.

TerKeurs: Numerically, zero, at the level of your numbers.

Linke: Ah, I now know what you are getting at. What preparation are you talking about? This would be the first issue.

TerKeurs: Trabeculae.

Linke: Next you would really have to know the species. It may very well vary with species like dog and rat.

I think, I wouldn't dare to do a comparison, a very close comparison, given those very different techniques with which you have measured forces. I would, for example, like to compare it when you have more similar conditions such that you do a complete set starting with mechanics on myofibrils, on cardiac cells, and on trabeculae, and then compare maybe even with the same experimenter. Because I think values can very well vary. I would not dare to compare data that have been obtained in a papillary muscle in the 80s with the data that I get now. For instance, we always have BDM in our solution, which gives a 9-10% difference. Maybe it is worth doing this comparison, but I would not dare to talk about it now considering the factor of 2 difference.

Granzier: You made this comparison in your '94 paper?

Linke: Yes, and I concluded that up to about 2.15 to 2.25 μm sarcomere length, the forces are similar, but you can see that there is high scatter in the data. And at that time we did not go into a statistical analysis. We didn't even calculate the mean plus/minus standard deviation, and I did not look closely at this data again.

Rief: I wanted to come back to your modeling of the PEVK segment, the wormlike chain thing. Did I get it right that the deviation that you model with the spring constant is used to explain the viscous response that you had during step relaxation?

Linke: No, we didn't really make that correlation.

Rief: Because if that were the case, then I think an additional spring constant that you add would not be the appropriate way to describe that.

Linke: I agree with that.

Rief: Also if you have that viscous response you should see a dependency on how fast you stretch.

Linke: Absolutely. Well you see, our approach was to fit the experimental data. We took the wormlike chain model and we couldn't fit the data in the high force regime. So we needed to modify the model, and we concluded something which made the most sense to us. The wormlike chain theory, as it is expressed in the Marko and Siggia model, does not consider stretch to beyond the predicted contour lengths. But in reality we think that this is not the correct model to apply to PEVK titin. It may work in the high extension range differently and that is why we used models that have these enthalpic stretch modules.

Rief: I totally agree, we faced similar problems, although with single molecule data, but I just wanted to point out if that is a viscous response, you shouldn't use an elastic factor.

Linke: No, we did not use it to describe viscosity.

Bullard: Muscle is operating at constant volume. When you stretch, the filaments are getting closer together and this may increase viscous drag.

Linke: Yes, we have thought about this, but when the filaments get closer together what would you expect? Maybe more viscous drag, but we see basically more compliance than expected from the wormlike chain theory. Since we find that at high stretch there is more compliance than we would predict from the wormlike chain model, we do not think that this can be explained by viscous drag.

Baatsen: There is an extra structure in parallel with the contractile apparatus. This is the collagen network. There is also the extracellular or the peripheral cytoskeleton which may also be a candidate for adding stiffness to the whole system. Collagen fibers, of course, are very stiff. But they may be oriented in such a way that at short muscle length they are almost perpendicular to the length of the muscle fiber. This may explain why the stiffness is not really noticed.

Linke: Is this a comment or a question?

Baatsen: A comment.

LINKS IN THE CHAIN:
THE CONTRIBUTION OF KETTIN TO
THE ELASTICITY OF INSECT MUSCLES

Belinda Bullard, David Goulding,
Charles Ferguson, and Kevin Leonard
European Molecular Biology Laboratory, Heidelberg, Germany

Abstract: Asynchronous flight muscle fibers are activated by periodic stretches and need to be stiff for strain to be transmitted to the contractile system. Kettin associated with thin filaments and projectin with thick filaments contribute to fiber stiffness. Kettin extends along thin filaments with the N-terminus in the Z-disc and the C-terminus outside. C filaments connecting thick filaments to the Z-disc contain projectin but not kettin. Insect flight myofibrils have a titin PEVK epitope which is only exposed on stretch, suggesting it is short and inaccessible. It is concluded that kettin stiffens thin filaments near the Z-disc and projectin and titin provide elasticity to C filaments.

Introduction

Insect flight muscle has proved to be an excellent system for investigating many aspects of muscle function. The exceptional regularity of the lattice of thick and thin filaments in the indirect flight muscle has been particularly useful in studying the structure by electron microscopy and X-ray diffraction. In striated muscle fibers, thick filaments containing myosin interdigitate with thin filaments containing actin. Thin filaments are anchored in the Z-disc. Fine strands connecting the ends of thick filaments to the Z-disc were first observed in insect flight muscle. Auber and Couteaux (1963) saw the filaments in electron micrographs of *Calliphora* flight muscle, and it was twelve years before similar filaments were described in vertebrate muscle (Locker and Leet, 1975).

Elastic Filaments of the Cell, edited by Granzier and Pollack
Kluwer Academic/Plenum Publishers, 2000

These connecting filaments, or C filaments, are largely responsible for the passive elasticity of both insect flight muscle and vertebrate striated muscle fibers.

Insect flight muscle may be synchronous, in which case each contraction follows a nervous impulse, or asynchronous when contraction frequency is independent of the frequency of nervous stimulation. Asynchronous fibers are activated by periodic stretches during oscillatory contraction, and for stress to be transmitted to the thick and thin filaments in the sarcomere, the fibers must be stiff. The importance of stiffness in the stretch activation of flight muscle led to a study of the mechanics and structure of C filaments in Ernst's laboratory in Pecs, Hungary and Pringle's laboratory in Oxford, England in the 1970s.

The rapid pace of recent work on elastic proteins in muscle has led to the identification of proteins in C filaments and on thin filaments in the Z-disc region of the insect flight muscle sarcomere. These proteins have a modular structure and different elastic properties depending on their function. Elastic proteins that are likely to contribute to the mechanical properties of insect muscle are projectin, kettin and titin (also called connectin).

Results and Discussion

Properties of kettin

Dr. Charles Trombitás, while in Ernst's laboratory, published remarkable electron micrographs of bee flight muscle showing that if actin filaments were broken in the region of the Z-disc by stretching myofibrils in rigor, and the myofibrils subsequently activated, the sarcomere was held together by the C filaments, and broken actin filaments could move into the other half of the sarcomere (Trombitás and Tigyi-Sebes, 1984). The inappropriate polarity of the broken actin filaments did not prevent their translocation into a half sarcomere already occupied by actin filaments of correct polarity. In a comprehensive study of the effect of stretching bee fibers in rigor, Trombitás and Tigyi-Sebes (1977) found that actin filaments are broken cleanly from the Z-disc and are uniform in length. Comparison of the lengths of intact actin filaments (in the right hand side of the sarcomere) and the broken filaments in Figure 1, shows that the broken filaments are shorter by about 55 nm and therefore break, not at the edge of the Z-disc, but at a short distance into the I-band. This suggests that there is a reinforcing component which stabilizes a finite length of the actin filaments in the region of the Z-disc, so that this part of the filaments remains attached to it. The stabilizing element is likely to be kettin.

Kettin is a modular protein in insect muscles with a molecular weight of 500 kDa or 700 kDa, depending on the muscle type. The sequence is a chain of immunoglobulin-like (Ig) domains separated by a 35 residue linker sequence that is not present in the other modular muscle proteins (Lakey *et al.*, 1993).

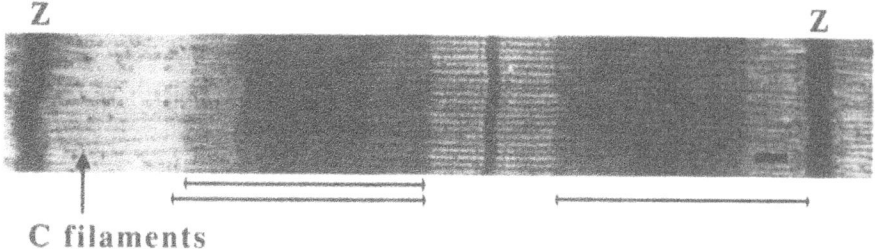

C filaments

Figure 1. *Thin filaments broken from the Z-disc in a bee myofibril in rigor. Fibers were extended before glycerination and then stretched in rigor conditions. Thin filaments in the left hand side of the sarcomere are broken cleanly from the Z-disc, exposing stretched C filaments. The broken thin filaments are about 55 nm shorter than the unbroken filaments in the right hand side of the sarcomere. The short section of thin filaments that stays attached to the Z-disc on the left side may be stabilized by kettin bound to actin in this region of the sarcomere. Long two-headed arrows show lengths of broken and unbroken thin filaments. Scale bar 110 nm. After Trombitás and Tigyi-Sebes (1977).*

The sequence obtained so far has 35 consecutive Ig-linker modules (Kolmerer *et al.*, 2000). We have also identified a highly homologous sequence in the *Caenorhabditis elegans* genome data base which has the same pattern of Ig domains and linkers as *Drosophila* kettin. Kettin differs from other modular proteins of the same family in binding to actin rather than myosin. The protein from the giant waterbug, *Lethocerus*, binds strongly to F-actin with a stoichiometry of one molecule of kettin to about 30 actin protomers in the filament (van Straaten *et al.*, 1999). This stoichiometry favors a model in which each Ig domain in the 500 kDa protein binds to a protomer in F-actin. Projectin, an 800 kDa protein associated with insect thick filaments, contains a regular repeating pattern of Ig and fibronectin-like (Fn) domains (Ayme-Southgate *et al.*, 1991; Fyrberg *et al.*, 1992; Daley *et al.*, 1998). Although both proteins contain Ig domains, only kettin binds to actin (Fig. 2) and the linker sequence may be needed for specific actin binding. The Fn domains in projectin, like those in the A-band region of titin, are characteristic of a thick filament-associated protein.

Kettin is found in the Z-disc region of the sarcomere. The position and orientation of kettin in *Drosophila* flight muscle have been determined by labeling with antibodies to epitopes towards the N- and C-termini of the molecule (van Straaten *et al.*, 1999). Figure 3 shows the position of these epitopes and an additional epitope approximately in the middle of the molecule. The labeling pattern is consistent with a model in which the N-terminal regions of kettin molecules from opposite sarcomeres overlap in the Z-disc, the middle of

Kettin   Projectin
S   P   S   P   S   P   S   P

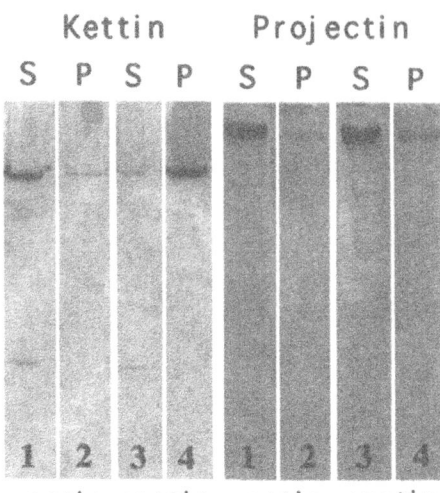
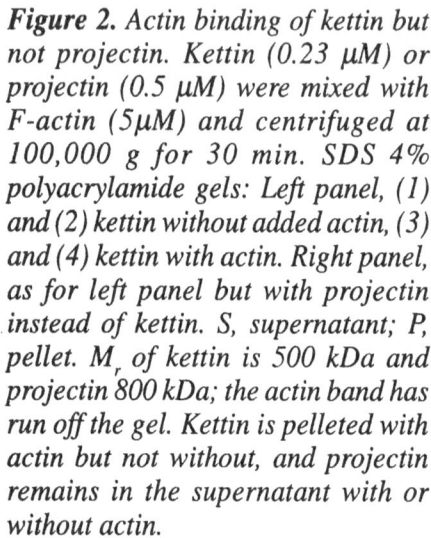

1   2   3   4   1   2   3   4
-actin +actin   -actin +actin

Figure 2. *Actin binding of kettin but not projectin. Kettin (0.23 μM) or projectin (0.5 μM) were mixed with F-actin (5μM) and centrifuged at 100,000 g for 30 min. SDS 4% polyacrylamide gels: Left panel, (1) and (2) kettin without added actin, (3) and (4) kettin with actin. Right panel, as for left panel but with projectin instead of kettin. S, supernatant; P, pellet. M_r of kettin is 500 kDa and projectin 800 kDa; the actin band has run off the gel. Kettin is pelleted with actin but not without, and projectin remains in the supernatant with or without actin.*

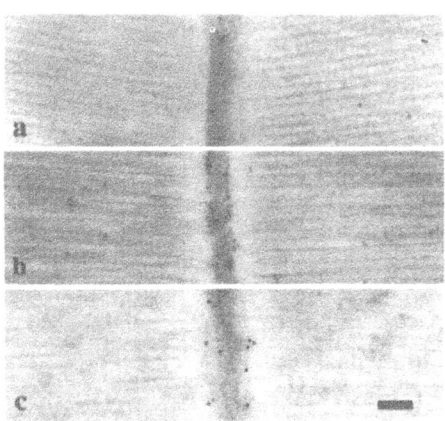

Figure 3. *Position of kettin in the sarcomere of* Drosophila *flight muscle. Electron micrographs of cryosections of* Drosophila *flight muscle fibers labeled with antibodies to sequence: (a) near the N-terminus (KET1), (b) near the middle of the molecule (KET3) and (c) near the C-terminus (KET2). Primary antibodies were followed by protein A tagged with 10 nm gold particles. The position of epitopes in the sequence is shown in Figure 4. The N-terminus of kettin is within the Z-disc and the C-terminus is outside. Scale bar 100 nm.*

the molecule is at the border of the Z-disc and the C-terminal region extends into the I-band (Fig. 4). In *Drosophila* flight muscle, kettin extends 54 nm along the thin filament outside the Z-disc (van Straaten *et al.*, 1999). The degree of overlap in the Z-disc is uncertain because the N-terminal sequence is not complete and the structure may differ from the regular repeating modules in the rest of the molecule. The structure of kettin in the Z-disc differs from that of vertebrate titin which spans the Z-disc (Young *et al.*, 1998; Gregorio *et al.*, 1998). This is not surprising because kettin binds to actin, whereas titin is linked to actin indirectly through binding of Z-repeats to α-actinin (Fig. 4).

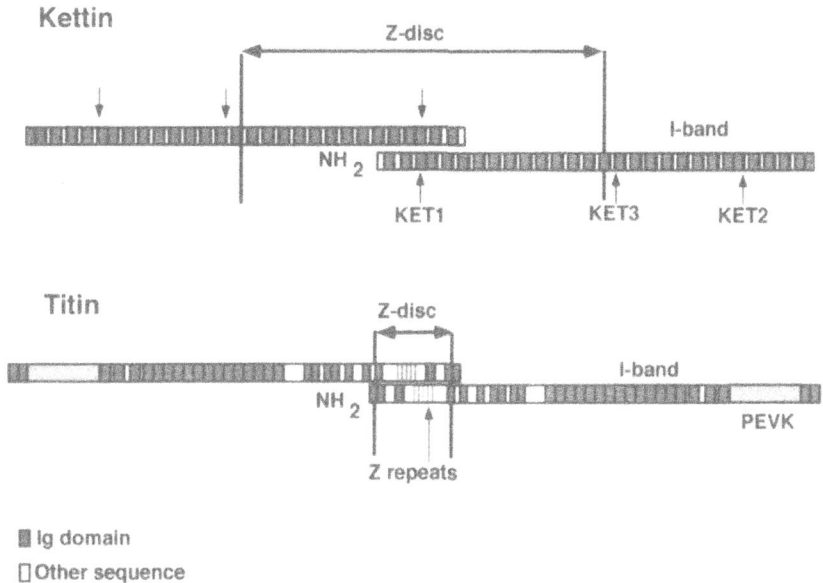

Figure 4. Plan of the layout of kettin in the Z-disc region of the sarcomere of Drosophila *flight muscle and comparison with the layout of titin in the vertebrate skeletal muscle Z-disc. KET1, KET2 and KET3 show the positions in the kettin sequence of the epitopes labeled with antibody in Figure 3. The N-terminus of kettin is inside the Z-disc but the exact position is not known. The widths of the insect and vertebrate Z-discs are not to scale; kettin would be compressed by association of Ig-linker modules with actin protomers. PEVK, extensible region of titin.*

The model that best fits both the stoichiometry of kettin binding to actin and the length of thin filament spanned by kettin, is one in which the molecule follows the genetic helix of the actin filament. Tropomyosin and myosin subfragment 1 (S1) compete with kettin for binding to actin (van Straaten *et al.*, 1999). These proteins may bind to the same site on actin as kettin, or interference in binding may be due to cooperative changes along the actin filament (Orlova *et al.*, 1995) which prevent a second protein binding. α-Actinin and kettin bind simultaneously to actin and therefore probably occupy different binding sites. The likely position of kettin on actin, based on competition with tropomyosin and S1 but not α-actinin, is on subdomain 1 on the outside of the actin filament, overlapping subdomain 3 (Fig. 5).

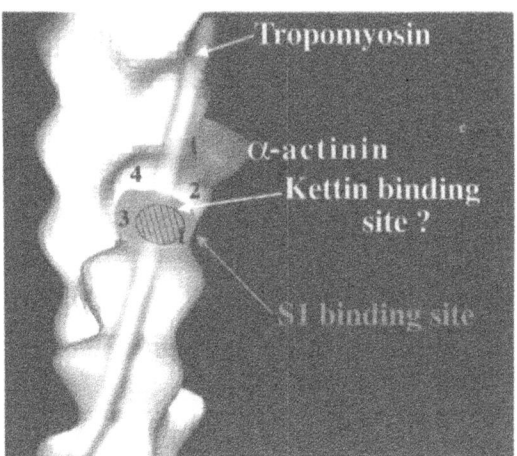

Figure 5. *Model of the thin filament to show the likely binding site of kettin. Tropomyosin and myosin S1 compete with kettin for binding to actin but α-actinin and kettin bind simultaneously to actin. The kettin binding site is shown on subdomain 1, the S1 binding site (gray), overlapping the tropomyosin binding position which is shown schematically. Subdomains 1 to 4 on an actin protomer are marked.*

Function of kettin

Kettin occurs in both synchronous and asynchronous insect flight muscle and therefore does not seem to have a function peculiar to stretch activated muscle. It has also been found in crayfish claw muscle (Maki *et al.*, 1995). The high affinity binding of kettin to actin would stabilize thin filaments axially in the region of the Z-disc. Crosslinks of α-actinin and other proteins anchor flight muscle thin filaments in the Z-disc (Cheng and Deatherage, 1989; Deatherage *et al.*, 1989). When the sarcomere is stretched considerably beyond the physiological limit, thin filaments break uniformly at the end of the region containing kettin, while C filaments stay attached to the Z-disc and are extended (Fig. 1). The length of thin filament that stays attached to the Z-disc in bee myofibrils is remarkably similar to the length of thin filament spanned by kettin outside the Z-disc in *Drosophila* myofibrils (about 55 nm).

In cardiac myofibrils of vertebrates, a short region of titin adjacent to the Z-disc is inextensible at physiological degrees of stretch and is bound to actin, which is thought to strengthen the link between titin and the Z-disc (Trombitás and Pollack, 1993; Linke *et al.*, 1997; Trombitás and Granzier, 1997; Granzier *et al.*, 1997). The length of the inextensible region of cardiac titin is about the same as the length of thin filament spanned by *Drosophila* kettin (90-100 nm from the center of the Z-disc), but the association of this part of titin with the thin filament differs from that of kettin. The affinity for actin of a titin peptide from this region is low (Linke *et al.*, 1997); the part of titin that binds to thin filaments does not have the repeating domain structure of kettin and is unlikely to bind to successive actin protomers. In the case of cardiac titin, the inextensible region outside the Z-disc is approximately 1100 amino acid residues or about 120 kDa (Linke *et al.*, 1997), whereas in *Drosophila* flight muscle, the part of kettin outside the Z-disc is about 290 kDa. Thus, the association of kettin with thin filaments is considerably more condensed than that of titin and this region of the thin filament would be expected to have greater rigidity in flight muscle than in cardiac or other vertebrate striated muscles.

Elasticity of the flight muscle sarcomere

Mechanical measurements on insect muscle are usually made using *Lethocerus* because of the conveniently large size or, more recently, *Drosophila* because of the genetic possibilities. In both insects, the relaxed stiffness of asynchronous flight muscle fibers is about six times that of synchronous muscle fibers from the same insect (Peckham *et al.*, 1990). Both insect fiber types are considerably stiffer than vertebrate skeletal muscle fibers (*Lethocerus* asynchronous fibers are 28 times as stiff as rabbit psoas fibers (Peckham and White, 1991). The high relaxed stiffness of asynchronous fibers is due to the short I-bands and to the elasticity of C filaments (White, 1983). Weak crossbridges may also contribute to relaxed stiffness (Granzier and Wang, 1993). However, these authors found that passive tension in *Lethocerus* flight muscle fibers is solely due to C filaments and independent of the presence of thin filaments.

The regularity of the lattice of thick and thin filaments in asynchronous flight muscle and the fact that the overlap region and the Z-disc have hexagonal symmetry, means that it is possible to follow C filaments in stretched fibers from their origin on the thick filament, across the I-band. Transverse sections of fibers stretched in rigor so that thin filaments are broken from the Z-disc, exposing C filaments, show a regular array of C filaments in the position of the thick filament lattice (White and Thorson, 1973; White, 1983). Such a clear demonstration that C filaments follow the thick filament lattice position is not possible in vertebrate striated muscle where the symmetry of the overlap region is threefold and the Z-disc has a square lattice, so that the C filament lattice must change as it enters the Z-disc.

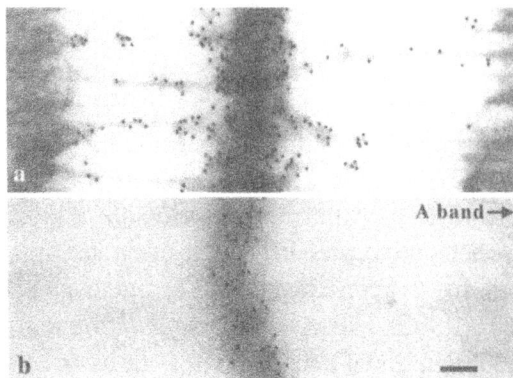

Figure 6*. Composition of C filaments.*
Lethocerus flight muscle fibres were stretched
in rigor; thin filaments are broken from the Z-
disc and C filaments span the gap between Z-
disc and A-band. (a) Electron micrograph of
cryosection labeled with anti-projectin, (b) cryo-
section labeled with anti-kettin. Primary
antibodies were followed by protein A tagged
with 10 nm gold particles. Projectin is in C
filaments but kettin is confined to the Z-disc.
Scale bar 100 nm.

Composition of C filaments

The protein composition of C filaments will determine their elasticity. In asynchronous flight muscle, projectin is in the part of the sarcomere containing C filaments (Saide *et al.*, 1990; Nave and Weber, 1990; Lakey *et al.*, 1990) and in electron micrographs of fibers stretched in rigor, labeled anti-projectin is seen on the extended filaments (Fig. 6a). Kettin is not in the C filaments and therefore would not contribute to their elasticity (Fig. 6b).

The widely differing stiffness of asynchronous and synchronous insect muscle fibers is likely to be due to the difference in the width of the I-band and to the proteins in the C filaments. In synchronous fibers, projectin is not in the C filament region of the sarcomere but confined to the A-band (Saide *et al.*, 1990; Lakey *et al.*, 1990). Therefore, the high resting elasticity of asynchronous fibers is likely to be due, at least partly, to projectin in the C filaments. During oscillatory contraction, asynchronous fibers are stretched by less than 5% of resting length. In *Drosophila* fibers, the length change *in vivo* is estimated to be 3.5 % (Chan and Dickinson, 1996). It is not clear what element in C filaments could stretch by this amount. The *Drosophila* projectin sequence

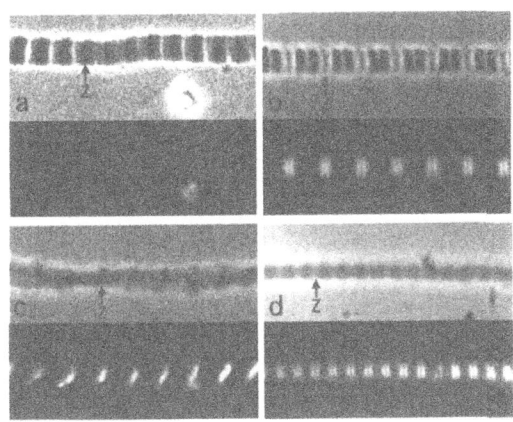

Figure 7. A stretch sensitive titin epitope in Lethocerus *and* Drosophila myofibrils. *Phase (top) and fluorescence (bottom) micrographs of: (a) unstretched* Lethocerus *flight myofibril, (b) stretched* Lethocerus *flight myofibril, (c) unstretched* Lethocerus *leg myofibril, (d) stretched* Drosophila *flight myofibril. Myofibrils in (b) and (d) were stretched in rigor by drawing a coverslip across myofibrils adhering to a slide. Myofibrils were labeled with antibody to the PEVK sequence of titin followed by a fluorescent second antibody. The antibody does not label unstretched flight myofibrils in* Lethocerus *or* Drosophila *(not shown) but labels an epitope in the region of C filaments in stretched myofibrils. Unstretched* Lethocerus *leg myofibrils are labeled close to the Z-disc.*

has no elastic PEVK region (Daley *et al.*, 1998). In vertebrate fibers, the tandem Ig region of I-band titin straightens out on moderate stretch (Gautel and Golding, 1996; Linke *et al.*, 1996, 1998; Trombitás *et al.*, 1998), but straightening the five tandem Ig domains near the N-terminus of projectin, could not account for the length change. Therefore, the extensible element in C filaments is unlikely to be projectin for length changes within the physiological range. In C filaments stretched beyond physiological limits (Fig. 6a), it is likely that the molecule is extended by unfolding Ig domains and recruitment of domains from the region of projectin bound to the thick filament, as occurs in overstretched titin (Granzier and Wang, 1993; Granzier *et al.*, 1996; Linke *et al.*, 1996).

The presence of titin in the C filaments of insect muscle fibers has not been shown. However, antibodies to vertebrate titin and to recombinant peptides from regions of a gene thought to be *Drosophila* titin, label the Z-disc region of *Drosophila* myofibrils (Machado *et al.*, 1998). We have found that an antibody to the extensible PEVK region of vertebrate titin labels an epitope in flight muscle myofibrils that is only exposed when the myofibrils are stretched to create an I-band, although the antibody labels unstretched leg myofibrils that have a clear I-band (Fig. 7). The stretch-sensitive PEVK epitope is therefore probably in C filaments in the short I-band of asynchronous fibers. As projectin does not have a PEVK sequence, the epitope is probably in the PEVK sequence identified in titin (Machado *et al.*, 1998). Since the stiffness of asynchronous fibers is high, the extensible PEVK region is probably short. Although C filaments have not been demonstrated in synchronous muscle, it is likely they exist and, lacking projectin, contain titin. The greater accessibility of the PEVK epitope in synchronous leg myofibrils compared to flight muscle myofibrils may be due both to the longer I-band and to the length of PEVK sequence, which is greater in titin molecules in fibers that are less stiff and more extensible (Labeit and Kolmerer, 1995).

The function of titin in vertebrate muscle is to determine sarcomere length, to act as a template for binding other proteins and to provide elasticity (Trinick, 1996; Maruyama, 1997; Gregorio *et al.*, 1999). Projectin and kettin in insect muscle could not fulfil the first two of these functions, which need a molecule extending across half the sarcomere. The sarcomere length in *Drosophila* flight muscle is 3.4 μm, so titin in these fibers would be 1.7 μm long and over 4 MDa, compared to around 3 MDa in vertebrate striated muscle.

Stretch activation

A rigid sarcomere structure is required for stretch activation during oscillatory contraction. In active insect flight muscle, both thick and thin filaments are under tension. Thick filaments are strained during oscillations due to the stiffness of C filaments attached to the Z-disc, and this strain is transmitted to thin filaments through crossbridges. Projectin and titin are likely to be responsible for the high stiffness of C filaments in asynchronous fibers. Opposing forces across the Z-disc from crossbridge activity in neighboring half sarcomeres would tend to pull thin filaments out of the Z-disc. The flight muscle Z-disc is an electron dense compact structure, wider than typical striated muscle Z-discs. Connecting links within the lattice hold actin filaments in place and contribute to the stiffness of active flight muscle fibers. Kettin, which effectively encases thin filaments, stabilizes the filaments in the Z-disc and may, in addition, strengthen α-actinin crosslinks between filaments. The elastic proteins in flight muscle maintain and transmit strain to thick and thin filaments. How this strain stimulates active tension development is still unknown.

Acknowledgment

The antibody used in Figure 7 was developed by Dr. M. Greaser and obtained from the Developmental Studies Hybridoma Bank, Department of Biological Sciences, University of Iowa, Iowa City, IA 52242.

References

Auber J, Couteaux R. Ultrastructure de la strie Z dans des muscles de Diptères. *J Microscopie* 1963;2:309-24.

Ayme-Southgate A, Vigoreaux JO, Benian GM, Pardue ML. *Drosophila* has a twitchin/titin related gene that appears to encode projectin. *Proc Natl Acad Sci USA* 1991;88:7973-77.

Chan WP, Dickinson MH. *In vivo* length oscillation of indirect flight muscles in the fruit fly *Drosophila melanogaster. J Exp Biol* 1996;199:2767-74.

Cheng N, Deatherage JF. Three-dimensional reconstruction of the Z-disk of sectioned bee flight muscle. *J Cell Biol* 1989;108:1761-74.

Daley J, Southgate R, Ayme-Southgate A. Structure of the *Drosophila* projectin protein: isoforms and implication for projectin filament assembly. *J Mol Biol* 1998;279:201-10.

Deatherage JF, Cheng N, Bullard B. Arrangement of filaments and cross-links in the bee flight muscle Z-disk by image analysis of oblique sections. *J Cell Biol* 1989;108:1775-82.

Fyrberg CC, Labeit S, Bullard B, Leonard K, Fyrberg EA. *Drosophila* projectin: relatedness to titin and twitchin and correlation with *lethal (4) 102 CDa* and *bent-Dominant* mutants. *Proc Roy Soc* ser B 1992;249:33-40.

Gautel M, Goulding D. A molecular map of titin/connectin elasticity reveals two different mechanisms acting in series. *FEBS Lett* 1996;385:11-14.

Granzier H, Wang K. Interplay between passive tension and strong and weak binding crossbridges in insect indirect flight muscle. *J Gen Physiol* 1993;101:235-70.

Granzier H, Helmes M, Trombitás K. Nonuniform elasticity of titin in cardiac myocytes: a study using immunoelectron microscopy and cellular mechanics. *Biophys J* 1996;70:430-42.

Granzier H, Kellermayer M, Trombitás K. Titin elasticity and mechanism of passive force development in rat cardiac myocytes probed by thin filament extraction. *Amer J Physiol* 1997;73:2043-53.

Gregorio CC, Trombitás K, Centner T, Kolmerer B, Stier G, Kunke K, Suzuki K, Obermayr F, Herrman B, Granzier H, Sorimachi H, Labeit S. The NH_2 terminus of titin spans the Z-disc: its interaction with a novel 19-K ligand (T-cap) is required for sarcomeric integrity. *J Cell Biol* 1998;143:1013-27.

Gregorio CC, Granzier H, Sorimachi H, Labeit, S. Muscle assembly: a titanic achievement? *Curr Opin Cell Biol* 1999;11:18-25.

Kolmerer B, Clayton J, Benes V, Allen T, Ferguson C, Leonard K, Weber U, Knekt M, Ansorge W, Labeit S, Bullard B. Sequence and expression of the kettin gene in *Drosophila melanogaster* and *Caenorhabditis elegans. J Mol Biol* 2000;296:435-48.

Labeit S, Kolmerer B. Titins: giant proteins in charge of muscle ultrastructure and elasticity. *Science.* 1995;270:293-96.

Lakey A, Ferguson C, Labeit S, Reedy M, Larkins A, Butcher G, Leonard K, Bullard B. Identification of high molecular weight proteins in insect flight and leg muscles. *EMBO J* 1990;9:3459-67.

Lakey A, Labeit S, Gautel M, Ferguson C, Barlow D, Leonard K, Bullard B. Kettin, a large modular protein in the Z-disc of insect muscles. *EMBO J* 1993;12:2863-71.

Linke WA, Ivemeyer M, Oliveri N, Kolmerer B, Rüegg JC, Labeit S. Towards a molecular understanding of the elasticity of titin. *J Mol Biol* 1996;261:62-71.

Linke WA, Ivemeyer M, Labeit S, Hinssen H, Rüegg JC, Gautel M. Actin-titin interaction in cardiac myofibrils: probing a physiological role. *Biophys J* 1997;73:905-19.

Linke WA, Stockmeier MR, Ivemeyer M, Hosser H, Mundel P. Characterizing titin's I-band region as an entropic spring. *J Cell Sci* 1998;111:1567-74.

Locker RH, Leet NG. Histology of highly stretched beef muscle. I. The fine structure of grossly stretched single fibers. *J Ultrastruct Res* 1975;52:64-72.

Machado C, Sunkel CE, Andrew DJ. Human autoantibodies reveal titin as a chromosomal protein. *J Cell Biol* 1998;141:321-33.

Maki S, Ohtani Y, Kimura S, Maruyama K. Isolation and characterization of a kettin-like protein from crayfish claw muscle. *J Muscle Res Cell Motil* 1995;16:579-85.

Maruyama K. Connectin/titin, giant elastic protein of muscle. *FASEB J* 1997;11:341-45.

Nave R, Weber K. A myofibrillar protein of insect muscle related to vertebrate titin connects Z band and A band: purification and molecular characterizations of an invertebrate mini-titin. *J Cell Sci* 1990;95:535-44.

Orlova A, Prochniewitz E, Egelman EH. Structural dynamics of F-actin: II. Cooperativity in structural transitions. *J Mol Biol* 1995;245:598-607.

Peckham M, Molloy JE, Sparrow JC, White DCS. Physiological properties of the dorsal longitudinal flight muscle and the tergal depressor of the trochanter muscle of *Drosophila melanogaster. J Muscle Res Cell Motil* 1990;11:203-15.

Peckham M, White, DCS. Mechanical properties of demembranated flight muscle fibers from a dragonfly. *J exp Biol* 1991;159:135-47.

Saide J, Chin-Bow S, Hogan J, Busquets-Turner L. Z-band proteins in the flight muscle and leg muscle of honeybee. *J Muscle Res Cell Motil* 1990;11:125-36.

Trinick J. Cytoskeleton: titin as scaffold and spring. *Curr Biol* 1996;6:258-60.

Trombitás K, Tigyi-Sebes A. Fine structure and mechanical properties of insect muscle. In *Insect Flight Muscle*, RT Tregear, ed. Elsevier, Amsterdam 1977;79-90.

Trombitás K, Tigyi-Sebes A. Crossbridge interaction with oppositely polarized actin filaments in double-overlap zones of insect flight muscle. *Nature* 1984;309:168-70.

Trombitás K, Pollack GH. Elastic properties of the titin filament in the Z-line region of vertebrate striated muscle. *J Muscle Res Cell Motil* 1993;14:416-22.

Trombitás K, Granzier H. Actin removal from cardiac myocytes shows that near the Z-line titin attaches to actin while under tension. *Amer J Physiol* 1997;273:C662-70.

Trombitás K, Greaser M, Labeit S, Jin J-P, Kellermeyer M, Helmes M, Granzier H. Titin extensibility *in situ*: entropic elasticity of permanently folded and permanently unfolded molecular segments. *J Cell Biol* 1998;140:853-59.

van Straaten M, Goulding D, Kolmerer B, Labeit S, Clayton J, Leonard K, Bullard B. Association of kettin with actin in the Z-disc of insect flight muscle. *J Mol Biol* 1999;285:1549-62.

White DCS. The elasticity of relaxed insect fibrillar flight muscle. *J Physiol* 1983;343:31-57.

White DCS, Thorson J. The kinetics of muscle contraction. *Prog Biophys* 1973;27:175-255.

Young P, Ferguson C, Bañuelos S, Gautel, M. Molecular structure of the sarcomeric Z-disk: two types of titin interactions lead to an asymmetrical sorting of α-actinin. *EMBO J* 1998;17:1614-24.

Discussion

Vigoreaux: Belinda, as you probably know, John Sparrow has a fly with an actin mutation in the 93 residue, E93K, which prevents the tropomyosin from sliding over. It looked to me like you mapped the kettin binding site very close to that. Is that where it is?

Bullard: Offhand I don't know where 93 is in the actin sequence.

Vigoreaux: It is right where the tropomyosin sits in the off state.

Bullard: Tropomyosin and kettin compete if tropomyosin is in the off or the on state.

Vigoreaux: The reason I am asking is because in the original E93K paper where they describe the structure of the sarcomere in the mutant, they show it had no Z-bands. In fact it was a high molecular weight protein that was missing from those myofibrils, which is probably kettin, if that is the binding site.

Bullard: Did they use sufficiently loose gels to see it?

Vigoreaux: I can't remember; we could probably look it up.

Pollack: Belinda, could you give an overall summary of what you think in insect flight muscle the connecting filament is actually composed of?

Bullard: I think it is composed of titin and projectin.

Pollack: OK, in what ratio? For those of us that are interested in the mechanics of that it would be interesting to know something more about the structure.

Bullard: Well, I think the only person who has quantitated titin on the thick filament is John Trinick, and what he is saying is that 6 titin molecules would give threefold symmetry. There could be more on insect thick filaments since the symmetry of the insect filament is fourfold. There could be 8 titins, and equally there could be 8 projectins.

Pollack: In electron micrographs of vertebrate muscle the connecting filament looks rather robust, rather thick relative to, for example, invertebrate muscle.

Bullard: Well, I think that when you normally see connecting filaments in flight muscle you would have unraveled at least the projectin because the projectin is not going to be extensible. There is no

PEVK region. So you must extend the Ig domains or the fibronectin domains because that is all there is apart from the kinase domain. And it looks as though the titin PEVK region is rather short, so again in C-filaments I think you are pulling the molecules out.

Greaser: This is just a comment about the 9D10, because it is an IgM. There is a difficulty in terms of penetration. Even at very short sarcomere lengths with rabbit psoas or cardiac muscle if you get down to a sarcomere length of about 1.7 nm, the staining is very weak.

Bullard: But even in cryosections where you would expect the epitope to be available, you still don't get any staining unless you stretch.

Greaser: Interesting.

Ayme-Southgate: If titin and projectin are together in the connecting filament, how do you think that these filaments stretch when the sarcomere becomes extended?

Bullard: I think the answer is, physiologically neither of them would stretch much. In *Drosophila* the length change is about 3%. But of course we see projectin in the C-filament, so it is definitely there. But I think physiologically projectin wouldn't stretch but maybe make the C-filaments rigid. There are 5 Ig domains at the N-terminus and that would not give you much to unravel.

TITIN AS A CHROMOSOMAL PROTEIN

Cristina Machado and Deborah J. Andrew
Department of Cell Biology and Anatomy,
The Johns Hopkins University School of Medicine, Baltimore, MD

Abstract: We identified titin as a chromosomal protein using a human autoimmune scleroderma serum. We cloned the corresponding gene in the fruitfly, *Drosophila melanogaster*. We have demonstrated that titin is not only expressed and localized in striated muscle but is also distributed uniformly on condensed mitotic chromosomes using multiple antibodies directed against different domains of both Drosophila and vertebrate titin. Titin is a giant sarcomeric protein responsible for the elasticity of striated muscle. Titin may also function as a molecular scaffold during myofibril assembly. We hypothesize that titin is a component of chromosomes that may function to determine chromosome structure and provide elasticity, playing a role similar to that proposed for titin in muscle. We have identified mutations in *Drosophila Titin* (*D-Titin*) and are characterizing phenotypes in muscle and chromosomes.

Introduction

Titin, also known as connectin, is a giant muscle protein that forms one of the three known filament systems of the sarcomere (with actin and myosin forming the other two). Titin is the third most abundant myofibrillar protein after actin and myosin (Wang *et al.*, 1979; Wang *et al.*, 1989). Titin is responsible for the elasticity of striated muscle and may function as a molecular scaffold for the assembly of myofibrils (for reviews, see Keller, 1995; Labeit and Kolmerer, 1995; Trinick, 1996; Labeit *et al.*, 1997; Maruyama, 1997; Gregorio *et al.*, 1999). Individual filamentous titin molecules span half a sarcomere from the Z-disk to the M-line, a distance of approximately 1.2 µm in relaxed skeletal muscle. The different isoforms of vertebrate titin exhibit a range of molecular weights, from 2993 to 3700 kDa, due to tissue-specific alternative

Elastic Filaments of the Cell, edited by Granzier and Pollack
Kluwer Academic/Plenum Publishers, 2000

splicing in regions of titin that localize to the Z-disk, I-band and M-line (Labeit and Kolmerer, 1995; Kolmerer *et al.*, 1996; Sorimachi *et al.*, 1997). Nearly 90% of titin's mass is comprised of immunoglobulin (Ig)-like and fibronectin type III (FN3)-like repeats (Labeit *et al.*, 1990; Maruyama *et al.*, 1993; Labeit and Kolmerer, 1995). In human skeletal titin there are 297 copies of these 100-residue repeats, the repeat number being somewhat smaller in human cardiac muscle. The I-band region of vertebrate titin also contains a 163 to 2200 residue PEVK domain rich in proline (P), glutamic acid (E), valine (V) and lysine (K). The PEVK domain and the Ig/FN3 domains confer elasticity to the titin filament (Gautel and Goulding, 1996; Linke *et al.*, 1996). Recent studies using optical tweezers and atomic force microscopy reveal that the application of physiological stretch forces induces elastic transitions in titin that involve Ig/FN3 domain straightening and PEVK domain unfolding. Application of more extreme, non-physiological, stretch forces induces Ig/FN3 unfolding (Kellermayer *et al.*, 1997; Rief *et al.*, 1997; Tskhovrebova *et al.*, 1997). Refolding of the Ig/FN3 domains, however, is slow and occurs only in the absence of stretch force. The number of Ig/FN3 repeats and the number of PEVK residues correlate with the elasticity of specific muscle types, the higher the number, the greater the elasticity of the muscle type (Labeit and Kolmerer, 1995; Labeit *et al.*, 1997). In addition to titin's Ig/FN3 repeats and elastic PEVK domain, the sequence of titin contains phosphorylation sites (Sebastyén *et al.*, 1995), recognition sites for muscle-specific calpain proteases (Sorimachi *et al.*, 1995; Kinbara *et al.*, 1997), and a serine/threonine kinase domain near the C-terminus (Labeit *et al.*, 1992; Takano-Ohmuro *et al.*, 1992).

Titin is hypothesized to function as the scaffold upon which the myofibrils are assembled (Keller, 1995; Trinick, 1996). Titin mRNA is expressed in myoblasts prior to cell fusion (Colley *et al.*, 1990) and titin mRNA and protein are among the earliest molecules to localize in the developing sarcomere (Fulton and Alftine, 1997; van der Ven and Fürst, 1997). Within each region of the sarcomere, titin binds to different proteins. In the Z-disk, the N-terminus of titin binds to the C-terminal region of α-actinin, an actin-binding protein that crosslinks titin to the thin actin filaments (Ohtsuka *et al.*, 1997a; Ohtsuka *et al.*, 1997b; Sorimachi *et al.*, 1997; Turnacioglu *et al.*, 1997). In cardiac muscle, the N-terminus of titin may bind to actin in the Z/I junction, and in short regions within the I-band (Trombitas *et al.*, 1997). The A-band region of titin binds to MyBP-C (C-protein), MyBP-H (H-protein) and myosin II (Itoh *et al.*, 1988; Fürst *et al.*, 1989; Soteriou *et al.*, 1993; Eilertsen *et al.*, 1994; Houmeida *et al.*, 1995; Freiburg and Gautel, 1996; Trombitas *et al.*, 1997). In the M-line, titin binds to M-protein and phosphorylated myomesin, which are myosin-binding proteins that crosslink titin to the thick myosin filaments (Eppenberger *et al.*, 1981; Obermann *et al.*, 1996; Obermann *et al.*, 1997). Evidence that titin functions as a myofibrillar scaffold comes from experiments using an N-terminal fragment of titin fused to green fluorescent protein (Turnacioglu *et al.*, 1997). This fusion protein localizes to the Z-disc in contracting myofibrils in culture; when overexpressed, this fusion protein causes myofibrillar disassembly. Direct evidence that titin is required for myofibrillar assembly during normal development awaits the identification and characterization of loss-of-function mutations in the titin gene.

Chromosomes are highly ordered, elastic structures that must maintain their integrity even during the physically strenuous process of cell division. Each human chromosome contains a single DNA molecule of between 50 to 250×10^6 base pairs, which would extend from 1.7 to 8.5 cm if uncoiled. Chromosomes are packaged into much more compact units in order to fit inside the cell nucleus, which measures only about 10-20 μm. During mitosis, the chromosomes are further compacted 5- to10-fold, reaching an overall 10,000-fold compaction. The high level of compaction that occurs during mitosis ensures (1) that tangles between sister chromatids or between neighboring nonhomologous chromosomes are eliminated and (2) that chromosomes are sufficiently compact so that they are away from the cleavage furrow at the end of mitosis. The compact structure of mitotic chromosomes is likely to arise by a highly regulated and organized process (1) each mitotic chromosome within a given cell type has a fixed axial diameter and length; (2) each chromosome gives a reproducible banding pattern when stained with DNA dyes; and (3) fluorescence *in situ* hybridization (FISH) experiments with specific DNA probes localize to reproducible positions on the chromosomes (for review, see Koshland and Strunnikov, 1996).

The mechanism by which chromosome architecture is established remains one of the most fundamental questions in cell biology. For the last two decades, scientists have been trying to learn how chromosomes achieve their regular hierarchical structure, but only a few key players involved in this assembly have been discovered. Several models predict the existence of a scaffold, composed of one or more proteins that could serve as a template upon which the chromosomes would be organized. However, the proteins known to be required for chromosome condensation during mitosis are likely to have active, enzymatic roles in the compaction process, rather than functioning as structural templates for higher-order assembly of chromatin.

The known proteins required for chromosome compaction include the histones H2A, H2B, H3 and H4, which package DNA into 10 nm fibers known as nucleosomes, histone H1, which is involved in further packaging of DNA into 30 nm fibers, topoisomerase II (topo II), SMC (structural maintenance of chromosomes) proteins and their associated subunits that comprise the condensins, and cohesins, molecules that link condensation and sister chromatid cohesion (for review see Earnshaw and Mackay, 1994; Hirano, 1995; Koshland and Strunnikov, 1996; Heck, 1997; Warburton and Earnshaw, 1997). Histones are bound to chromosomes and provide a basal level of chromatin compaction throughout the cell cycle. Although histone modifications (acetylation or phosphorylation) may trigger changes in chromatin condensation (Wei *et al.*, 1999 and for review, Wolffe, 1998), modified histones are unlikely to provide structural/scaffolding activity in mitotic chromosome architecture. Although topo II is also present throughout the cell cycle, it does appear to have an active role in chromosome condensation during mitosis (for review, see Earnshaw and Mackay, 1994; Hirano, 1995; Heck, 1997; Warburton and Earnshaw, 1997). Mutations in yeast topo II fail to condense chromosomes in prophase, depletion of topo II in Xenopus mitotic extracts or early Drosophila embryos prevents condensation, inhibitors of topo II block mammalian chromosome condensation both *in vivo* and *in vitro*, and topo II is found associated with mitotic chromosomes prepared under certain fixation conditions. The enzymatic activity of topo II is to create double strand breaks that allow the passage of one DNA strand through another, an activity that can resolve catenated DNA circles *in vitro* and reduce the superhelicity of DNA. Thus topo II may function during mitosis by either relaxing superhelical domains formed during the condensation of chromatin or by decatenating sister chromatids during replication and thus eliminating steric problems during condensation. Topo II may play a separate structural role in mitotic chromosomes based largely on its initial biochemical identification as a chromosomal "scaffold" protein (scaffold proteins have been operationally defined as the nonsoluble fraction from isolated chromosome preparations) and subsequent immunolocalization to metaphase chromosomes (Adolf *et al.*, 1977; Paulson and Laemmli, 1977). However, the presence of topo II in mitotic scaffolds may simply be a consequence of its earlier requirements for DNA and RNA synthesis.

SMC proteins, of which there are at least four family members, were identified as important in mitotic chromosome structure and function through several divergent approaches (for review, see Peterson, 1994; Hirano, 1995; Saitoh *et al.*, 1995; Koshland and Strunnikov, 1996; Heck, 1997). SMCs were first found in yeast as mutations that resulted in defects in chromosome condensation and segregation (Strunnikov *et al.*, 1993; Saka *et al.*, 1994; Strunnikov *et al.*, 1995). Sc2, a chicken SMC protein, was identified as a major component of the chromosome scaffold (Saitoh *et al.*, 1994). Finally, XCAP-C and XCAP-E, which are homologous to yeast SMC4 and SMC2, respectively, were isolated from mitotic extracts from Xenopus eggs (Hirano and Mitchison, 1994). Chromatin condensation fails when antibodies to XCAP-C protein are added to Xenopus mitotic extracts further supporting a role for SMCs in chromosome condensation. Moreover, SMC proteins accumulate in foci near the geometrical center of condensed chromatin supporting a potential role in both the condensation process and the maintenance of condensation. SMC proteins are relatively large (~1000-2000 residues) and share a common head-rod-tail structural organization. The N-terminal head contains an NTP-binding domain, which is followed by two central coiled-coil regions and a C-terminal conserved domain with a helix-loop-helix motif. In Xenopus, XCAP-C and XCAP-E form part of a larger protein complex which is required for condensation based on immunodepletion and rescue experiments with Xenopus mitotic extracts (Hirano *et al.*, 1997). This larger complex, known as the 13S condensin complex, contains XCAP-C, XCAP-E, XCAP-D2, XCAP-G and XCAP-H. XCAP-H is homologous to the Drosophila *barren* gene product. BARREN is a chromosomal protein required for mitotic chromosome segregation, again consistent with a potential function in chromosome condensation. The 13S condensin complex has recently been shown to induce positive supercoils into closed circular DNA in an ATP-dependent manner (Kimura and Hirano, 1997). This activity is proposed to use ATP-hydrolysis to wrap DNA into a right-handed supercoil and may represent the mechanism underlying compaction of chromatin during mitosis.

Cohesins were independently identified by two groups as yeast mutations that cause premature sister chromatid separation (Guacci *et al.*, 1997; Michaelis *et al.*, 1997). Cohesins interact with SMC proteins both biochemically and genetically. Mutations in one of the two cohesin molecules, known as MCD1 or SCC1, result not only in premature sister chromatid separation but also in chromosome condensation defects, potentially linking chromatid cohesion and condensation (Guacci *et al.*, 1997). Since little is known of the biochemistry of cohesins, it is difficult to determine how they mediate these roles in chromosome behavior.

The limited number of proteins known to be required for chromosome condensation, topo II, SMC/condensins and cohesins, were identified either by biochemical assays or clever genetic screens in yeast. The isolation of the same or related proteins by multiple independent methods suggests either that there are very few proteins required for this process or that alternative approaches may be necessary to find novel components required for chromosome condensation. We have found a protein, titin, that may serve not only a scaffolding role in the assembly of mitotic chromosomes but may also provide the elasticity necessary to prevent chromosome breakage and subsequent loss of genetic material during the physically strenuous process of cell division. We identified titin as a chromosomal protein using an approach already proven successful in the identification of other proteins associated with chromosomes (see below) and our preliminary genetic characterization fully supports a role for this protein in chromosome behavior during mitosis.

Results and Discussion

Identification of titin as a chromosomal component

Human autoimmune sera have been used successfully to identify a wide range of nuclear antigens, including both single- and double-stranded DNA, RNA, histones, small nuclear RNA-binding proteins, transcription factors, nuclear lamins, topo I and II, heterochromatin-associated proteins and centromere proteins (Tan, 1982; Earnshaw and Rothfield, 1985; Tan *et al.*, 1987; Tan, 1989; Mole-Bajer *et al.*, 1990; Earnshaw and Rattner, 1991; Tan, 1991; Fritzler, 1997; Saunders *et al.*, 1993; Shibata *et al.*, 1993; Bejarano and Valdivia, 1996; Spain *et al.*, 1997; Sugiura *et al.*, 1997). We recently identified titin as a chromosomal component using a human autoimmune scleroderma serum (Machado, 1998; Machado *et al.*, 1998). The scleroderma serum recognized an epitope on condensed mitotic chromosomes from both human cultured cells and early Drosophila embryos. Using the scleroderma serum to directly screen a Drosophila expression library, we isolated the Drosophila homologue of vertebrate titin (*D-Titin*) based on sequence similarity, the size of the corresponding protein, developmental expression and subcellular localization (Fig. 1). We generated antisera against two different domains of Drosophila Titin and showed that the antisera not only stain the Z-disks of Drosophila sarcomeres, but also give uniform staining of condensed human and Drosophila mitotic chromosomes. Moreover, monoclonal and polyclonal antibodies directed against several different vertebrate titin epitopes also stained condensed chromosomes in human epithelial cells (Machado *et al.*, 1998). These results suggest an additional, novel role for titin in chromosome structure and/or chromosome dynamics during mitosis.

Titin and chromosome structure

Based on the uniform distribution of titin along condensed chromosomes, we envision a role for titin in the assembly of higher-order condensed mitotic chromosomes, a role similar to that proposed for titin as a scaffolding element in myofibrils (Trinick, 1994; Trinick, 1996). Mitotic chromosome condensa-

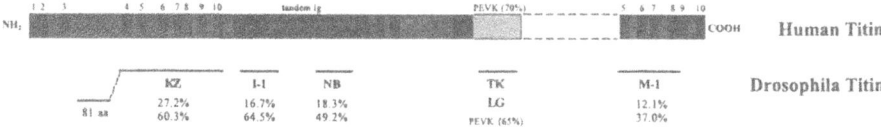

Figure 1. *Comparison of the Human Titin protein with the available Drosophila Titin protein sequence. Numbers 1-10 refer to the first ten Ig repeats from the region of titin that was initially localized to the Z disc. The "Ig" domains are in dark gray and the "PEVK" domain is in light gray.*

tion is essential for proper chromosome segregation, and for ensuring that chromosomes are away from the cleavage furrow during cytokinesis (for reviews see Hirano, 1995; Koshland and Strunnikov, 1996). Chromosome condensation is believed to be a highly regulated process: Chromosome length and banding patterns are uniform in a given cell type. Specific sequences are always found at the same position on the chromosome. The axial diameter of mitotic chromosomes is also invariant in a given cell type (Koshland and Strunnikov, 1996). The regularity of chromosome structure suggests the involvement of a protein that functions as a "molecular ruler," a function already ascribed to titin in muscle (Trinick, 1994; Trinick, 1996).

Titin and chromosome elasticity

Titin may provide elasticity to chromosomes, and resistence to chromosome breakage during mitosis. The elastic properties of purified muscle titin (Kellermayer *et al.*, 1997; Rief *et al.*, 1997; Tskhovrebova *et al.*, 1997) correspond well to the elastic properties of chromosomes in living cells and of chromosomes assembled *in vitro* using *Xenopus* egg extracts (Houchmandzadeh *et al.*, 1997; Houchmandzadeh and Dimitrov, 1999). Studies using optical tweezers and atomic force microscopy reveal that the PEVK domain and the Ig/FN3 repeats constitute a two-spring system acting in series to confer reversible extensibility to titin. Similarly, metaphase chromosomes from living cells also show two levels of extensibility (Houchmandzadeh *et al.*, 1997). Metaphase chromosomes from cultured newt lung cells can be stretched up to 10 times their normal length and return to their native shape. Further non-physiological extensions of chromosomes from 10- to 100-fold are irreversible. Measurements of longitudinal deformability and bending rigidity of *in vitro* assembled chromosomes reveal properties of chromosomes that are entirely consistent with models developed for titin elasticity (Houchmandzadeh and Dimitrov, 1999). Our discovery of titin on chromosomes integrates the mechanical properties of purified titin with the elastic properties of eukaryotic chromosomes.

Identification of mutations in D-Titin

The most direct method to determine the role of *D-Titin* in muscles and chromosomes is to isolate and characterize loss-of-function mutations in the gene. As a first step toward the identification of *D-Titin* mutations, we carried out *in situ* hybridization to polytene chromosomes using genomic clones and cDNAs from the *D-Titin* gene. These experiments localized *D-Titin* to cytological interval 62C1-2, a fortuitous position since a saturation mutagenesis has been done in the region, and three P-elements have also been localized there. Both EMS- and γ-ray-induced alleles of fifteen different complementation groups (genes) in the region had been made and mapped with respect to each other using a series of small deficiencies that delete one or more adjacent genes in the region (Sliter *et al.*, 1989; Wang *et al.*, 1994). By genomic Southerns and *in situ* hybridizations to polytene chromosomes, we found that *D-Titin* is deleted in *Df(3L)-1* and *Df(3L)-2* but not in *Df(3L)-3*, *Df(3L)-4* or *Df(3L)-5* (Fig. 2). These experiments localize *D-Titin* to a genetic interval known to contain only a single gene. We have obtained five EMS-induced *D-Titin* alleles, four γ-ray-induced *D-Titin* alleles, and have shown that a lethal P-element insertion that maps to 62C1-2 fails to complement *D-Titin*. Viable excisions of the P-element insertional mutation fully complement *D-Titin*.

We are in the process of examining *D-Titin* protein accumulation in embryos homozygous for each of the *D-Titin* alleles and characterizing the mutant phenotypes. Preliminary results fully support a role for titin in myofibrillar assembly and in chromosome structure and function. The identification of such a large structural protein on chromosomes will be a useful tool for isolating and characterizing other chromosomal proteins, and for understanding the molecular basis for the higher-order chromosome architecture and behavior at different stages throughout the cell cycle.

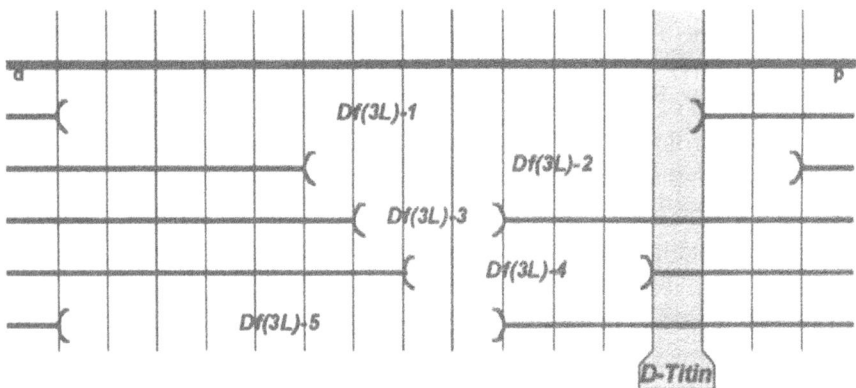

Figure 2. Genetics of cytological region 62B/C. Fifteen complementation groups (genes) map to the region. Df(3L)-1, Df(3L)-2, Df(3L)-3, Df(3L)-4 and Df(3L)-5 were used to localize D-Titin by genomic Southerns and in situ hybridization to polytene chromosomes. The dark lines indicate regions not deleted by the deficiencies and breaks represent regions deleted by the deficiencies. D-Titin is deleted in Df(3L)-1 and Df(3L)-2 but not in Df(3L)-3, Df(3L)-4 or Df(3L)-5. Thus, D-Titin maps to an interval containing only a single gene for which there are more than ten extant alleles, including five EMS-induced, four γ-ray-induced, one insertion allele and several additional excision alleles.

References

Bejarano LA, Valdivia MM. Molecular cloning of an intron-less gene for the hamster centromere antigen CENP-B. *Biochimica et Biophysica Acta* 1996;1307:21-25.

Colley NJ, Tokuyasu KT, Singer SJ. The early expression of myofibrillar proteins in round post mitotic myoblasts of embryonic skeletal muscle. *J Cell Sci* 1990;95:11-22.

Earnshaw WC, Mackay AM. Role of nonhistone proteins in the chromosomal events of mitosis. *FASEB J* 1994;8:947-956.

Earnshaw WC, Rattner JB. The use of autoantibodies in the study of nuclear and chromosomal organization. *Methods in Cell Biology* 1991;35:135-175.

Earnshaw WC, Rothfield N. Identification of a family of human centromere proteins using autoimmune sera from patients with scleroderma. *Chromosoma* 1985;91:313-321.

Eilertsen KJ, Kazmierski ST, Keller TCS III. Cellular titin localization in stress fibers and interaction with myosin II filaments in vitro. *J Cell Biol* 1994;126:1201-1210.

Eppenberger HM, Perriard JC, Rosenberg UB and Strehler EE. The Mr 165,000 M-protein myomesin: a specific protein of cross-striated muscle cells. *J Cell Biol* 1981;89:185-193.

Freiburg A, Gautel M. A molecular map of the interaction between titin and myosin-binding protein C. Implications for sarcomeric assembly in familial hypertrophic cardiomyopathy. *Eur J Biochem* 1996;235:317-323.

Fritzler MJ. Autoantibodies: diagnostic fingerprints and etiological perplexities. *Clin Invest Med* 1997;20:103-115.

Fulton AB, Alftine C. Organization of protein and mRNA for titin and other myofibril components during myofibrillogenesis in cultured chicken skeletal muscle. *Cell Struc Funct* 1997;22:51-58.

Fürst DO, Nave R, Osborn M, Weber K. Repetitive titin epitopes with a 42 nm spacing coincide in relative position with known A band striations also identified by major myosin-associated proteins. An immunoelectron-microscopical study on myofibrils. *J Cell Sci* 1989;94:119-125.

Gautel M, Goulding D. A molecular map of titin/connectin elasticity reveals two different mechanisms acting in series. *FEBS Lett* 1996;385:11-14.

Gregorio CC, Granzier H, Sorimachi H and Labeit S. Muscle assembly: a titanic achievement? *Curr Opin Cell Biol* 1999;11:18-25.

Guacci V, Koshland D, Strunnikov A. A direct link between sister chromatid cohesion and chromosome condensation revealed through the analysis of *MCD1* in S. cerevisiae. *Cell* 1997;91:59-70.

Heck MMS. Condensins, cohesins and chromosome architecture: How to make and break a mitotic chromosome. *Cell* 1997;91:5-8.

Hirano T, Kobayashi R, Hirano M. Condensins, chromosome condensation protein complexes containing XCAP-C, XCAP-E and a Xenopus homolog of the Drosophila Barren protein. *Cell* 1997;89:511-521.

Hirano T, Mitchison TJ. A heterodimeric coiled-coil protein required for mitotic chromosome condensation in vitro. *Cell* 1994;79:449-458.

Hirano T, Mitchison TJ, Swedlow JR. The SMC family: from chromosome condensation to dosage compensation. *Curr Opin Cell Biol* 1995;7:329-336.

Houchmandzadeh B, Dimitrov S. Elasticity measurements show the existence of thin rigid cores inside mitotic chromosomes. *J Cell Biol* 1999;145:215-223.

Houchmandzadeh B, Marko JF, Chatenay D, Libchaber A. Elasticity and structure of eukaryote chromosomes studied by micromanipulation and micropipette aspiration. *J Cell Biol* 1997;139:1-12.

Houmeida A, Holt J, Tskhovrebova L, Trinick J. Studies of the interaction between titin and myosin. *J Cell Biol* 1995;131:1471-1481.

Itoh Y, Suzuki T, Kimura S, Ohashi K, Higuchi H, Sawada H, Shimizu T, Shibata M, Maruyama
    K. Extensible and less-extensible domains of connectin filaments in stretched vertebrate
    skeletal muscle as detected by immunofluorescence and immunoelectron microscopy using
    monoclonal antibodies. *J Biochem* 1988;104:504-508.
Keller TCS. Structure and function of titin and nebulin. *Curr Opin Cell Biol* 1995;7:32-3.
Kellermayer MSZ, Smith SB, Granzier HL, Bustamante C. Folding-unfolding transitions in single
    titin molecules characterized with laser tweezers. *Science* 1997;276:1112-1116.
Kimura K, Hirano T. ATP-dependent positive supercoiling of DNA by 13S condensin: A
    biochemical implication for chromosome condensation. *Cell* 1997;90:625-634.
Kinbara K, Sorimachi H, Ishiura S, Suzuki K. Muscle-specific calpain, p94, interacts with the
    extreme C-terminal region of connectin, a unique region flanked by two immunoglobulin
    C2 motifs. *Arch Biochem Biophys* 1997;342:99-107.
Kolmerer B, Olivieri N, Witt C, Herrmann BG, Labeit S. Genomic organization of the M-line
    titin and its tissue-specific expression in two distinct isoforms. *J Mol Biol* 1996;256:556-
    563.
Koshland D, Strunnikov A. Mitotic chromosome condensation. *Ann Rev Cell Dev Biol*
    1996;12:305-333.
Labeit S, Barlow DP, Gautel M, Gibson T, Gibson M, Holt J, Hsieh C-L, Francke U, Leonard K,
    Wardale J, Whiting A, Trinick J. A regular pattern of two types of 100-residue motif in the
    sequence of titin. *Nature* 1990;345:273-276.
Labeit S, Gautel M, Lakey A, Trinick J. Towards a molecular understanding of titin. *EMBO J*
    1992;11:1711-1716.
Labeit S, Kolmerer B. Titins, giant proteins in charge of muscle ultrastructure and elasticity.
    Science 1995;270:293-296.
Labeit S, Kolmerer B, Linke WA. The giant protein titin. *Circul Res* 1977;80:290-294.
Linke WA, Ivemeyer M, Olivieri N, Kolmerer B, R'egg JC, Labeit S. Towards a molecular
    understanding of the elasticity of titin. *J Mol Biol* 1996;261:62-71.
Machado C. Isolating the *D-Titin* gene in *Drosophila melanogaster* using a human autoimmune
    serum. *Ph.D. Thesis, University of Porto,* 1998.
Machado C, Sunkel CE, Andrew DJ. From muscles to chromosomes: human antibodies reveal
    titin as a chromosomal protein. *J Cell Biol* 1998;141:321-323.
Maruyama K. Connectin/titin, giant elastic protein of muscle. *FASEB J* 1997;11:341-345.
Maruyama K, Endo T, Kume H Kawamura Y, Kanzawa N, Nakavchi Y, Kimura S, Kawashima
    S. A novel domain sequence of connectin localized at the I-band of skeletal muscle
    sarcomeres: homology to neurofilament subunits. *Biochem Biophys Res Commun*
    1993;194:1288-1291.
Michaelis C, Ciosk R, Nasmyth K. Cohesins: Chromosomal proteins that prevent premature
    separation of sister chromatids. *Cell* 1997;91:47-58.
Mole-Bajer J, Bajer AS, Zinkowski RP, Balczon RD, Brinkley BR. Autoantibodies from a patient
    with scleroderma CREST recognized kinetochores of the higher plant Haemanthus. *Proc
    Natl Acad Sci USA* 1990;17:1627-1631.
Obermann WM, Gautel M, Steiner F, van der Ven PF, Weber K, Fürst DO. The structure of the
    sarcomeric M band: localization of defined domains of myomesin, M-protein, and the 250
    kD carboxy-terminal region of titin by immunoelectron microscopy. *J Cell Biol*
    1996;134:1441-1453.
Obermann WM, Gautel M, Weber K, Fürst DO. Molecular structure of the sarcomeric M band:
    mapping of titin and myosin binding domains in myomesin and the identification of a potential
    regulatory phosphorylation site in myomesin. *EMBO J* 1997;16:211-220.
Ohtsuka H, Yajima H, Maruyama K, Kimura S. Binding of the N-terminal 63 kDa portion of
    connectin/titin to α-actinin as revealed by the yeast two-hybrid system. *FEBS Lett*
    1997a;401:65-67.
Ohtsuka H, Yajima H, Maruyama K, Kimura S. The N-terminal Z repeat 5 of connectin/titin
    binds to the C-terminal region of a-actinin. *Biochem Biophys Res Commun* 1997b;235:1-3.

Paulson JR, Laemmli UK. The structure of histone-depleted chromosomes. *Cell* 1977;12:817-828.

Peterson CL. The SMC family: novel motor proteins for chromosome condensation? *Cell* 1994;79:389-392.

Rief M, Gautel M, Oesterhelt F, Fernandez JM, Gaub HE. Reversible unfolding of individual titin immunoglobulin domains by AFM. *Science* 1997;276:1109-1112.

Saitoh N, Goldberg I, Earnshaw WC. The SMC proteins and the coming of age of the chromosome scaffold hypothesis. *Bioessays* 1995;17:759-766.

Saitoh N, Goldberg IG, Wood ER, Earnshaw WC. ScII: an abundant chromosome scaffold protein is a member of a family of putative ATPases with unusual predicted tertiary structure. *J Cell Biol* 1994;127:303-318.

Saka Y, Sutani T, Yamashita Y, Saitoh S, Takeuchi M, Nakaseko Y, Yanagida M. Fission yeast cut3 and cut14, members of the ubiquitous protein family are required for chromosome condensation and segregation in mitosis. *EMBO J* 1994;13:4938-4952.

Saunders WS, Chue C, Goebl M, Craig C, Clark RF, Powers JA, Eissenberg JC, Elgin SC, Rothfield NF, Earnshaw WC. Molecular cloning of a human homologue of Drosophila heterochromatin protein HP1 using anti-centromere autoantibodies with anti-chromo specificity. *J Cell Sci* 1993;104:573-582.

Sebastyén MG, Wolff JA, Greaser ML. Characterization of a 5.4 kb cDNA fragment from the Z-line of rabbit cardiac titin reveals phosphorylation sites for proline-directed kinases. *J Cell Sci* 1995;108:3029-3037.

Shibata S, Muryos T, Saitoh Y, Brumeanu TD, Bona CA, Kasturi KN. Immunochemical and molecular characterization of anti-RNA polymerase I autoantibodies produced by tight skin mouse. *J Clin Invest* 1993;92:984-992.

Sliter TJ, Henrich VC, Tucker RL, Gilbert LI. The genetics of the *Dras3-Roughened-ecdysoneless* chromosomal region (62B3-4 to 62D3-4) in *Drosophila melanogaster*: Analysis of recessive lethal mutations. *Genetics* 1989;123:327-336.

Sorimachi H, Freiburg A, Kolmerer B, Ishiura S, Stier G, Gregorio C, Labeit D, Linke WA, Suzuki K, Labeit S. Tissue-specific expression and a-actinin binding properties of the Z-disc titin. Implications for the nature of vertebrate Z-discs. *J Mol Biol* 1997;271:1-8.

Sorimachi H, Kinbara K, Kimura S, Takahashi M, Ishiura S, Sasagawa N, Sorimachi N, Shimada H, Tagawa K, Maruyama K, Suzuki K. Muscle-specific calpain, p94, responsible for limb girdle muscular dystrophy type 2A, associates with connectin through IS2, a p94-specific sequence. *J Mol Biochem* 1995;270:31158-31162.

Soteriou A, Gamage M, Trinick J. A survey of the interactions made by the giant protein titin. *J Cell Sci* 1993;104:119-123.

Spain TA, Sun R, Gradzka M, Lin SF, Craft J, Miller G. The transcriptional activator Sp1, a novel autoantigen. *Arthritis and Rheumatism* 1997;40:1085-1095.

Strunnikov AV, Hogan E, Koshland D. *SMC-2*, a *Saccaromyces cerevisiae* gene essential for chromosome segregation and condensation defines a subgroup within the SMC-family. *Genes and Dev* 1995;9:587-599.

Strunnikov AV, Larionov VL, Koshland D. *SMC1*: an essential yeast gene encoding a putative head-rod-tail protein is required for nuclear division and defines a new ubiquitous protein family. *J Cell Biol* 1993;123:1635-1648.

Sugiura K, Muro Y, Nagai Y, Kamimoto T, Wakabayashi T, Ohashi M, Hagiwara M. Expression cloning and intracelular localization of a human ZF5 homologue. *Biochimica et Biophysica Acta* 1997;1352:23-26.

Takano-Ohmuro H, Nakauchi Y, Kimura S, Maruyama K. Autophosphorylation of β-connectin (titin 2) *in vitro*. *Biochem Biophys Res Commun* 1992;183:31-35.

Tan EM. Autoantibodies to nuclear antigens (ANA): Their immunobiology and medicine. *Advances Immunol* 1982;33:167-240.

Tan EM. Antinuclear antibodies: diagnostic markers for autoimmmune diseases and probes for cell biology. *Advances Immunol* 1989;44:93-151.

Tan EM. Autoantibodies in pathology and cell biology. *Cell* 1991;67:841-842.

Tan EM, Reimer G, Sullivan K. *Intracellular autoantigens: diagnostic fingerprints but aetiological dilemmas.* Chichester, John Wiley & Sons, 1987.

Trinick J. Titin and nebulin: protein rulers in muscle? *TIBS* 1994;19:405-408.

Trinick J. Titin as a scaffold and spring. *Cytoskeleton Curr Biol* 1996;6:258-260.

Trombitas K, Greaser ML, Pollack GH. Interaction between titin and thin filaments in intact cardiac muscle. *J Muscle Res Cell Motil* 1997;18:345-351.

Tskhovrebova L, Trinick J, Sleep JA, Simmons RM. Elasticity and unfolding of single molecules of the giant muscle protein titin. *Nature* 1997;387:308-312.

Turnacioglu KK, Mittal B, Dabiri GA, Sanger JM, Sanger JW. An N-terminal fragment of titin coupled to green fluorescent protein localizes to the Z-bands in living muscle cells: overexpression leads to myofibril disassembly. *Mol Biol Cell* 1997;8:705-717.

van der Ven PF, Fürst DO. Assembly of titin, myomesin and M-protein into the sarcomeric M band in differentiating human skeletal muscle cells in vitro. *Cell Struct Funct* 1997;22:163-171.

Wang K, Fanger BO, Guyer CA, Staros JV. Electrophoretic transfer of high-molecular weight proteins for immunostaining. *Meth Enzymology* 1989;172:687-696.

Wang K, McClure J, Tu A. Titin: major myofibrillar components of striated muscle. *Proc Natl Acad Sci USA* 1979;76:3698-3702.

Wang M, Champion LE, Biessmann H, Mason JM. Mapping a mutator, mu2, which increases the frequency of terminal deletions in *Drosophila melanogaster. Mol Gen Genet* 1994;245:598-607.

Warburton PE, Earnshaw WC. Untangling the role of DNA topoisomerase II in mitotic chromosome structure and function. *Bioessays* 1997;19:97-99.

Wei Y, Yu L, Bowen J, Gorovsky MA, Allis CD. Phorphorylation of histone H3 is required for proper chromosome condensation and segregation. *Cell* 1999;97:99-109.

Wolffe AP. *Chromatin: structure and function.* San Diego, CA: Academic Press, 1998.

Discussion

Pollack: I think it was a beautiful and fascinating presentation. At the end, one thing that struck me is that you are anticipating that titin is a template for the organization of DNA, just as it might be a template for the laying down of myosin. Is that true?

Andrew: That is certainly what we would like to think. The next set of experiments that we would like to do, is to get antibodies directed against different domains of titin, and do the immuno-electron microscopy on chromosomes, so that we can see if there is a regular array of titin or whether it is more of an isotropic arrangement. Based on the fact that these mutant chromosomes are de-condensed, it is certainly involved in packaging of the chromatin. The fact that one of the phenotypes is premature sister chromatid separation, [there are other proteins for which mutations in the corresponding gene have sister chromatid separation defects] suggests that titin might be involved in how these other proteins assemble on the chromosomes as well. So yes, we like to think that. Chromosomes have a very regular structure. If you look in the same cell type, from cell to cell, the diameter of the chromosomes is always the same. And if you look at particular sequences by FISH, you can see where those sequences are found on the chromosome. They are always found at approximately the same position from chromosome to chromosome. So although chromosomes at the gross level don't look like very regular structures, I think they are.

Labeit: Recent data from Christian Witt indicate that the antibody to the immunogenetic titin region detects a different gene locus in the genome. So putting everything together, my question is, why is there such variation in results of different scientists? You get most antibody to stain chromosomes, Michael Wray doesn't get most antibodies to stain. Why is this?

Andrew: Yes, Dimitrov has also reported difficulty seeing staining. I can't address exactly what we might be doing differently. We have given the Drosophila antibody out to two other labs and the other labs have used multiple fixation techniques and get staining, and have looked at multiple cell types. They do detect staining with our Drosophila antibody.

Labeit: So one possibility would be that at this stage we should be careful whether we talk about titin, or titin-like proteins? Is it possible that it could be a different gene?

Andrew: I think this one is eerily like titin, given the size. One thing that we will be doing is making antibodies to the little domain of the protein that maps by homology to the M-line, and see whether we also see staining with that.

Linke: That would be the perfect chance to name the protein trombitin.

Trinick: Are you suggesting that kettin may be a proteolitic fragment of titin?

Andrew: Either that or a splice variant.

Trinick: Well, perhaps I could ask Belinda then, if you just take Drosophila and dissolve it in SDS at very early times without any chance of degredation, is there a kettin band?

Bullard: Yes.

Trinick: So it's not a proteolitic fragment, it can't be.

Bullard: I hate to commit myself at this stage. I think it is a very nice story. We have found there is a 7 kb gap between the end of the kettin sequence and your NB clone. Your PEVK sequence is much further downstream, so there is something in the middle there, that is not kettin and not titin.

Andrew: An intron?

Bullard: No, it is not an intron. Then there is the fact that the nematode has kettin but no titin sequence. Now we've discussed that earlier and you said you thought that nematode genomic sequence might be missing bits, I suppose that is a possibility. I am fairly convinced kettin is a genuine protein. But there is the story of zeugmatin, which turned out to be titin.

Bullard: One other question I wanted to ask you is about the N-terminal antibodies. They reacted with the muscle but not with the chromosome?

Andrew: To vertebrate titin, yes.

Bullard: To vertebrate titin. So why do you think that was?

Andrew: Our hypothesis is that there are alternative splice forms of titin for the chromosomes, muscles and to encode kettin. So all that is required is to have some alternative splice form that doesn't encode the epitopes that are not detected on chromosomes.

Bullard: The KZ antibody used with the chromosomes is to a sequence that we would call kettin. So there is kettin in chromosomes and kettin has completely different properties from titin. It binds to actin not myosin, it has no fibronectin domains, it is not extensible.

Andrew: Well, we are going to use your antibodies as soon as you send them.

Bullard: You already have anti-kettin because you have anti-KZ.

Andrews: KZ is overlapping. The one we don't have is the other.

Bullard: That also reacts with kettin.

Andrew: And that [KZ] definitely reacts with chromosomes.

Bullard: So you've got kettin in chromosomes then?

Andrew: Or titin in chromosomes. I would like to comment that when we grind up *Drosophila* embryos and run the extracts out on gels, we see that mega-dalton size polypeptide. We don't detect something in the 500 kDa range where kettin would be.

Labeit: So next Charles wants to keep the option for Trombitin alive?

Trombitás: No, I just would like to ask a question. Do you think that in the Drosophila muscle titin may play a template role for the thick filament, as it may do in the vertebrate muscle?

Andrew: Well, we can look, that is all I can say. We have the means now; we have some escapers, so we can look in some of the escapers that make it to later stages. I think so, but nothing like the experiment.

Ayme-Southgate: Deb, this is a geneticists question. If the gene is really titin and if it has a role in chromosomes, wouldn't you expect some mutation to be female sterile?

Andrew: Yes, and we haven't looked at that yet. Another group that we have provided the antibody to, looked at another species and they detect the protein in meiotic chromosomes.

Vigoreaux: I do not understand what happens to titin during interphase.

Andrew: When we use the *Drosophila* antibodies to the *Drosophila* cells, we detected signal in the nucleus, with the exception of the nucleoli.

Vigoreaux: So homogeneous distribution?

Andrew: Yes, a homogeneous distribution. Much like what we see with the scleraderma serum, and the vertebrate antibodies in vertebrate cells.

Can I ask a question? What would you like me to do to demonstrate that this is titin and not some relative?

Greaser: Get the whole sequence.

Labeit: I am just wondering, if my basic biology doesn't leave me behind, erythroblasts have nuclei, erythrocytes do not. Then maybe you could do Western blots on erythrocytes with your antibodies?

Andrew: Yes, that would work.

Labeit: In general terms I wonder if what you see could be a titin-like gene, and not titin. I wonder if this is still a possibility.

Andrew: So I can say that the genome project in *Drosophila* is still far behind that of worms and if you take vertebrate titin and put it through the *Drosophila* genomic sequence database, you can pull up another related molecule, and it scores quite high on a search. But it scores quite high, I think, because of its size, because they have a contig through the region. If you take fragments of that gene, if you guess where the open reading frame might be, and you put it back, it won't pull up vertebrate titin, whereas this one will. So right now I would argue that this is the most related thing in the Drosophila genome that has been sequenced. I agree that the complete sequence would be the answer. I have only one person in my lab working on this; the rest of my lab does something very different. I do do some work on this, but I also have to keep my other grants going. So we decided not to sequence the entire gene. We hope that the genome project gets it done, and we hope they get it done right. My concern with the worm project is that we already know that we can take some of the fragments from the gene, and pull out regions from a contig, but other fragments are missing, as well as genes that should be in that contig. So I think regions like where titin is found present a special problem because of the repetitive sequence. In fact, some of the cDNAs we got were only obtained by using special *E. coli* strains that are recombination defective, etc. You must know the problem. So we are not doing it now, but we hope the genome project does it soon and right.

Labeit: Maybe when we come to the issue of cellular titins, we will have similar questions. What is really needed to establish the existence of titin.

Greaser: Your scleroderma antibody, where does it stain on a muscle myofibril?

Andrew: We actually see staining in the Z-disc, in the M-line, and we can see some staining around the myofibril. So it is not as clean as the N-terminal antibody that we made.

Greaser: Does it Western blot? Does it react with muscle titin on a Western blot?

Andrew: We didn't have enough of the serum to actually do Westerns with it.

ROLE OF THE ELASTIC PROTEIN PROJECTIN IN STRETCH ACTIVATION AND WORK OUTPUT OF *DROSOPHILA* FLIGHT MUSCLES

Jim O. Vigoreaux[1a], Jeffrey R. Moore[2a], and David W. Maughan[2a]

[1]*Department of Biology,* [2]*Department of Molecular Physiology and Biophysics,* [a]*Cell and Molecular Biology Program, The University of Vermont, Burlington, VT*

Abstract: We examine how the stretch activation response of the *Drosophila* indirect flight muscles (IFM) is affected by the projectin mutation *bent[Dominant]*. IFM from flies heterozygous for this mutation (*bent[D]* / +) produce ~85% full length projectin and ~15% truncated projectin lacking the kinase domain and more C-terminal sequences. Passive stiffness and power output of mutant fibers is similar to that of wild-type (+/+) fibers, but the amplitude of the stretch activation response (delayed tension rise) was significantly reduced. Measurement of actomyosin kinetics by sinusoidal analysis revealed that the apparent rate constant of the delayed tension rise ($2\pi b$) increased in proportion to the decrease in amplitude, accounting for the near wild-type levels of power output and nearly normal flight ability. These results suggest that projectin plays a crucial role in stretch activation, possibly through its protein kinase activity, by modulating crossbridge recruitment and kinetics.

Introduction

Most animals rely on muscles for locomotion and survival. The various forms of locomotion (e.g., running, hopping, flying, swimming) highlight the need to

Elastic Filaments of the Cell, edited by Granzier and Pollack
Kluwer Academic/Plenum Publishers, 2000

understand how a common actomyosin-based contractile mechanism has evolved to fulfill these specific activities. The diversity in muscle function arises in part from specialized adaptations in sarcomere design and contractile kinetics. Particularly interesting examples of variations on the contractile theme are the muscles that work in an oscillatory manner, two well-known examples being vertebrate cardiac muscle and insect IFM. The stretch activation response, an innate property of all striated muscles, is greatly enhanced to augment work production and power output in these muscle types. Stretch activation is manifested as a delayed rise in tension when an activated fiber is subjected to a quick stretch or release (Steiger, 1977; Pringle, 1978). Deficits in stretch activation lead to loss of flight ability in insects (Tohtong *et al.*, 1995), and cardiac hypertrophy and possibly death in vertebrates (Vemuri *et al.*, 1999).

The molecular basis of stretch activation remains elusive. A number of models have been invoked to explain the stretch-induced increase in actomyosin ATPase and delayed rise in tension. Two possibilities, not mutually exclusive, are that stretch induces a change in crossbridge kinetics (Thorson and White, 1969; Tawada and Kawai, 1990) and/or alters crossbridge recruitment (Dickinson *et al.*, 1997). Either one of these possibilities requires a mechanism by which the change in length can be sensed and stress transmitted to the myosin head.

One characteristic feature of insect IFM is its high resting stiffness that arises, at least in part, from connecting filaments that link thick filaments to Z bands (White, 1983; Granzier and Wang, 1993). This stiffness persists in the active muscle and acts mechanically in parallel with crossbridges (White, 1983). Thus it is conceivable that strain on the connecting filaments affects crossbridge cycling and/or recruitment such that stretch activation and oscillatory work are enhanced in IFM (Thorson and White, 1983).

Projectin, a member of the titin superfamily of elastic proteins, is the only identified protein component of the connecting filaments (Saide, 1981). Unlike titin, which extends from Z band to M line, IFM projectin is believed to extend from Z bands to the A/I junction (Saide *et al.*, 1989). To establish if projectin (and hence connecting filaments) play a role in stretch activation, we conducted mechanical tests on IFM fibers from *bent Dominant* heterozygotes (*bent^D*/+). *bent^D* is a recessive lethal projectin mutant allele in *Drosophila melanogaster* that contains a rearrangement breakpoint within, or just N-terminal to the protein kinase domain (Ayme-Southgate *et al.*, 1995; Fig. 1). Here we show that the *bent^D* allele encodes a truncated projectin, likely missing the kinase domain and other C-terminal sequences. Fibers from *bent^D*/+ flies have normal relaxed stiffness but reduced oscillatory work, demonstrating that projectin plays a fundamental role in the stretch activation response.

Results

The bent^D allele encodes a truncated projectin

Drosophila has a single projectin gene that encodes at least three muscle isoforms that differ in molecular weight, only one of which is expressed in IFM (Vigoreaux *et al.*, 1991). In contrast, IFM skinned fibers from *bent^D/+* show an additional, slightly faster migrating projectin band on SDS-PAGE not present in wild-type (+/+) IFM (Fig. 1). Scanning densitometry of coomassie stained gels revealed that the shortened projectin constitutes approximately 15-19% of the total projectin in IFM of heterozygous flies. These results suggest that the *bent^D* mutation prematurely terminates transcription or translation, but the resulting shortened protein is at least partially stable. The presence of this shortened projectin in skinned fibers suggests that it is incorporated into the myofibrillar lattice. However, the presence of the shortened projectin does not appear to have a detrimental effect on myofibrillar structure, because electron micrographs of *bent^D/+* IFM are essentially indistinguishable from +/+ (results not shown).

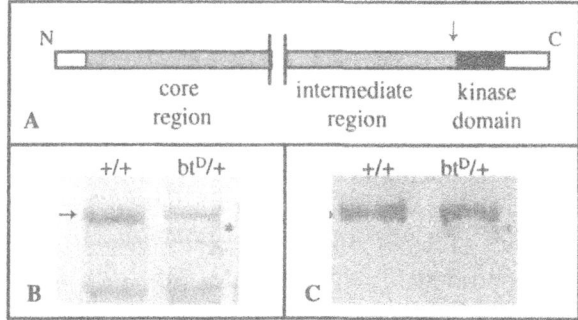

Figure 1. **A**. *Schematic representation of* Drosophila *projectin and location of the* bent^D *mutation (Ayme-Southgate et al., 1995). A breakpoint in the genomic sequence, most likely due to an inversion, is found close to, or within, the amino end of the kinase domain (arrow)* **B**. *Coomassie-stained 3-15% SDS-PAGE and (**C**) corresponding Western blot incubated with a mixture of anti-projectin monoclonal (Saide et al., 1989) and polyclonal (Maroto et al., 1992) antibodies. IFM fibers were dissected, skinned, and solubilized in a modified Laemmli sample buffer containing 8M urea and a cocktail of protease inhibitors (Vigoreaux et al., 1991). Note the presence of the shorter projectin polypeptide (*) in the* bent^D/+ *sample but not in the +/+ sample. Arrows point to the full length projectin.*

Skinned fibers from bent^D heterozygotes show an altered stretch activation response

We study stretch activation by sinusoidal analysis of skinned fibers. This technique enables the measurement of parameters that represent the stretch-activation response, namely magnitude and rate of myosin crossbridge attachment and detachment, force generation, work output and power output. At different levels of calcium activation ranging from pCa 8 to 5, sinusoidal length changes are applied (0.25% peak to peak) at frequencies ranging from 0.5 to 1000 Hz. At each frequency, signals representing muscle fiber tension and length are measured to obtain the dynamic stiffness modulus (Y(f)) of the fiber. Y(f) is defined as the ratio of stress to strain in the frequency domain, where stress is the force per cross-sectional area and strain is the fractional length change

$$Y(f) = (\Delta F/A) / (\Delta L/Lo)$$

The dynamic stiffness modulus can be resolved into two components: the elastic modulus (E_e), which is in-phase with the applied length change, and the viscous modulus (E_v), which is 90° out-of-phase with the applied length change. The latter provides a measure of the work-producing (phase lag) and work-absorbing (phase lead) properties of the fiber. On a Nyquist plot (a vector modulus plot), the work done by the fiber during each length oscillation is proportional to the amplitude of the negative viscous modulus (work $= 2\pi E_v (\Delta L/L)^2$). Fibers from *bent^D/+* have an E_v approximately half that of +/+ (92 ± 15 kNm^{-2} vs. 175 ± 29 kNm^{-2} in +/+; p <0.01) resulting in a decreased dynamic stiffness and reduced work output (Table 1). In addition, the frequency at which power output reaches its maximum (F_{max}, Table 1) was increased in *bent^D/+*. This frequency is near the resonance frequency of the flight system (i.e., near the wing beat frequency of the fly). Interestingly, the magnitude of the increase in F_{max} is roughly proportional to the magnitude of the decrease in E_v. Hence total power (Wm^{-3}), which is proportional to the product of E_v and F_{max}, is not significantly affected in *bent^D/+* (Table 1). As a result, *bent^D/+* flies have nearly normal flight ability (flight index of 4.2 ± 0.5 compared to 5.6 ± 0.05 for wild-type, p=0.15) despite the reduced work-producing properties.

We determined the effect of the *bent^D* mutation on actomyosin kinetics by sinusoidal length perturbation analysis and interpreted the results as the sum of three viscoelastic elements according to the equation

$$Y(f) = A(i2\pi f / \alpha)^k - Bif / (b+if) + Cif / (c+if)$$

where the term with A is a composite viscoelastic element, and terms with B and C are two exponential crossbridge processes (see Dickinson *et al.*, 1997 for full definition of terms and explanation of model). Element A represents the passive properties of several sarcomeric structures, including but not limited to connecting filaments, crossbridges, and thick and thin filaments. In relaxed

Table 1. **Mechanical performance of skinned IFM fibers**

	E_e (kNm^{-2})	E_v (kNm^{-2})	Dyn stiff (kNm^{-2})	F_{max} (Hz)	Power$_{total}$ (Wm^{-3})
+/+	490 ± 66	175 ± 29	522 ± 70	105 ± 16	78 ± 17
bent^D/+	$312 \pm 40^*$	$92 \pm 15^{**}$	$327 \pm 41^*$	$152 \pm 7^{**}$	64 ± 11

Values represent mean $\pm$ SEM. * p<0.05 ; **p<0.01.
Measurements done at 12°C.

Table 2. **Model parameters of Ca^{2+}-activated skinned IFM fibers**

	A (kNm^{-2})	B (kNm^{-2})	C (kNm^{-2})	b (Hz)	c (Hz)
+/+	622 ± 100	848 ± 121	801 ± 112	84 ± 15	640 ± 87
bent^D/+	$304 \pm 48^{**}$	$559 \pm 77^*$	$473 \pm 70^*$	$150 \pm 5^{**}$	622 ± 27

Values represent mean $\pm$ SEM. * p<0.05 ; **p<0.01.

muscle A is presumably dominated by the properties of connecting filaments (Dickinson *et al.*, 1997). The *bent^D* mutation has no discernible effect on A in relaxed muscle (*bent^D/+*: 141 ± 42 kNm^{-2}; +/+: 127 ± 25 kNm^{-2}; p=0.3). However, in active fibers model parameters A, B, and C are significantly reduced (Table 2). Process B corresponds to the delayed rise in tension or stretch activation demonstrated in step analysis (Huxley, 1974). Process C corresponds to phase 2, the rapid decay preceding force redevelopment (Kawai and Brandt, 1980). The reduction in process B parallels the reduction in E_v and decreased work output. The apparent rate constant of process B, $2\pi b$ (b in Table 2), is nearly doubled in *bent^D/+* fibers. $2\pi b$ reflects the kinetics of crossbridge transitions from a pre-force state to a force-producing state (Zhao and Kawai, 1993). Crossbridge power, which is proportional to the product of B and b, is not significantly changed (*bent^D/+*: 187 ± 27 Wm^{-3}; +/+: 141 ± 23 Wm^{-3}) since the increase in frequency b compensates for the decrease in magnitude of process B. The apparent rate constant of process C, $2\pi c$, is also unchanged.

Discussion

The results presented here show that the projectin mutation *bent^D* affects the stretch activation response of skinned fibers via alterations in crossbridge recruitment and kinetics of the force-producing step. The former is indicated by the reduction in processes A, B, and C, and the latter by the increase in $2\pi b$. On the other hand, we cannot ascribe any effects of the mutation to myofibrillogenesis or to sarcomere structure as *bent^D/+* myofibrils appear well-ordered with no apparent defects on myofilament length or lattice interdigitation.

The thick filaments and thin filaments of IFM have matching helix repeats of 38 nm, a property that may be critical for stretch activation. One proposal is that the probability of actomyosin interaction increases cyclically during extension as the optimal actin binding sites translate past myosin heads (Wray, 1979). A shorter projectin polypeptide may result in subtle axial displacement of myosin heads so they no longer encounter preferred actin sites upon stretch (Tregear *et al.*, 1998). Our electron microscopy analysis of *bent^D/+* does not offer sufficient resolution to discard this possibility. More likely, however, is that the mutation affects crossbridge function through a reduction in protein kinase activity.

The truncated projectin encoded by the *bent^D* allele appears to be stably incorporated in the myofibrillar lattice, although at much lower levels than full-length projectin. Since this mutant protein retains most of the modular repeated motifs I and II, it is not surprising that relaxed stiffness is unchanged in the mutant. On the other hand, the changes observed in active dynamic stiffness can be ascribed to reduced protein kinase activity since the mutant protein is most likely missing the kinase domain.

Insect projectin has been shown to autophosphorylate *in vitro* (Maroto *et al.*, 1992; Ayme-Southgate *et al.*, 1995; Weitkamp *et al.*, 1998) as well as to phosphorylate a 30 kD thin filament-associated protein identified as troponin I (Weitkamp *et al.*, 1998). In contrast, twitchin has been shown to phosphorylate itself as well as myosin regulatory light chain (Heierhorst *et al.*, 1994; Heierhorst *et al.*, 1995). Among *Drosophila* IFM myofibrillar proteins, only myosin regulatory light chain (RLC; Dickinson *et al.*, 1997), flightin (Vigoreaux and Perry, 1994), troponin T (Domingo *et al.*, 1998), and projectin (Maroto *et al.*, 1992) have been shown to be phosphorylated *in vivo*. Which, if any of these proteins, serves as the *in vivo* substrate of projectin kinase remains to be established. We do note, however, some interesting parallels between the mechanical properties of *bent^D* and those of RLC and flightin mutants (Fig. 2). Homozygous mutant *RLC^{S66A,S67A}*, in which both myosin light chain kinase substrate serines are substituted for nonphosphorylatable alanines, show a marked reduction in magnitude (i.e., number of force-producing crossbridges) but not in the frequency (i.e., crossbridge kinetics) at which work output is maximum (Tohtong *et al.*, 1995). This was interpreted as a decrease in crossbridge recruitment (X_d to X_p, Fig. 3 and Dickinson *et al.*, 1997). *bent^D* heterozygotes also show a reduction in magnitude and, in addition, an increase in frequency, while the

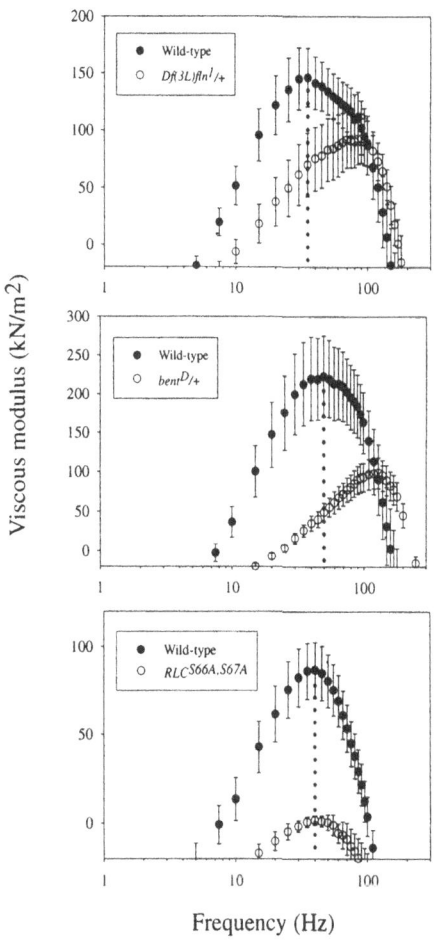

Figure 2. *Plot of viscous modulus vs. frequency for various muscle mutants.* **Top:** *+/+ vs. Df(3L)fln ¹/+. Note that both plots have similar form; however, Df(3L)fln ¹ /+ fibers achieved maximum E$_v$ at a higher frequency. Hence oscillatory work, which is proportional to E$_v$ is lower at the natural resonant frequency (broken line). The difference in E$_v$ amplitude is not significantly different from +/+ (from Vigoreaux et al., 1998).* **Middle:** *+/+ vs. bent ᴰ/ +. Note the reduction in E$_v$ amplitude and the shift in frequency.* **Bottom:** *+/+ vs. RLC [S66A,S67A] / RLC [S66A,S67A]. Amplitude of viscous modulus is significantly reduced but frequency is unchanged (from Tohtong et al., 1995).*

flightin deficiency heterozygote *Df(3L)fln ^{1}* shows an increase in frequency but no change in amplitude of work output (Vigoreaux *et al.*, 1998). Characteristic frequency b is increased by an almost identical 44% in both *bent^D* and *Df(3L)fln^1* heterozygotes. Furthermore, the amount of flightin in *Df(3L)fln^1* IFM is reduced by approximately 22% (Vigoreaux *et al.*, 1998), which closely matches the ~19% reduction in full-length projectin in *bent^D* IFM. Altogether, these results raise the intriguing possibility that projectin kinase may modulate crossbridge function through either RLC or flightin, or perhaps both.

Immunolocalization studies (Saide *et al.*, 1989; Nave and Weber, 1990) and length measurements of isolated projectin molecules (~0.25 μm; Nave and Weber, 1990) suggest that a single projectin anchored at the Z band cannot extend too far beyond the A/I junction of insect flight muscle. Hence it is unlikely that RLC or flightin, which are distributed throughout the A band, are direct substrates of projectin kinase. One possibility is that projectin is an upstream kinase in a pathway that leads to phosphorylation of RLC and/or flightin via a soluble protein kinase. Both RLC and flightin undergo extensive phosphorylation very early in adult life (i.e., during the few hours immediately following eclosion and before the first flight episode (Vigoreaux and Perry, 1994)). Changes in sarcomere length and width also occur during this period (Reedy and Beall, 1993). We speculate that these phosphorylations may serve to establish a state of high static tension and/or stiffness which prepares the muscle for quick mechanical activation, equivalent to a warm-up but with longer-lasting effects. Projectin kinase, perhaps activated by stress from hardening of the cuticle and/or sarcomere growth, may participate in establishing this functional state by initiating a mechanical signaling cascade that eventually results in a steady-state level of RLC and/or flightin phosphorylation. Phosphorylation of RLC is likely to affect crossbridge recruitment, while phosphorylation of flightin may affect crossbridge kinetics (Fig. 3).

These studies underscore the importance of the genetic approach to understanding the molecular basis of stretch activation. The *Drosophila* IFM model system is well suited for this research given its prominent stretch activation response and the availability of experimental techniques to generate muscle mutations and study their effects *in vivo* and *in vitro*.

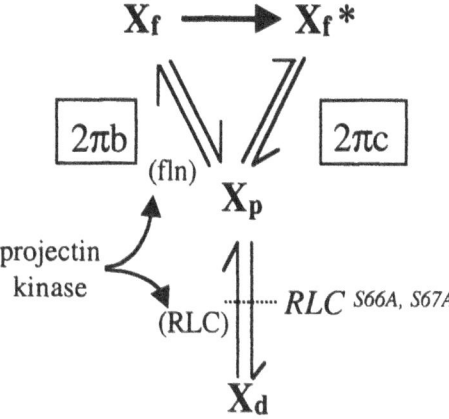

Figure 3. A model for the function of projectin kinase in modulating crossbridge activity. For clarity only three composite crossbridge states are shown. bent^D increases apparent rate constant $2\pi b$. Likewise, fibers from Df(3L)fln ^{1} flies show elevated $2\pi b$, suggesting that flightin (fln) may be a target for projectin kinase modulation of crossbridge activity. bent^D also decreases the population of crossbridges in putative force-producing (X_f) and/or force-maintaining (X_f^*) states (i.e., reduced process B). The mutant RLC ^{S66A,S67A}, which lacks two phosphorylatable serines, shows a marked reduction in force production (process B), possibly by reducing the number of pre-force crossbridges (X_p) that can advance to the force-producing state (Dickinson et al., 1997). X_d represent detached crossbridges.

References

Ayme-Southgate A, Southgate R, Saide J, Benian GM, Pardue ML. Both synchronous and asynchronous muscle isoforms of projectin (the Drosophila bent locus product) contain functional kinase domain. *J Cell Biol* 1995;128:393-403.

Dickinson MH, Hyatt CJ, Lehmann F-O, Moore JR, Reedy MC, Simcox A, Tohtong R, Vigoreaux JO, Yamashita H, Maughan DW. Phosphorylation-dependent power output of transgenic flies: an integrated study. *Biophys J* 1997;73:3122-3134.

Domingo A, Gonzalez-Jurado J, Maroto M, Diaz C, Vinos J, Carrasco C, Cervera M, Marco R. Troponin-T is a calcium-binding protein in insect muscle: in vivo phosphorylation, muscle-specific isoforms and developmental profile in Drosophila melanogaster. *J Muscle Res Cell Motil* 1998;19:393-403.

Granzier HLM, Wang K. Interplay between passive tension and strong and weak binding cross-bridges in insect indirect flight muscle. *J Gen Physiol* 1993;101:235-270.

Heierhorst J, Probst WC, Kohanski RA, Buku A, Weiss KR. Phosphorylation of myosin regulatory light chains by the molluscan twitchin kinase. *Eur J Biochem* 1995;233:426-431.

Heierhorst J, Probst WC, Vilim FS, Buku A, Weiss KR. Autophosphorylation of molluscan twitchin and interaction of its kinase domain with calcium/calmodulin. *J Biol Chem* 1994;269:21086-21093.

Huxley A. Muscular contraction. *J Physiol (Lond)* 1974;243:1-43.

Kawai M, Brandt PW. Sinusoidal analysis: a high resolution method for correlating biochemical reactions with physiological processes in activated skeletal muscles of rabbit, frog and crayfish. *J Muscle Res Cell Motil* 1980;1:279-303.

Maroto M, Vinos J, Marco R, Cervera M. Autophosphorylating protein kinase activity in titin-like arthropod projectin. *J Mol Biol* 1992;224:287-291.

Nave R, Weber K. A myofibrillar protein of insect muscle related to vertebrate titin connects Z band and A band: purification and molecular characterization of invertebrate mini-titin. *J Cell Science* 1990;95:535-544.

Pringle JWS. Stretch activation of muscle: function and mechanism. *Proc R Soc Lond B* 1978;201:107-130.

Reedy MC, Beall C. Ultrastructure of developing flight muscle in Drosophila. I. Assembly of myofibrils. *Develop Biol* 1993;160:443-465.

Saide JD. Identification of a connecting filament protein in insect fibrillar flight muscle. *J Mol Biol* 1981;153:661-679.

Saide JD, Chin-Bow S, Hogan-Sheldon J, Busquets-Turner L, Vigoreaux JO, Valgeirsdottir K, Pardue ML. Characterization of components of Z-bands in the fibrillar flight muscle of Drosophila melanogaster. *J Cell Biol* 1989;109:2157-2167.

Steiger GJ. Stretch activation and tension transients in cardiac, skeletal and insect flight muscle. In *Insect flight muscle,* RT Tregear, ed. North Holland: Amsterdam, 1977;221-268.

Tawada K, Kawai M. Covalent cross-linking of single fibers from rabbit psoas increases oscillatory power. *Biophys J* 1990;57:643-647.

Thorson J, White DCS. Distributed representations for actin-myosin interaction in the oscillatory contraction of muscle. *Biophys J* 1969;9:360-390.

Thorson J, White DCS. Role of cross-bridge distortion in the small-signal mechanical dynamics of insect and rabbit skeletal muscle. *J Physiol (Great Britain)* 1983;343:59-84.

Tohtong R, Yamashita H, Graham M, Haeberle J, Simcox A, Maughan D. Impairment of muscle function caused by mutations of phosphorylation sites in myosin regulatory light chain. *Nature* 1995;374:650-655.

Tregear RT, Edwards RJ, Irving TC, Poole KJ, Reedy MC, Schmitz H, Towns-Andrews E, Reedy MK. X-ray diffraction indicates that active cross-bridges bind to actin target zones in insect flight muscle. *Biophys J* 1998;74:1439-1451.

Vemuri R, Lankford EB, Poetter K, Hassanzadeh S, Takeda K, Yu ZX, Ferrans VJ, Epstein ND. The stretch-activation response may be critical to the proper functioning of the mammalian heart. *Proc Natl Acad Sci U S A* 1999;96:1048-1053.

Vigoreaux JO, Hernandez C, Moore J, Ayer G, Maughan D. A genetic deficiency that spans the flightin gene of *Drosophila melanogaster* affects the ultrastructure and function of the flight muscles. *J Exp Biol* 1998;201:2033-2044.

Vigoreaux JO, Perry ML. Multiple isoelectric variants of flightin in Drosophila stretch-activated muscles are generated by temporally regulated phosphorylations. *J Muscle Res Cell Motil* 1994;15:607-616.

Vigoreaux JO, Saide JD, Pardue ML. Structurally different Drosophila striated muscles utilize distinct variants of Z band-associated proteins. *J Muscle Res Cell Motil* 1991;12:340-354.

Weitkamp B, Jurk K, Beinbrech G. Projectin-thin filament interactions and modulation of the sensitivity of the actomyosin ATPase to calcium by projectin kinase. *J Biol Chem* 1998;273:19802-19808.

White DCS. The elasticity of relaxed insect fibrillar flight muscle. *J Physiol (Great Britain)* 1983;343:31-57.

Wray JS. Filament geometry and the activation of insect flight muscles. *Nature* 1979;280:325-326.

Zhao Y, Kawai M. The effect of the lattice spacing change on cross-bridge kinetics in chemically skinned rabbit psoas muscle fibers. *Biophys J* 1993;64:197-210.

Discussion

Bullard: There is a myosin light chain kinase. How do you think that links up with the projectin kinase?

Vigoreaux: Well, it turns out that the titin kinase has homology to the myosin light chain kinase. It has been shown that the titin kinase and the projectin kinase are both in the MLCK family of kinases. Now it has been shown that projectin autophosphorylates, and I think there is some evidence that it may be phosphorylating its kinase domain. So it may be possible that MLCK is a substrate of projectin kinase. The other thing that is also possible is that, as I mentioned, myosin light chain is hyperphosphorylated in *Drosophila*. There are multiple phosphovariants that suggest there must be at least 15 phosphorylated residues on the protein. So there is likely to be more than one kinase that phosphorylates MLC.

TerKeurs: Very exciting, elegant presentation, Jim. You say projectin kinase is involved in the phosphorylation at two levels, the crossbridge and light chain. You also said that animals with a lack of phosphorylation at the regulatory light chain don't fly well, but apparently that means they are alive.

Vigoreaux: They are, yes.

TerKeurs: OK. So why are the homozygotes of the *bent^D* mutant dead?

Vigoreaux: Why are they dead? The mutation removes the kinase domain plus other things. Everything else that is behind the kinase domain. So there might be other sequences in there that may be important for the function of non-flight muscle. The reason these animals are dead is because their embryonic muscles don't work very well. We are only looking at the flight muscle. There might be differences, isoform differences for example, or something of that matter. There might be sequences deleted in the *bent^D* that are significant in the non-flight muscle but maybe not so significant in the flight muscle.

TerKeurs: Does that mean, that you implicitly assume that this model is complete for the flight muscle?

Vigoreaux: I am only talking about the flight muscle, yes.

Labeit: Debbie, does D-titin have a kinase domain?

Andrew: We don't have it yet. We probably have around 5% of the sequence. Genomic sequence even we have difficulty getting.

Labeit: It seems to be an important question in the field, whether *Drosophila* titin has a kinase domain. If you would have one, it would be exciting to get mutations in it, and see how it affects muscle structure and chromosome structure.

Greaser: How do you deal with the problem that projectin is strapped down and so are the light chains. How are you going to get the kinase to the light chains so that you can get phosphorylation?

Vigoreaux: I think it is a cascade initiated by the projectin kinase, perhaps it phosphorylates a soluble kinase that then goes on to phosphorylate other proteins.

Greaser: OK, so there's something in between here?

Vigoreaux: There must be because physically it is not possible.

Granzier: Just as a follow-up, what kind of soluble kinase would it be? You are using skinned fibers. Anything soluble would have diffused out.

Vigoreaux: Correct, correct. But my point is that all of these phosphorylations are established early in adult life. OK. It is sort of the equivalent of a warm-up. It prepares the muscle for high physical activity. So all of that takes place early in adult life and those phosphorylations are pretty much steady-state; they persist throughout the life of the adult. I mean a fly has to be ready to take off as soon as you lift your hand, you know, to swat them. They don't have time to sit there and warm up while you go like this. They have to be on the go all the time, ready to take off immediately. So maybe these muscles are in a high state of static tension which allows them to be activated very quickly. I'm not proposing that this kinase is active during cyclical contractions. This is a state that is established early in adult life and persists through the life of the animal.

Sanger: Jim, I just want to follow up with Belinda Bullard's observation on the first part where the Z-bands are not aligned. Is this *bent^D* mutant? Does it affect all the muscles in the fly?

Vigoreaux: Yes.

Sanger: Well, has anyone looked at the visceral muscles or the cardiac muscles to see if the Z-bands, where they are composed of dense bodies, whether they are in position or they get misaligned?

Vigoreaux: Maybe Agnes could talk about that because she has looked at embryonic muscle.

Sanger: I just wonder if you have some evidence that this projectin may also be playing a role in Z-band connections.

Vigoreaux: Between myofibrils you mean? I hope not.

Ayme-Southgate: You are almost telling the beginning of my talk. But that's OK, I'll answer the question now. The one thing that is really peculiar with projectin is that it's localized in the I-band of flight muscle. But if you look at any other muscle in the adult in the lower view of the embryo, it's actually over the A-band. So I'll tell you more later.

Vigoreaux: The point is, projectin is not likely to be in dense bodies in non-flight muscle?

Trombitás: I think that is not crictical; the Z-lines are in register. I prepared, I don't know how many thousand bees, and I checked the Z-line interconnection by scanning electron microscope. I showed that many times in normal conditions, the Z-lines are not aligned very well.

Linke: You never talked about regulatory proteins. Is there any particular reason you excluded them completely? I remember there was some idea that troponin H has something to do with stretch activation. Why didn't you talk about it?

Vigoreaux: I didn't talk about those because this is a titin meeting, so I didn't want to stir the pot too much. But obviously the picture is very complex. There is thin filament regulation. As I mentioned, the phosphorylation of myosin light chain, plays a modulatory role in stretch activation. So there is likely to be some important contribution from the thin filament regulatory complex, yes. But it's hard to encompass all of that in 25 minutes.

Baatsen: Would it be at all possible that another protein might be affected, like for instance, desmin?

Vigoreaux: Yes, it is very likely in fact. A lot of mutations cause secondary effects. For example, if you eliminate myosin from the muscle, you not only eliminate myosin, you eliminate a whole bunch of other proteins. It could be that the mutation is only affecting the myosin gene, but by affecting the expression of the myosin gene, you affect the expression of all the other proteins that are down stream in the assembly pathway. That's always a possibility. That is why we look at the structure of the muscle before we do any functional studies. We want to be sure that the structure is somewhat preserved. The best that we can do is run gels and try to identify all of the proteins in the myofibril and make sure that all of them are there, and that they're in roughly the same amounts as in wild-type flies. For the *bent^D*, we haven't seen any reduction or absence of known myofibrillar proteins. But you have a good point. That is something that we always have to keep in mind.

DROSOPHILA PROJECTIN:
A LOOK AT PROTEIN STRUCTURE
AND SARCOMERIC ASSEMBLY

Agnes Ayme-Southgate[1], Richard Southgate[1], and Michelle Kulp McEliece

Department of Biological Sciences, Lehigh University, Bethelehem, PA
[1]present: Department of Biology, College of Charleston, Charleston, SC

Abstract: The large projectin protein is found in all *Drosophila* muscles; however, it shows a dual sarcomeric localization depending on the muscle type. In larval and adult synchronous muscles, projectin is found localized over the A-band. Initial *in vitro* binding assays indicate interactions of several projectin regions with themselves and myosin heavy chain. These interactions might be critical for the assembly of projectin over the myosin filament during embryonic myofibrillogenesis and larval growth. On the other hand, projectin localizes over the I-Z-I region in indirect flight muscles. Correspondingly, projectin is found in association with forming Z-bands during pupation and colocalizes with α-actinin and kettin.

Introduction

The basic sarcomeric organization of both invertebrate and vertebrate striated muscles is very similar, with two interdigitated filaments, called thin and thick filaments, anchored at the Z-band and M-line respectively (Squire, 1981). Beyond this basic organization, striated muscles show a wide range of muscle-specific variations aimed at physiological adaptations, from the length of the individual filaments to the speed of contraction/relaxation. How the complex myofibrillar structure is built "to specification" during myofibrillogenesis and maintained through cycles of contraction-relaxation is still partly a mystery.

Elastic Filaments of the Cell, edited by Granzier and Pollack
Kluwer Academic/Plenum Publishers, 2000

251

To explain some of the properties of muscle structure and function, early models suggested the existence of a third filament system with elastic properties. In vertebrate striated muscles, such elastic filament has been shown to be composed of the gigantic protein, titin (Wang *et al.*, 1984; 1991; Horowits *et al.*, 1989; see other chapters in this book and references therein). Electron microscopy studies of insect indirect flight muscles also have revealed the presence of fine connections between the Z-bands and the myosin thick filaments, called C- (connecting) filaments (Auber *et al.*, 1963; Saide and Ullrick, 1974; Ashhurst, 1977; Trombitas and Tigyi-Sebe, 1977; Candia-Carnevali *et al.*, 1980; Deatherage *et al.*, 1989). These connecting filaments were proposed as elastic components involved in the indirect flight muscle property known as "stretch-activation" (Pringle, 1978).

Results and Discussion

Elastic filaments in insect indirect flight muscles

The highly specialized indirect flight muscles (abbreviated as IFM) are powerful muscles adapted for the rapid repeated contractions necessary for flight. These muscles are stretch activated, undergoing multiple rounds of contraction per nerve impulse, and because they lack the 1:1 relationship between nerve stimulation and contraction, the IFM are also referred to as asynchronous muscles (Crossley, 1978). The stretch-activation mechanism is explained as a "delayed increase in tension due to stretch," which activates the muscle and results in contraction (Pringle, 1978). The stretch activation mechanism is an intrinsic property of the myofibrillar apparatus (Jewell and Ruegg, 1966).

Electron microscopy of total stretched myofibrils and of purified Z-disks have shown the presence of a filament extending or "projecting" from the Z-band toward the myosin filament and just overlapping the tip of the A-band (Bullard *et al.*, 1977; Saide, 1981). Saide analyzed the composition of the connecting C-filament of honeybee flight muscles by antibody staining and biochemical analysis and characterized a high molecular weight polypeptide called projectin (Saide, 1981).

In honeybee IFM, connecting filaments can be extended to well over ten times their rest length. When muscles are stretched in rigor and then released, the recoil forces of the connecting filaments cause the sarcomere to shorten, leading to the crumpling of thin filaments held in rigor (Trombitás and Tigyi-Sebes, 1977). Some models suggest the projectin filaments play an important role in the stretch-activation mechanism and in this hypothesis, projectin would be an elastic IFM protein conferring high resting stiffness to the IFM and/or capable of transferring stress to the thick filament during stretching.

Arthropod projectin

Projectin or mini-titin proteins have been identified from several arthropod species. Four different insect species were used for these studies (Lethocerus, Apis, *Drosophila* and Locusta) and in each case, a high molecular weight polypeptide between 700 and 1200 kiloDaltons (kDa) estimated size was characterized. The localization of these large proteins was reported within the I-Z-I region of insect flight muscles (Saide, 1981; Hu *et al.*, 1986; Locker and Wild, 1986; Saide *et al.*, 1989; Lakey *et al.*, 1990; Nave and Weber, 1990; Maroto *et al.*, 1992).

Crayfish muscles were also used for purification of an equivalent protein, and antibodies raised against the crayfish protein were shown to cross-react with *Drosophila* muscles (Hu *et al.*, 1990; Maroto *et al.*, 1992; Vibert *et al.*, 1993). Some antibodies, raised against projectin from one insect species show cross-reactivity to the equivalent protein in other insects (Hu *et al.*, 1990; Nave and Weber, 1990). Some of these purified mini-titin proteins were also recognized by some antibodies directed against the vertebrate protein, titin (Nave and Weber, 1990; Vibert *et al.*, 1993). Cross reactivity between *Drosophila* projectin and the large nematode protein, twitchin, was also demonstrated (Vigoreaux *et al.*, 1991).

From the antibodies' cross-reactivity and the biochemical profile of these arthropod proteins, it seems clear these proteins belong to the same family and are related to titin and twitchin. More recently, another protein, different from projectin, was characterized from *Drosophila* by its crossreactivity with titin antibodies and was called D-titin (Machado *et al.*, 1998).

Drosophila *projectin protein structure*

The isolation of the *Drosophila* projectin gene allowed the determination of its amino acid sequence (Ayme-Southgate *et al.*, 1991; 1995; Fyrberg *et al.*, 1992; Daley *et al.*, 1998). This revealed the presence of two repeated amino acid motifs, referred to as the Fn and Ig domains. The Fn motif has homology to fibronectin type II domain, while the Ig motif resembles most closely the immunoglobulin I set (Benian *et al.*, 1989; Pfuhl and Pastore, 1995). These two domains are also found in the other family members, titin and twitchin in particular (Benian *et al.*, 1989; Labeit and Kolmerer, 1995 and references therein). To date, the projectin protein is composed of a total of 39 Fn domains and 31 Ig domains with a molecular weight of approximately 831 kDa (Daley *et al.*, 1998).

The central "core" region contains 28 Fn and 14 Ig motifs, arranged in the Fn-Fn-Ig pattern repeated 14 times (Fig. 1). This modular structure suggests many possible projectin interactions, which will be discussed below. Two additional patterns were identified, an Fn-Fn-Fn-Ig pattern and an Fn-Fn-Ig-Ig pattern, both found in the "intermediate region," just before the kinase domain

Figure 1. *Domain organization in the projectin protein. This figure presents a schematic representation of the projectin domain structure. NH_2 and COOH termini contain only Ig domains and unique sequences, while the core and intermediate regions are composed of both Fn and Ig domains.*

(Ayme-Southgate *et al.*, 1995). The same change of pattern is also found at the same position in titin and twitchin and its importance for the positioning of the kinase domain on its substrate has been suggested.

The COOH- and the NH_2-termini are composed of arrays of Ig domains interspersed with unique sequences. The COOH terminal unique regions do not show any amino acid sequence conservation with sequences in twitchin or titin. However, the size of each unique sequence in the projectin COOH terminus is very close to the size of the corresponding regions in twitchin. It could be that the length, not the sequence, of these unique regions is important for correct positioning or assembly of projectin in the sarcomere (Daley *et al*, 1998; see below). The NH_2-terminal unique sequences, in contrast to the ones found at the COOH-terminus show homology with unique regions from both twitchin and titin NH_2-termini. The NH_2-terminal region of projectin is still not completed, and other homologies might be characterized, in particular PEVK sequences, which have been found in titin and involved with titin's elastic properties (Linke *et al.*, 1996).

Projectin muscle specificity

Based on its proposed involvement with stretch-activation, projectin was first thought of as an IFM-specific protein. Further studies on the localization of projectin in different *Drosophila* muscle types revealed, however, that projectin is also present in all larval and adult muscles other than IFM (Vigoreaux *et al.*, 1991). These muscles do not display stretch-activation and the ratio nerve impulse-contraction is 1:1; hence, these muscles are often referred to as synchronous muscles (Crossley, 1978).

A single gene, located on chromosome four (location 102CD), encodes *Drosophila* projectin, and transcripts from this gene are detected in both flight and synchronous muscles (Ayme-Southgate *et al.*, 1991). Western analysis reveals, on the other hand, the presence of multiple projectin isoforms within the various muscle types. The IFM-specific projectin is significantly smaller than the isoforms found in synchromous muscle types (Vigoreaux *et al.*, 1991). Although there may be differences related to post-translational modification, the different isoforms are most probably, generated through alternative splicing of a primary transcript from a single gene.

Recent evidence confirms the presence of alternative splicing at the 3' end of the gene (Daley *et al.*, 1998; Southgate and Ayme-Southgate, unpublished observations). This alternative pathway would generate two isoforms with different COOH-termini (Fig. 2). In one pathway, the protein COOH-terminus would include five Ig motifs and all three unique regions. The alterna-

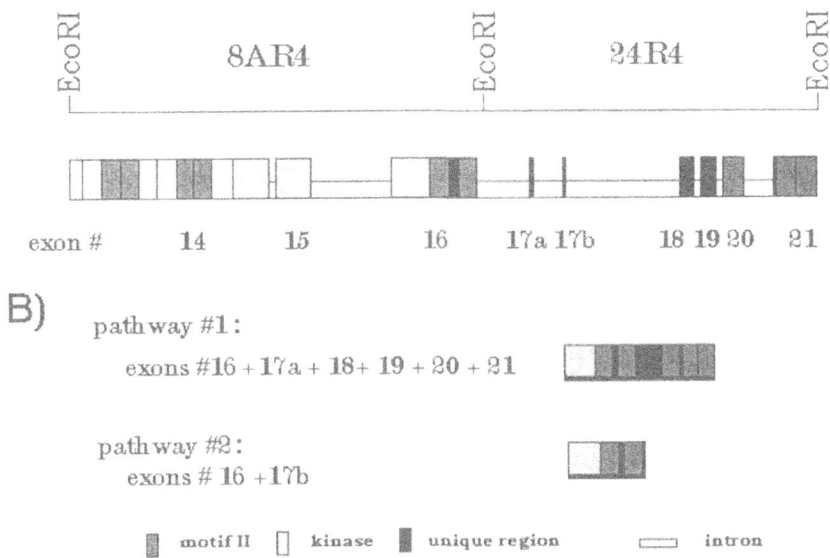

Figure 2. Alternative splicing at the 3'end of the projectin gene. (A) The exon-intron structure for the last 8 exons at the 3'end of the projectin gene is presented. Exons 17a and 17b are alternatively spliced. 8AR4 and 24R4 refer to the two subclones used in the in situ *hybridization studies. 8AR4 probe hybridizes to both IFM and synchronous muscles while 24R4 probe only recognizes transcripts in the synchronous muscles. (B) The projected protein domain structure for the two alternatively spliced forms is diagrammed. The smaller form arises from an early termination codon in exon 17b. The shorter form only includes 2 of the 5 terminal Ig domains.*

tive splicing pathway generates a protein shorter by about 401 amino acids (or 40 kDa) due to the presence of an earlier termination codon. The COOH-terminus of the shorter isoform would only include 2 Ig domains and one short unique sequence (see Fig. 2). The muscle-specificity (if any) of these two spliced forms still need to be confirmed but indirect evidence from *in situ* hybridization points to the possibility the longer form would be synchronous muscle-specific (A. Ayme-Southgate, unpublished observations).

Projectin Assembly in Embryonic Muscles

In all larval and adult synchronous muscles, anti-projectin antibodies detect projectin localization over the A-band only, with the exclusion of the M line, reciprocal of the distribution seen in IFMs (Vigoreaux *et al.*, 1991). One of the main questions is how the dual localization is achieved during myofibrillogenesis in the different muscle types.

The first stage of *Drosophila* myofibrillogenesis occurs during embryonic development (Crossley *et al.*, 1978). Around 6.0 hours into embryonic development (i.e., hours after egg laying, abbreviated a.e.l.), the somatic mesoderm is segregated and is composed of myoblasts. Between 9.0 and 13.0 hours a.e.l. the myoblasts fuse and form polynucleated cells. After 14.0 hours a.e.l. small amounts of organized myofibrillar material are present. There is, still, no real sarcomeric organization at this time, with the Z bodies appearing only around 15.0 hours. Between 16.0 and 18.0 hours a.e.l., the typical sarcomere structure takes shape.

Projectin's accumulation and assembly occurs through specific steps during embryonic myofibrillogenesis. Projectin is detectable on embryonic squashes around 9-10 hours a.e.l. with an appearance of unordered dots. Around 13 hours a.e.l., most projectin protein now displays staining in longitudinal continuous filaments. Projectin pattern finally develops, around 15-16 hours a.e.l. into a striated, discontinuous A-band-like pattern (Kulp McEliece, unpublished observations). This pattern of assembly is somewhat similar to the one described for titin, in particular the transition from dots to filaments might represent an unfolding of the protein (Tokuyasu and Maher, 1987; Colley *et al.*, 1990; Isaacs *et al.*, 1992; van der Ven *et al.*, 1993; Rhee and Sanger, 1994; van der Loop *et al.*, 1996).

The A-band localization of projectin is representative of the location assumed by the related C. elegans protein, twitchin. Twitchin has been implicated in regulation of muscle contraction and is thought to interact with the components of the thick filament, in particular myosin heavy chain (Moerman *et al.*, 1982; Benian *et al*, 1989). Interaction between myosin heavy chain and the central region of the titin molecule has also been demonstrated through multiple *in vitro* binding assays (reviewed in Gautel, 1996; Trinick, 1996). The interactions between titin and myosin are thought to play a critical role in organizing the two proteins with respect to one another, and titin has been proposed both as a scaffold and a ruler (reviewed in Fürst and Gautel, 1995; Gautel, 1996; Trinick, 1996).

Initial data from *in vitro* binding assay experiments are consistent with the interaction of the central core region of the projectin protein with myosin heavy chain (Ayme-Southgate, unpublished observation). Interestingly, during *Drosophila* embryogenesis, myosin assembles in continuous filaments concomitantly with projectin, around 13 hours a.e.l. (Kulp McEliece, unpublished observations). Interaction between projectin and myosin probably takes place at that time, but their importance in the assembly process remains to be determined. It is noteworthy that at that same time point, there is still no evidence for the presence of forming Z bodies, even though α-actinin is already present in the cells.

Projectin in larval muscles and assembly

It has been proposed that titin (Trinick, 1984; Horowits *et al.*, 1989), projectin (Nave and Weber, 1990; Hu *et al.*, 1990) and twitchin (G. Benian, personal communication) are thin, elongated molecules. Titin, with a maximum 3.7 MDa size (Labeit and Kolmerer, 1995), has been shown to reach the length of 1 μm. Projectin is presently estimated at a molecular weight of 831 kDa, but the protein may well reach 1 MDa in size (Daley *et al.*, 1998). Projectin's length has been estimated by electron microscopy at 0.2 to 0.3 μ (Nave and Weber, 1990). The length of projectin is sufficient for two projectin molecules to encompass the A-band length of an embryonic sarcomere but as the sarcomeres elongate during larval growth, more projectin molecules must be recruited and assembled over the A-band. How this is accomplished is unknown, but recent evidence suggests a possible mechanism. The assembly of new projectin molecules over the A-band is thought to occur in two basic ways: 1) The projectin molecules assemble directly onto the growing myosin thick filaments without interacting with one another, 2) the projectin molecules interact with myosin heavy chain and one another, forming a staggered projectin arrangement (Fig. 3). Initial results from *in vitro* binding assays demonstrate interactions between specific regions of the projectin protein (Fig. 3). This would support the idea of a staggered arrangement of projectin molecules along the myosin filament. In that respect, the phenotype of one projectin mutant allele is interesting. The mutant bt [1-2k] is truncated, missing the last 3 Ig domains (Ayme-Southgate *et al.*, 1995) at the COOH terminus. This allele dies at the larval stage with the main lethal period occurring around the second instar. Further analysis should reveal the nature of the sarcomere defect.

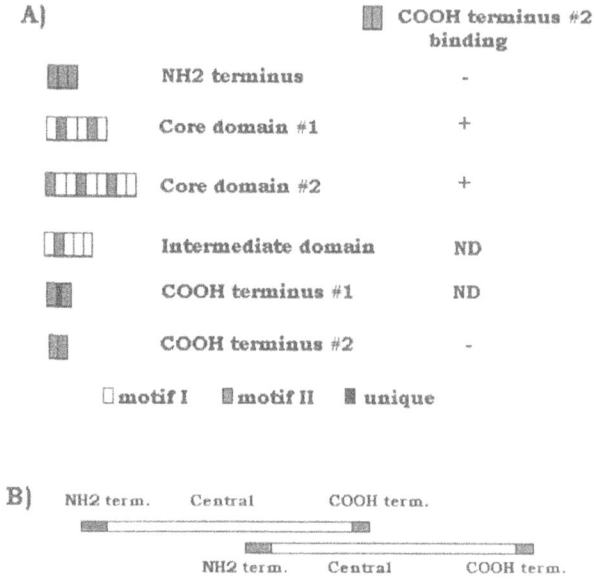

Figure 3. Model for projectin staggered arrangement. (A) This presents the position and extent of the various projectin fusion proteins used in the in vitro *binding assays. Binding interaction between these fusion proteins and the projectin COOH terminus fusion peptide is indicated by "+" for binding and "–" for no binding. ND: not determined. (B) This schematic depicts the possible self-association of two projectin molecules in a staggered arrangement, consistent with the* in vitro *binding assays. This type of arrangement could account for the distribution of projectin over the entire A-band in larval and adult muscles. Association with myosin thick filaments is not presented but is assumed to occur concomitantly.*

Projectin in IFM muscles and assembly

During pupation, all the larval muscles are first histolyzed and new adult muscles are formed. Of all the adult muscles, the IFMs have been the most studied (Fyrberg and Beall, 1990). In the IFMs, within 42 hours after the start of pupation, initial fibrils occur inside "sleeves" of microtubules. Z bodies are present but irregular, and there are very few "stress fiber-like structures" (abbreviated as SFLS) present. By 50 hours, striated narrow myofibrils are found throughout the muscle and at 60 hours, the typical adult Z-band is organized (Reedy and Beall, 1993).

In IFMs, projectin localizes over the I-Z-I region, and the question has arisen of how projectin can assume two such different localization in the two main muscle types. Based on its I-Z-I localization, projectin is very probably anchored into the Z-band (Bullard *et al.*, 1977; Saide, 1981; Lakey *et al.*, 1993). It has been proposed this anchoring may occur through association with some Z-band proteins, possibly alpha-actinin. The association of projectin with forming Z bodies occurs very early during IFM myofibrillogenesis. As early as 30 hours after the start of pupation (abbreviated as a.s.p.), projectin is found associated with aggregates which also stained with alpha-actinin and kettin antibodies (Ayme-Southgate, unpublished observations). These aggregates may represent forming Z bodies. The region of the projectin protein involved with Z-band association is still unknown.

The projectin molecule is long enough (0.2-0.3 μ) to cover the I-Z distance in the IFMs (0.1-0.2 μ; Crossley, 1978), even though it is still unclear how much of the projectin protein enters the Z-band and overlaps with the thick filaments. A single projectin molecule could span the distance and there would be no need to invoke the staggering of projectin molecules like in synchronous muscles. How the projectin molecule is directed to associate with the forming Z-band in IFM is unknown. Several scenarios are possible: 1) The presence of an alternative, IFM-specific exon could be responsible for projectin interactions with other Z-band proteins (alpha-actinin/ kettin); 2) The timing of Z-band assembly could also be a critical factor; Z bodies form relatively late during embryogenesis, while they appear early during pupation.

Conclusion

Much remains to be learned in understanding the process of projectin assembly and the related issue of how the different isoforms can assume different localization within the sarcomere. The elucidation of the complete protein structure will allow an indepth study of the structure-function relationship. The availability of several mutant alleles of the projectin gene also allows a genetic analysis of projectin assembly and function. Initial data on mutant analysis indicate that projectin is an essential component and that the presence of defective projectin leads to unorganized sarcomeres (Kulp McEliece, unpublished observations).

References

Auber J, Couteaux R. Ultrasructure de la strie Z dans des muscles de Dipteres. *J Microscopie* 1963;2:309-324.

Ashhurst DE. The Z-line: its structure and evidence for the presence of connecting filaments. In *Insect flight Muscle: Proceedings of the Oxford Symposium*, RT Tregear, ed. Elsevier, Amsterdam, North Holland 1977;57-73.

Ayme-Southgate A, Vigoreaux JO, Benian GM, Pardue ML. *Drosophila* has a twitchin/titin-related gene that appears to encode projectin. *Proc Natl Acad Sci USA* 1991;88:7973-7977.

Ayme-Southgate A, Southgate R, Saide J, Benian G, Pardue ML. Both synchronous and asynchronous muscle isoforms of projectin (the *Drosophila bent* locus product) contain functional kinase domains. *J Cell Biol* 1995;128:393-403.

Benian GM, Kiff JE, Neckelmann N, Moerman DG, Waterston, RH. Sequence of unusually large protein implicated in regulation of myosin activity in *C. elegans. Nature* 1989;342:45-50.

Bullard B, Hammond KS, Luke BM. The site of paramyosin in insect flight muscle the presence of an unindentified protein between myosin filaments and Z line. *J Mol Biol* 1977;115:417-440.

Candia-Carnevali MD, De Eguileor M, Valvassori R. Z line morphology of functionally diverse insect skeletal muscles. *J Submicrosc Cytol* 1980;12:427-446.

Colley NJ, Tokuyasu KT, Singer SJ. The early expression of myofibrillar proteins in round postmitotic myoblasts of embryonic skeletal muscle. *J Cell Sci* 1990;95:11-22.

Crossley AC. The morphology and development of the *Drosophila* muscular system. In *Genetics and Biology of Drosophila* vol 2B, M Asburner, TRF Wright, eds. Academic Press, London, 1978;499-599.

Daley J, Southgate R, Ayme-Southgate A. Structure of the projectin isoforms and implications for projectin assembly and functions. *J Mol Biol* 1998;279:201-210.

Deatherage JF, Cheng N, Bullard B. Arrangement of filaments and cross-links in the bee flight muscle Z-disk by image analysis of oblique sections. *J Cell Biol* 1989;108:1775-1782

Furst DO, Gautel M. The Anatomy of a Molecular Giant: How the Sarcomer Cytoskeleton is Assembled from Immunoglobulin Superfamily Molecules. *J Mol Cell Cardiol* 1995;27:951-959.

Fyrberg EA, Beall C. Genetic approaches to myofibril form and function in *Drosophila. Trends in Genet* 1990;6:126-131.

Fyrberg CC, Labeit S, Bullard B, Leonard K, Fyrberg EA. *Drosophila* projectin: Relatedness to titin and twitchin and correlation with *lethal (4) 102CDa* and *bent-dominant* mutants. *Proc R Soc Lond B* 1992;249:33-40.

Gautel M. The super-repeats of titin/connectin and their interactions: glimpses at sarcomeric assembly. *Adv Biophys* 1996;33:27-37.

Horowits R, Maruyama K, Podolsky RJ. Elastic behavior of connecting flaments during thick filament movement in activated skeletal muscle. *J Cell Biol* 1989;109:2169-2176.

Hu DH, Kimura S, Maruyama K. Sodium dodecyl sulfate-gel electrophoresis of connectin-like high molecular weight proteins of various types of vertebrate and invertebrate muscles. *J Biochem* 1986;99:485-1492.

Hu DH, Matsuno A, Terakado K, Matsuura T, Kimura S, Maruyama K. Projectin is an invertebrate connectin (titin): Isolation from crayfish claw muscle and localization in crayfish claw muscle and insect flight muscle. *J Muscle Res Cell Motil* 1990;11:497-511.

Isaacs WB, Kim IS, Struve A, Fulton, AB. Association of titin and myosin heavy chain in developing skeletal muscle. *Proc Natl Acad Sci USA* 1992;89:7496-7500.

Jewell BR, Ruegg C. Oscillatory contraction of insect fibrillar muscle after glycerol extraction. *Proc R Soc Lond B* 1966;164:428-459.

Labeit S, Kolmerer, B. Titins: giant proteins in charge of muscle ultrastructure and elasticity. *Science* 1995;270:293-296.

Lakey A, Ferguson C, Labeit S, Reedy M, Larkins A, Butcher G, Leonard K, Bullard B. Identification and localization of high molecular weight proteins in insect flight and leg muscles. *EMBO J* 1990;9:3459-3467.

Lakey A, Ferguson C, Labeit S, Reedy M, Larkins A, Butcher G, Leonard K, Bullard B. Kettin, a large modular protein in the Z-disc of insect muscles. *EMBO J* 1993;12, 2863-2871.

Linke WA, Ivemeyer M, Olivieri N, Kolmerer B, Ruegg JC, Labeit S. Towards a molecular understanding of the elasticity of titin. *J Mol Biol* 1996;261, 62-71.

Locker RH, Wild DJC. A comparative study of high molecular weight proteins in various types of muscle across the animal kingdom. *J Biochem* 1986;99:1473-1484.

Machado C, Sunkel CE, Andrew DJ. Human autoantibodies reveal titin as a chromosomal protein. *J Cell Biol* 1998;141:321-333.

Maroto M, Vinos J, Marco R, Cervera M. Autophosphorylating protein kinase activity of titin-like arthropod projectin. *J Mol Biol* 1992;224:287-291.

Moerman DG, Plurad S, Waterston RH, Baillie DL. Mutations in the *unc54* myosin heavy chain gene of *Caenorhabditis elegans* that alter contractility but not muscle structure. *Cell* 1982;29:773-781.

Nave R, Weber K. A myofibrillar protein of insect muscle related to vertebrate titin connects Z-band and A-band: Purification and molecular characterization of invertebrate mini-titin. *J Cell Sci* 1990;95:535-544.

Pfuhl M, Pastore A. Tertiary structure of an immunoglobulin-like domain from the giant muscle protein titin: a new member of the I set. *Structure* 1995;3:391-401.

Pringle FRS. Stretch activation of muscle: Function and mechanism. *Proc R. Soc Lond B* 1978;201:107-130.

Reedy MC, Beall C. Ultrastructure of developing flight muscle in *Drosophila*. I. Assembly of myofibrils. *Dev Biol* 1993;160:443-465.

Rhee D, Sanger JM, Sanger JW. The premyofibril: evidence for its role in myofibrillogenesis. *Cell Motil Cytoskeleton* 1994;28:1-24.

Saide JD, Ullrick WC. Purification and properties of the isolated honeybee Z-disc. *J Mol Biol* 1974;87:671-683.

Saide JD. Identification of a connecting filament protein in insect fibrillar flight muscle. *J Mol Biol* 1981;153:661-679.

Saide JD, Chin-Bow S, Hogan-Sheldon J, Busquets-Turner L, Vigoreaux JO, Valgeirsdottir K, Pardue ML. Characterization of domponents of Z-bands in the fibrillar flight muscle of *Drosophila melanogaster. J Cell Biol* 1989;109:2157-2167.

Squire J (ed). *The Structure and Basis of Muscular Contraction*. Plenum Press, New York, 1981.

Tokuyasu KT, Maher PA. Immunocytochemical studies of cardiac myofibrillogenesis in early chick embryos. I: Presence of immunofluorescent titin spots in premyofibril stages. *J Cell Biol* 1987;105:2781-2793.

Trinick J. Titin and nebulin: Protein rulers in muscle? *Trends in Bio Sci* 1984:19:405-409.

Trinick J. Interactions of titin/connectin with the thick filament. *Adv Biophys* 1996;33:81-90.

Trombitas C, Tigyi-Sebes A. Fine structure and mechanical properties of insect muscle. In *Insect Flight Muscle: Proceedings of the Oxford Symposium*, RT Tregear, ed. Elsevier, Amsterdam, North Holland 1977;79-90.

van der Loop FT, van der Ven PF, Furst DO, Gautel M, van Eys GJ, Ramaeckers FC. Integration of titin into the sarcomeres of cultured differentiating human skeletal muscle cells. *Eur J Cell Biol* 1996;69:301-307.

van der Ven PF, Schaart G, Croes HJ, Jap PH, Ginssel LA, Ramaekers FC. Titin aggregates associated with intermediate filaments align along stress fiber-like structures during human skeletal muscle cell differentiation. *J Cell Sci* 1993;106, 749-759.

Vibert P, Edelstein SM, Castellani L, Elliot BW. Mini-titins in striated and smooth molluscan muscles: Structure, location and immunological crossreactivity. *J Muscle Res Cell Motil* 1993;14:598-607.

Vigoreaux JO, Saide JD, Pardue M-L. Structurally different *Drosophila* striated muscles utilize distinct variants of Z-band associated proteins. *J Musc Res Cell Motil* 1991;12:340-354.

Wang K, Ramirez-Mitchell R, Palter D. Titin is an extraordinarily long, flexible and slender myofibrillar protein. *Proc Natl Acad Sci USA* 1984;81:3685-3689.

Wang K, McCarter R, Wright J, Beverly J, Ramirez-Mitchell R. Regulation of skeletal muscle stiffness and elasticity by titin isoforms: a test of the segmental extension model of resting tension. *Proc Natl Acad Sci USA* 1991;88:7101-7105.

Discussion

Sanger: You said that you wanted a mutant that disrupts the Z-band. I thought that White, Sparrow, and Freiberg had one where α-actinin wasn't there and the Z-bands were disrupted.

Ayme-Southgate: That is an actin mutant. The problem is that there are α-actinin mutations available in *Drosophila*, some of which are flightless. But the problem with those mutations is that they are not null. We come back to the same problem. α-actinin is made, it is the wrong form of α-actinin, but it is made. I have looked at them, and there is no obvious visible effect. I can still even pick up α-actinin in those mutants. So they are really not useful. I really need something where the Z-band disappears. Maybe a kettin mutant.

Bullard: How about an actin null mutant?

Ayme-Southgate: I am just a little bit worried about using an actin null mutant because then I am dealing with actin missing and the Z-band. I have not looked at interactions between projectin and actin, but I think after some of the science I have heard here, I cannot totally exclude that possibility. Then if projectin does not assemble correctly, I won't be able to tell if it is because it is missing its actin interaction or if it is because it is missing the Z-band. I will probably end up doing it because it will be better than not doing anything at all. But it will be difficult to interpret the data.

Trombitás: I would like to know where is the N-terminal and where is the C-terminal end. In your earlier publication it was not clear. Now you mention that probably the N-terminal end is at the Z-disk?

Ayme-Southgate: We don't know that for sure.

Trombitás: But is it possible?

Ayme-Southgate: We always assume that the N-terminal is at the Z-band, just because of analogy with titin. But we have no evidence for that.

Linke: You showed this picture where you said that several possibilities of projectin and myosin interaction exist. You said that it could be four projectin molecules per myosin, but then you said it could be staggered, it could be heads to tails. You didn't answer that question.

Ayme-Southgate: The only data I have at this point is this interaction between the N and the C-terminal parts of projectin. If that is true *in vivo*, of course, you have to take *in vitro* binding assay with a grain of salt. If this is true, that would favor the staggered arrangement. But that is why I am really interested in looking at the mutant, because I think the mutant will give me an idea of what is happening *in vivo*.

Linke: Is anything known about the ratio?

Ayme-Southgate: The ratio between myosin and projectin? I don't know. Belinda, do you want to comment?

Bullard: Yes, we did look into that question with Lethocerus muscle. Projectin is 0.3 microns long, and the half-length in the Lethocerus leg muscle fitted 3 projectin molecules end to end.

Ayme-Southgate: If we do the same calculation, if we assume that projectin is about 0.1 μm, and that the whole A-band is about 2.1 μm, that gives us as many as 10 molecules. Am I calculating right?

Bullard: Isolated projectin is 280 nm long.

Ayme-Southgate: OK, so that comes up to about 3. But if they're staggered you may be able to put in more.

Greaser: Just a comment. If you have an antibody against a defined region, you should be able to quickly sort out some of your models.

Ayme-Southgate: All the antibodies that we have at this point, stain the entire A-band. We don't have anything that will stain small domain regions. We have fusion proteins now, we can send them out and get antibodies against them. But the problem is that I am not sure all of those are going to be useful because the fusion protein contains Ig and Fn domain. I am worried that the antibody will recognize related domains as well.

Greaser: You shouldn't worry. This is not too likely. And also I would guess that if it is like the titin story, those domains that are near the ends are less likely to be of the repetitive form.

Ayme-Southgate: That is correct. To raise antibodies I was going to use the kinase domain because that is obviously unique, and the N and the C terminals regions.

ROLE OF TITIN IN NONMUSCLE AND SMOOTH MUSCLE CELLS

Thomas C. S. Keller, III[1,2], Kenneth Eilertsen[2],
Mark Higginbotham[1], Steven Kazmierski[1],
Kyoung-Tae Kim[1], and Michaella Velichkova[2]

*Department of Biological Science[1] and
Program of Molecular Biophysics[2]
Florida State University, Tallahassee, FL*

Abstract: Extensive investigation of vertebrate striated muscle titin has yielded significant insight into its structure and function in striated muscle. We have begun to investigate other members of the titin protein family found in vertebrate smooth muscle and nonmuscle cells. Smooth and nonmuscle titins resemble striated muscle titin in molecular size and morphology but differ in their interactions with myosin II filaments and in the structural contexts in which they exist *in vivo*. Divergence of these titins from the muscle titin paradigm demonstrates the versatility of this remarkable family of giant proteins.

Introduction

The muscle titin paradigm

Titin (also known as connectin) was discovered more than two decades ago as the third major filament in the contractile units of striated muscle known as sarcomeres (Wang *et al.*, 1979; Maruyama *et al.*, 1977). The other two major types of filaments in sarcomeres – actin thin filaments and myosin II (myosin) bipolar thick filaments – interact to produce contractile force. Titin interacts with these filaments and other components to establish and maintain the highly ordered sarcomere structure that is crucial for its function (for recent review, see Maruyama, 1997).

Elastic Filaments of the Cell, edited by Granzier and Pollack
Kluwer Academic/Plenum Publishers, 2000

265

Specific properties of the unusually long and elastic titin molecule sup-port its function in striated muscle. In developing muscle, titin assembles na-scent myosin filaments into the sarcomeres (Gregorio *et al.*, 1999) and regu-lates overall sarcomere morphogenesis through the activity of its kinase do-main (Mayans *et al.*, 1998). In adult muscle sarcomeres, each titin molecule spans half a sarcomere (Maruyama, 1997; Keller, 1995). The titin N-terminal region interacts with one Z-line boundary of the sarcomere. The titin C-termi-nal region interacts with the M-line in the center of the sarcomere and one pole of a myosin II bipolar thick filament. This anchorage and the elastic properties of the titin molecules maintain the central position of the myosin thick fila-ments in contracting sarcomeres and provide the passive resistance to exten-sion exhibited by relaxed muscle (Trombitás and Pollack, 1993).

Results and Discussion

Nonmuscle (cellular) and smooth muscle titins
Smooth muscle and nonmuscle cells also use actin-myosin structures to produce force for important cell functions. These structures share some but not all of the structural properties of muscle sarcomeres. Nonmuscle cells assemble actin and myosin filaments into a variety of stable and transient cytoskeletal structures including stress fibers, circumferential rings, and cleavage furrows. As in muscle sarcomeres, periodically distributed myosin filaments interdigi-tate with antiparallel actin filaments in these structures. Nonmuscle myosin filaments and muscle thick filaments, however, differ in size and spacing in their respective structures. Likewise, smooth muscle contains sarcomere-like arrangements of actin and myosin filaments, but these also differ in myosin filament form and other properties of organization from both striated muscle sarcomeres and nonmuscle cytoskeletons.

We have found titins in both smooth muscle and nonmuscle cells. Our investigations of these titins suggest that they interact with myosin filaments and participate in structural organizations in ways not readily explained by the striated muscle titin paradigm.

We first discovered a nonmuscle or cellular titin (c-titin) in the brush bor-der cytoskeleton (Eilertsen and Keller, 1992). This stable cytoskeleton sup-ports the array of microvilli projecting from the apical ends of intestinal epithe-lial cells. We subsequently found c-titin associated with transient cytoskeletal structures such as the stress fibers in fibroblasts (Eilertsen *et al.*, 1994) and those formed in platelets following activation (M. Velichkova and T. Keller, unpublished observations). More recently, we have begun to characterize a smooth muscle titin that we found in chicken gizzard smooth muscle.

Both c-titin isolated from chicken intestinal epithelial cells brush borders and human blood platelets and smooth muscle titin isolated from chicken giz-

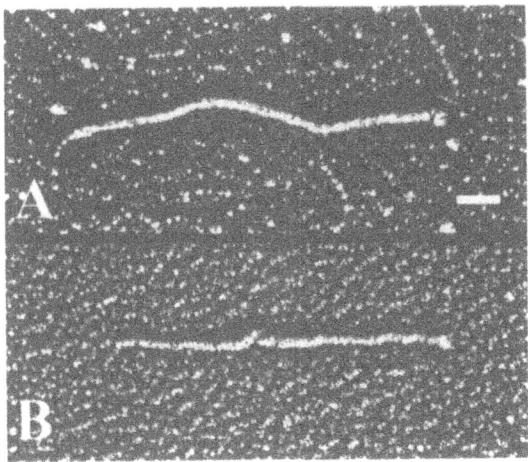

Figure 1. *Metal replicas of human blood platelet and chicken smooth muscle titins. (A) platelet titin. (B) gizzard smooth muscle titin. Molecules were rotary-replicated with platinum. Bar, 100 nm.*

zard resemble striated muscle titin in molecular size and morphology. All of the titins migrate at similarly slow rates on SDS-PAGE, indicating that they all are extremely high molecular weight molecules (Eilertsen and Keller, 1992; Eilertsen *et al.*, 1994; platelet and smooth muscle titin data not shown). Likewise, the brush border c-titin (Eilertsen and Keller, 1992) as well as the platelet and smooth muscle titins are ~900 nm long molecules, with a globular region at one end (Fig. 1). This globular end is known to be the N-terminal region of striated muscle titin, but it has yet to be confirmed which end it represents in c-titin and smooth muscle titin.

Coassembly of c-titin and nonmuscle myosin filaments in vitro

Despite these molecular similarities, the striated muscle, smooth muscle, and cellular titins associate with myosin filaments of different sizes and assembly states, and they self-assemble different structural organizations *in vitro*. C-titin purified from brush borders and coassembled with brush border myosin organizes highly ordered linear arrays of myosin bipolar filaments (Eilertsen *et al.*, 1994). The ~300 nm long myosin bipolar filaments in these arrays most likely contain ~30 myosin molecules.

Clearly, the packing state of nonmuscle myosin in these 'minifilaments' differs significantly from that of striated muscle myosin in the muscle thick filaments, which are ~ 1 μm long and composed of >300 myosin molecules. We have found that there are as few as one or two c-titin molecules associated with each nonmuscle myosin minifilament (Eilertsen *et al.*, 1994), whereas 6-

12 titins are estimated to be associated with striated muscle myosin thick filaments (Whiting *et al.*, 1989). Although the structural organizations are different, the ratio of titins to myosins (most likely ~ 1:25) is similar in each case.

C-titin isolated from human blood platelets also coassembles with nonmuscle myosin filaments *in vitro*. We have found three different structural organizations formed from platelet c-titin and myosin *in vitro*. In one type of structure, platelet c-titin aligns myosin bipolar minifilaments in a linear arrangement similar to that formed by the brush border proteins (not shown). In the second type of organization, the myosin bipolar filaments form geometric patterns, mostly triangles, with their ends abutting at nodes in the structure. (Fig. 2A). The third structural variation also contains geometric patterns of myosin filaments (Fig. 2B). The myosin filaments in these structures, however, appear to be small side-polar filaments. Side-polar filaments, which are known to exist in smooth muscle, have myosin heads distributed along their length and lack the central bare zone of bipolar filaments. The side-polar minifilaments formed from platelet c-titin and myosin range in length from ~150 to ~300 nm between nodes in the *in vitro* structure.

Interaction of platelet c-titin with both bipolar and side-polar myosin minifilaments and the different structural organizations formed by their self-assembly indicates that this c-titin and myosin can interact in multiple configu-

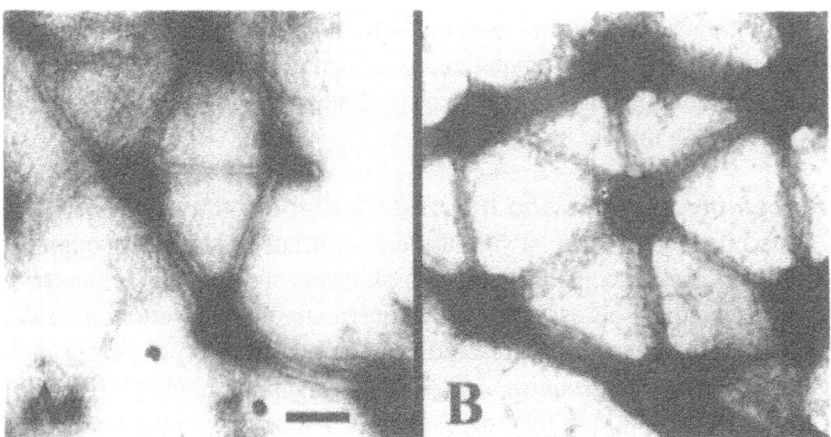

Figure 2. Coassemblies of human c-titin and myosin II. (A) Geometric array of c-titin and myosin bipolar filaments. (B) Geometric array of c-titin and myosin side-polar minifilaments that lack bare zones and have myosin heads distributed along their length. Coassemblies were negatively stained with uranyl acetate. Bar, 100 nm.

rations. This interaction flexibility may increase the versatility of c-titin-myosin structures participating in the assembly and dynamic changes of the platelet cytoskeleton following activation.

Interaction of smooth muscle titin with bipolar and side-polar filaments

Smooth muscle titin also displays a certain degree of flexibility in how it interacts with smooth muscle myosin. At physiological ionic strength, the titin and myosin form aggregates of large side-polar filaments *in vitro*, similar to those that exist in smooth muscle *in vivo* (Fig. 3A). This coassembly state so far appears to be unique to the smooth muscle proteins both *in vivo* and *in vitro*. Some of the side-polar filaments in coassembly preparations exceed the ~900 nm length of the smooth muscle titin. It will be interesting to learn how the 900 nm long smooth muscle titin molecules associate with these filaments.

Under low ionic strength *in vitro* conditions, smooth muscle titin and myosin coassemble into geometric arrays of minifilaments, most of which appear to be side-polar minifilaments similar to those formed from platelet titin and myosin (Fig. 3B). Although these structures have yet to be found in smooth muscle *in vivo*, the ability of these proteins to form two different organizations *in vitro* demonstrates that the interaction between smooth muscle titin and myosin also is not rigidly proscribed by the molecular structure of either protein.

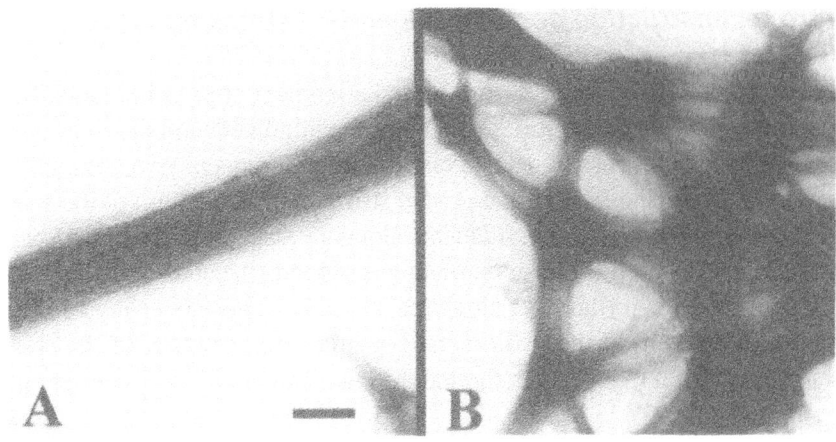

Figure 3. *Coassemblies of chicken gizzard smooth muscle titin and myosin II. (A) A long aggregate of smooth muscle titin and smooth muscle myosin side-polar filaments assembled in physiological ionic strength conditions. (B) Geometric array of smooth muscle titin and smooth muscle myosin side-polar minifilaments assembled at low ionic strength. Bar, 100 nm.*

Specificity of titin-myosin interaction

Consistent with the differences in structures formed by striated muscle, smooth muscle, and nonmuscle titins and myosins, we have found that interaction between the striated muscle and nonmuscle proteins is specific with respect to their source. For example, striated muscle titin interacts with muscle myosin *in vitro* but not with nonmuscle myosin (Eilertsen *et al.*, 1994). Likewise, c-titin interacts with nonmuscle myosin but not with muscle myosin. This specificity of interaction could account for the observation that muscle myosin introduced into the cytoplasm of nonmuscle cells fails to incorporate into cellular structures such as stress fibers. Smooth muscle titin incorporates into cytoskeletal structures when injected into nonmuscle cells and therefore may interact with c-titin.

The region of striated muscle titin that interacts with the myosin thick filament molecular is composed of superrepeat patterns of FN3 and Ig structural domains (Labeit *et al.*, 1992). The molecular dimensions of this muscle titin region and the pattern of the superrepeat structure are consistent with the structural features of the thick filament. Cloning and sequencing of the analogous myosin filament-binding region of c-titin will reveal the molecular basis for its interaction with nonmuscle myosin. Failure to interact with muscle myosin may indicate that c-titin may differ from muscle titin in the existence and/or specific sequence of Ig and FN3 domains in its myosin-binding region. It seems more likely, however, that this region of c-titin contains similar structural domains, but differs in the superrepeat pattern of those domains.

In vitro *reconstruction of c-titin-dependent stress fiber-like structures*

The linear side-by-side and end-to-end alignment of myosin bipolar filaments in the *in vitro* coassemblies resembles the alignment and distribution of myosin filaments in stress fibers and circumferential rings *in vivo*. The major difference between the *in vitro* and *in vivo* structures is the length of the space between the ends of the myosin filaments along the structures. In stress fibers, I-band-like spaces spanned by actin filaments separate the ends of the side-by-side groups of myosin filaments along the structure. These inter-myosin filament spaces shorten during actin-myosin-mediated contraction of the stress fibers in the manner that I-bands shorten during contraction of muscle sarcomeres.

In the *in vitro* c-titin-myosin coassemblies, the myosin filaments abut end-to-end along the length of the structure. How the ~900 nm long titin molecules are laid out in this structure remains unclear. If completely extended, each titin molecule is long enough to span three of the ~300 nm myosin filament striations (Fig. 4A). Interaction of a single titin molecule with three myosin fila-

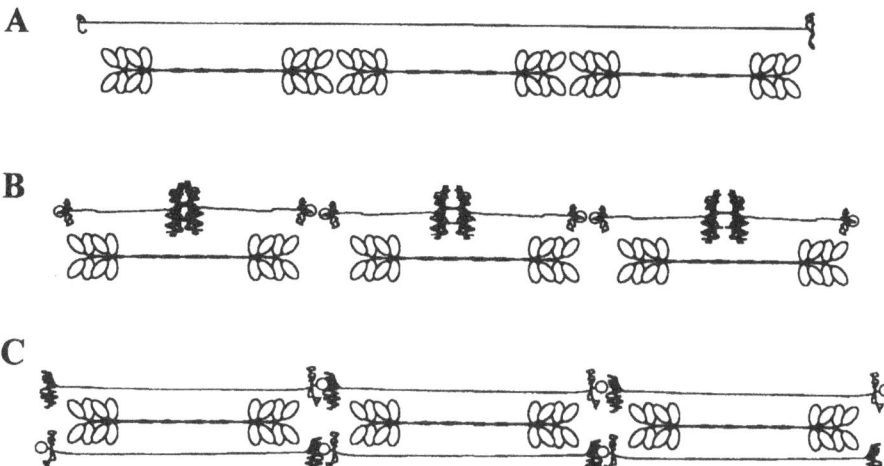

Figure 4. *Three possibilities for the arrangement of c-titin and myosin filaments in linear coassemblies. (A) each 900 nm long titin molecule associates with three myosin bipolar filaments and spans three striations. (B) each c-titin molecule associates with one pole of a myosin bipolar filament. In this possibility, the globular C-terminal ends of two c-titins associate in a position analogous to the M-line, and the highly folded N-terminal domains interact in a position analogous to the Z-line position in a striated muscle sarcomere. (C) each c-titin molecule interacts with a myosin bipolar filamen such that both ends of the c-titin overhang the myosin filament and interact with other highly folded c-titin ends interact end-to-end and side-by-side along the structure.*

ments along its length, however, seems unlikely. It is much more likely that each titin molecule interacts with one myosin bipolar filament and that one or both ends of the titin are highly folded in the region between the ends of the abutting myosin filaments. This possibility is most consistent with observations that ribbons of side-by-side myosin filaments get recruited into stress fibers near the margins of cells *in vivo* (Verkhovsky *et al.*, 1995).

If the layout of c-titin in these structures is analogous to that in striated muscle, then the C-terminus of each c-titin molecule would be located in the middle of the bare zone of the bipolar filament (Fig. 4B). We see no evidence, however, for the density of protein that would be expected if the globular regions of two c-titin molecules interact near the center of the bare zone of each myosin bipolar filament.

Another possibility is that each titin molecule interacts with one myosin filament along its length and overhangs the myosin either on one or perhaps on both ends (Fig. 4C). We have obtained rare images in which myosin bipolar dimers appear to be associated near the middle of the extended region the c-titin molecule (Fig. 5). The c-titin molecule overhangs both ends of the myosin dimer. This observation supports the possibility that both ends of the c-titin in

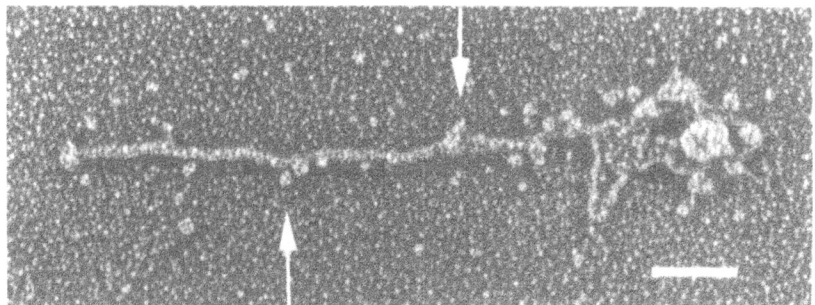

Figure 5. *Metal replica of myosin bipolar dimer interaction with a c-titin molecule. The myosin bipolar dimer appears to associate near the middle of the c-titin molecule. Bar, 100 nm.*

the *in vitro* structures project into the regions between the ends of the myosin filaments. Determining the actual position and folding pattern of c-titin will require mapping the positions of epitopes along the c-titin molecule in the coassemblies. We currently are developing monoclonal antibodies to human platelet c-titin for this purpose.

Regardless of the exact position of c-titins in the coassemblies, the c-titin molecules are likely to be highly folded. If the integral c-titins are highly folded, then the structure of the *in vitro* coassembly might represent the most collapsed form of the underlying structure of stress fibers. *In vivo*, the stress fiber structure might be extended by forces or other assembly parameters not replicated *in vitro*. Such a structure could maintain its structural integrity during normal extension by simply straightening the highly folded c-titin molecules. As in striated muscle titin, additional extension while maintaining the integrity of the structure then might involve elastic stretch of specific domains of the c-titin molecules.

Building a stress fiber in vitro

Despite the differences in myosin filament spacing, c-titin and nonmuscle myosin self-assemble *in vitro*, even in the absence of actin filaments, into the same underlying organization of myosin filaments that is present in stress fibers. This suggests that c-titin may play a greater role than actin in the morphogenesis of myosin-containing cytoskeletal structures *in vivo*. We investigated this possibility by comparing the structures of actin-myosin and c-titin myosin coassemblies. Actin-myosin coassemblies fail to replicate the stress fiber-like myosin filament distribution pattern of the titin-myosin coassemblies. Instead, the myosin filaments tend to interact through their heads with parallel actin filaments like skewed rungs on a ladder. Formation of the striated pattern of myosin filaments requires c-titin. Taken together, our results indicate that c-titin is necessary and sufficient to organize nonmuscle myosin filaments into stress fiber-like arrays *in vitro* and possibly *in vivo*.

The emergence of c-titin as an organizer of myosin structures raised the possibility that c-titin also might organize other stress fiber components. In stress fibers, α-actinin localizes in the spaces between the ends myosin filaments along the structure. Although the α-actinin clearly interacts with the actin filaments, more than this interaction is responsible for this distinct localization of α-actinin in stress fibers. When microinjected into cells, the actin-binding domain of α-actinin associates uniformly along the stress fiber actin filaments, whereas the intact protein or the central region lacking the actin-binding domains mimics endogenous α-actinin localization (Pavalko and Burridge, 1991).

The positions of α-actinin association with stress fibers are roughly analogous to those of the Z-lines in striated muscle. In the Z-line region of muscle, α-actinin associates with muscle titin through direct interaction with several amino acid motifs known as Z-repeat domains near the N-terminus of muscle titin (Sorimachi *et al.*, 1997). We have found that c-titin also interacts directly with α-actinin (Eilertsen *et al.*, 1997). We also have found that when coassembled with c-titin and myosin *in vitro*, α-actinin associates with the coassembly structures primarily in regions between the ends of the myosin filaments, as it does *in vivo* (Fig. 6). These findings suggest that c-titin plays a crucial role in organizing myosin II in cells and that interaction of the myosin-associated protein c-titin and the actin-binding protein α-actinin may be responsible for integrating actin and myosin filaments into functional cytoskeletal structures such as stress fibers.

Multiple structural organizations of titin and myosin in the brush border cytoskeleton

Our investigation of c-titin localization in brush borders first revealed the possibility of multiple titin-myosin organizational states within an integrated cytoskeletal structure. The brush border cytoskeleton contains three structural domains (Keller and Mooseker, 1991). One is composed of the core bundles of actin filaments that support the microvilli. This domain contains bridges of myosin I between the actin filaments and the membrane but lacks c-titin and myosin and thus will not be considered here.

The other two brush border structural domains contain myosin II (Keller *et al.*, 1985). One of these domains is composed of the circumferential ring of actin filaments that encircles the terminal web and is associated with the cytoplasmic face of the belt desmosome junction. This ring contains myosin bipolar minifilaments that are aligned side-by-side in clusters that are interdigitated with antiparallel actin filaments in a pseudo-sarcomeric fashion around the ring (Hirokawa *et al.*, 1983; Drenkhahn and Dermeitzel, 1988). With respect to

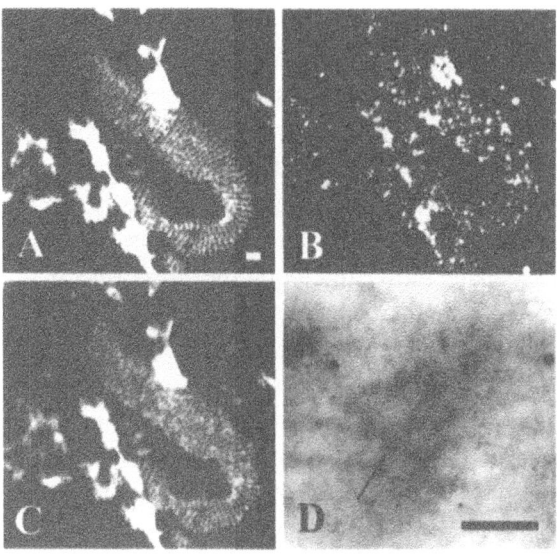

Figure 6. *In vitro coassemblies of c-titin, myosin II filaments, and α-actinin.(A) immuno-fluorescence localization of the myosin II filaments reveals periodic distri-bution along the structures. (B) immuno-fluorescence localization of α-actinin reveals a punctate distribution pattern along the structures. (C) overlay of the two localization patterns reveals that much of the α-actinin is localized in the spaces between the ends of the myosin filaments in positions analogous to those of α-actinin association in stress fibers in vivo. (D) Immunogold (5 nm gold) localization of α-actinin in the regions surrounding the heads of the myosin filaments in negatively stained in vitro coassemblies visualized with electron microscopy. The two 15 nm gold particles are labeling c-titin molecules. Bars, (A-C) 1 mm; (D) 100 nm.*

actin-myosin organization, the circumferential ring thus resembles a circular stress fiber. Interaction between the actin and myosin filaments can produce force for constriction of the circumferential ring (Keller *et al.,* 1985; Broschat *et al.*, 1983). Ring force production may regulate transcellular transport across the tight junctions of the cell with its neighbors. Constriction of similar rings in embryonic epithelial cells produces force for morphological remodeling of a variety of tissues including neural tube folding.

We have found that the brush border circumferential ring also contains c-titin (Figs. 7C and 7D). As such, the role c-titin may play in its morphogenesis and maintaining its structural stability is apparent from our understanding of c-titin-myosin interaction *in vitro.*

The third brush border domain is a specialized region of the cell cortex known as the terminal web, in which the microvillus core actin bundles are rooted and crosslinked by a web of fodrin molecules and myosin bipolar minifilaments. In this region, all of the actin filaments are bundled in the mi-crovillus rootlets. The myosin bipolar filaments span the distance between two microvillus rootlets. Adjacent rootlets are ~100-150 nm apart. This dis-tance is close enough for the heads of the myosins in the bipolar filaments,

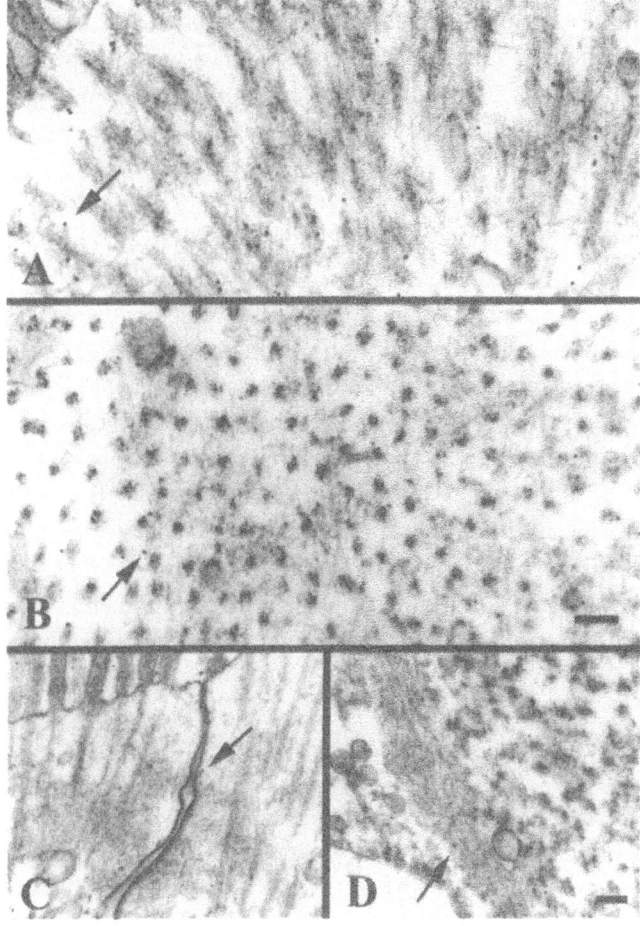

Figure 7. *Immunogold localization of c-titin in the brush border terminal web and circumferential ring domains. (A) Immunolocalization of c-titin with 5 nm gold reveals association with the microvillus rootlets and in spaces between the rootlets in longitudinal sections of the terminal web. (B) Immunogold localization of c-titin in the spaces between the microvillus rootlets in a cross section of the brush border terminal web domain. (C) Immunogold localization of c-titin in cross sections of the circumferential ring associated with the belt desmosome junction between neighboring intestinal epithelial cells. (D) Immunogold localization of c-titin in a tangential section of the brush border circumferential ring. Arrows indicate a few selected 5 nm gold particles. Bars, 100 nm.*

which are estimated to contain as few as four myosin molecules (Hirokawa *et al.*, 1982), to interact with the actin filaments in adjacent rootlets.

Although some of the myosin filament crossbridges are at oblique angles with respect to the actin filaments in the rootlets, others appear to be perpendicular to the rootlets. This unusual orientation of actin and myosin makes it difficult to deduce the role myosin plays in this domain, and the function of this myosin remains a mystery.

It also makes it difficult to predict the layout of c-titin in this domain. Our initial immunofluorescence localization investigations revealed that c-titin coexists with myosin in this domain (Eilertsen and Keller, 1992). We subsequently have confirmed c-titin localization between the microvillus rootlets by immunogold localization using an anti-c-titin antibody (Figs. 7A and 7B). If a single c-titin or at most two c-titin molecules are associated with each myosin filament, then a substantial portion of each terminal web c-titin is available to interact with other c-titins or with the vertical microvillus rootlets. Morphogenesis of this domain may involve formation of geometric arrangements of bipolar filaments similar to those formed from platelet c-titin and myosin and at the assembling ends of linear coassemblies of brush border c-titin and myosin. Perhaps the heads of the myosin filaments and the ends of associated c-titin molecules then interact with the forest of vertically oriented actin bundles in the developing terminal web to stabilize the terminal web organization.

Conclusion

Vertebrate striated muscle titin has served as a useful paradigm for understanding the structure and function of titins in general. Our investigations of smooth muscle and nonmuscle titins, however, have begun to reveal that smooth muscle and cellular titins diverge from the striated muscle paradigm. These differences include variations in the sizes and structures of the myosin filaments with which the titins interact, in the ultrastructure of the titin-myosin coassemblies formed *in vitro*, and in the cellular structures with which titin interacts *in vivo*. Consideration of these differences reveals that there is much left to learn about the specific features of members of the titin family of proteins.

Acknowledgments

The National Science Foundation and the Florida Affiliate of the American Heart Association support work in the authors' lab. We would like to thank Kim Riddle for help with the light and electron microscopy.

References

Broschat KO, Stidwell RP, Burgess DR. Phosphorylation controls brush border motility by regulating myosin structure and association with the cytoskeleton. *Cell* 1983;35:561-71.

Drenkhahn D, Dermeitzel R. Organization of the actin filament cytoskeleton in the intestinal brush border: a quantitative and qualitative immunoelectron microscope study. *J Cell Biol* 1988;107:1037-48.

Eilertsen KJ, Kazmierski ST, Keller TCS, III. Cellular titin localization in stress fibers and interaction with myosin II filaments in vitro. *J Cell Biol* 1994;126:1201-10.

Eilertsen KJ, Kazmierski ST, Keller TCS, III. Interaction of α-actinin with cellular titin. *Eur J Cell Biol* 1997;74:361-64.

Eilertsen KJ, Keller TCS, III. Identification and characterization of two huge protein components of the brush border cytoskeleton: Evidence for a cellular isoform of titin. *J Cell Biol* 1992;119:549-57.

Gregorio CC, Granzier H, Sorimachi H, Labeit S. Muscle Assembly: a titanic achievement? *Curr Opin Cell Biol* 1999;11:18-25.

Hirokawa N, Keller TCS, Chasan R, Mooseker MS. Mechanisms of Brush Border Contractility Studied by the Quick-freeze, Deep-etch Method. *J Cell Biol* 1983;96:1325-36.

Hirokawa N, Tilney LG, Fujiwara K, Heuser JE. Organization of actin, myosin, and intermediate filaments in the brush border of intestinal epithelial cells. *J Cell Biol* 1982;94:425-43.

Keller TCS, Conzelman KA, Chasan R, Mooseker MS. The role of myosin in terminal web contraction in isolated intestinal epithelial brush borders. *J Cell Biol* 1985;100:1647-55.

Keller TCS, Mooseker MS. "Enterocyte cytoskeleton: its structure and function." In *Handbook of Physiology. Section 6: The gastrointestinal system* 4th ed, M Field, RA Frizzell, eds. Bethesda, MD: American Physiological Society, 1991.

Keller TCS, III. Structure and function of titin and nebulin. *Curr Opin Cell Biol* 1995;7:32-38.

Labeit S, Gautel M, Lakey A, Trinick J. Towards a molecular understanding of titin. *EMBO J* 1992;11:1711-16.

Maruyama K. Connectin/titin, giant elastic protein of muscle. *FASEB J* 1997;11:341-45.

Maruyama K, Matsubara S, Natori R, Nonomura Y, Kimura S, Ohashi K, Murakami F, Handa S, Eguchi G. Connectin, an elastic protein of muscle: characterization and function. *J Biochem (Tokyo)* 1977;82:317-37.

Mayans O, van der Ven PF, Wilm M, Mues A, Young P, Furst DO, Wilmanns M, Gautel M. Structural basis for activation of the titin kinase domain during myofibrillogenesis. *Nature* 1998;395:863-69.

Pavalko FM, Burridge K. Disruption of the actin cytoskeleton after microinjection of proteolytic fragments of α-actinin. *J Cell Biol* 1991;114:481-91.

Sorimachi H, Freiburg A, Kolmerer B, Ishiura S, Stier G, Gregorio CC, Labeit D, Linke WA, Suzuki K, Labeit S. Tissue-specific expression and alpha-actinin binding properties of the Z-disc titin: implications for the nature of vertebrate Z-discs. *J Mol Biol* 1997;270:688-95.

Trombitás K, Pollack GH. Elastic properties of the titin filament in the Z-line region of vertebrate striated muscle. *J Muscle Res Cell Motil* 1993;14:416-22.

Verkhovsky AB, Svitkina TM, Borisy GG. Myosin II filaments assemblies in the active lamella of fibroblasts: their morphogenesis and role in the formation of actin filament bundles. *J Cell Biol* 1995;131:989-1002.

Wang K, McClure J, Tu A. Titin: major myofibrillar components of striated muscle. *Proc Natl Acad Sci USA* 1979;76:3698-702.

Whiting A, Wardale J, Trinick J. Does titin regulate the length of muscle thick filaments? *J Mol Biol* 1989;205:263-67.

Discussion

Pollack: What role do you see, what potential role do you see for titin in the terminal web area, located with myosin?

Keller: That's a really interesting question, so what I didn't take time to point out are the dimensions of that terminal web region. These bundles of actin filaments are about 100 nm apart. I left the slide out here, but we actually have immunogold labeling of the cross filaments between these things that are 100 nm apart. The dimensions of this molecule fit into that; it is really unusual how that is put together. The suggestion is that on either end of this molecule it might be interacting with actin filaments over a considerable length, because I don't know where all the excess length is in the system. Now what happens in the brush border is that, first of all, virtually all nutrient transport occurs through that membrane. We used to have the old joke, you are what you eat, well actually you are what you absorb. So these cells, to a large extent, control what you are, because this is where the absorption occurs. But probably all that absorption consists of small molecules. On the other hand, there are large vesicles, one of which was on that image that I showed. These are golgi-derived vesicles that are coming to replace membrane that sort of sluffs off this system. That has to make it through this terminal web region, which is so highly crossed linked. There were actually coated vesicles that pinch off right at the base of the microvilla, and they come down through that region. So there is actually a remarkable amount of the vesicular traffic through this region that is so highly crosslinked. So one of the fantasies, then, is that maybe an extensible filament would be able to squeeze its way out of the way or allow these things to squeeze in there and actually sort of stretch small parts of the system. Nobody knows what the myosin is doing there. If you think about the orientation of myosin, here the actin filaments are vertical with respect to the myosin, which is to a large extent horizontal. It may be at oblique angles here, but nobody knows what that myosin is doing there. One of the possibilities is that it is simply crosslinking and being a sort of a structural component. But there is a great crosslinker in there already, in the foldrin molecule, although from the earlier work that we heard from Mathias today, maybe titin is a better crosslinker than foldrin, if it doesn't stretch quite as well.

Pollack: Since titin is becoming so ubiquitous…

Keller: That's good, that's good.

Pollack: Why?

Keller: Why?

Pollack: Why is it good? The question is again sort of toward a general role. Now titin is being seen in a number of systems that look extremely interesting. So would you be willing to speculate further on a more general role as opposed to perhaps a role simply related to elasticity or a template?

Keller: Well, we think it is a major organizer. Our prejudice is that wherever myosin II is localized in cells, there will be a titin that organizes it. It is very clear that actin does not organize myosin II because there is actin in a lot of places where there is no myosin II; there are other myosins there. If you do co-assemblies with actin and myosin II, you get actin filaments, and then the myosin II are kind of rungs on the ladder, you don't get this nice striated structure that you see in the stress fibers. I don't think it is important only in terms of its elastic properties; I think it is a major organizer of myosin II activity in cells, and thereby where cells can produce contractile force. It is kind of a chicken and egg problem: (What then organizes the titin?). We have no idea about what is doing that. But we have evidence for titin localization in stress fibers and cleavage furrows, so it is clearly a basic housekeeping protein in cells. It was not invented for muscle.

Trinick: Have you tried any of the various monoclonal antibodies to the vertebrate molecule? Because, particularly if you had antibodies that were at opposite ends of the molecule, then that might tell you something about the distribution.

Keller: Right. We've tried a couple of the available monoclonals. Not enough and we haven't done it systematically, but we are developing monoclonals against the human platelet titin. We have 3 or 4 of these that look like they will be very useful in terms of mapping the layout in *in vivo* structures and particular in these *in vitro* structures. I was talking to Wolfgang about this earlier. One of our fantasies is being able to grab on to the ends of these *in vitro* structures and stretching them. Because really what we are assembling *in vitro*, where the myosins are abutted up against each other, is most likely the most collapsed version of that structure, which *in vivo* is normally extended, probably by other forces in the cell. So what would be neat to do is to grab onto the ends of one of these highly striated structures and stretch it, see how much it stretches and see what forces are actually produced. They are going to be all passive forces there (there is no actin in these structures), and see how that can be better extended. The real mystery is, why is this titin so big when the rest of the elements are mini with respect to the sarcomere structure?

Sanger: Stress fibers have particular striations and spacings. What was the myosin spacings verses the titin spacings in your stress fibers that were stained in those human epithelial cells?

Keller: Actually those were chicken skin fibroblasts. We didn't map closely the striations there; that was a polyclonal antibody and I may be picking up more than one epitope. If you actually look at the staining, it wouldn't be easy to measure those distances.

Sanger: But they did look periodic, so they have a measurement between the densities.

Keller: They are periodic. They are periodic in such a way that they look to be exactly complementary to the α-actinin. I have not done that measurement, but whatever the α-actinin measurement is, that is the periodicity.

Sanger: In a wide variety of fibroblasts, the sarcomeric spacing with α-actinin is on average, about 1.1 microns. That has been done on several cell types. That would mean that your titin also has a periodicity of 1.1 μm if it matches.

Keller: That wouldn't surprise me. That is why I am saying that what we are generating *in vitro* is the most collapsed version of this. So I think there's great potential that this is actually going to look very much like a sarcomere organization. The other proteins seem to be distributed appropriately, and that the reason the titin is so big, is because actually in terms of the real dimensions, that is much more in line with the real dimensions of what this may take on in the *in vivo* context.

Baatsen: Is the actomyosin interaction in the brush border comparable to a rigor state?

Keller: Well, you know, there is no reason to think that ATP would be excluded from that region. So I don't think you can look at it as a rigor state.

Baatsen: Because you mentioned that it had a linking function possibly. Possibly it has to be a strong bond in many conditions. And perhaps certain ionic conditions may allow for some more dynamic rebuilding of that network?

Keller: In the isolated condition?

Baatsen: No, actually, *in vivo*. I am more interested in that, really.

Keller: I'm not sure what you mean by rebuilding. It's a very stable structure in the cell. In fact, what I was talking about earlier is that there must be some dynamics there because there are vesicles that squeeze through these regions. So either there is an unbinding of the proteins to allow these things to go through and then a rebinding, relatively rapid rebinding, to maintain this structure, or there is something in there that is stretching, and that is where the titin might come in.

I also forgot to mention that the circumferential ring is organized very much like a circular stress fiber. It actually produces contractile force; you can do that *in vitro*. So that is a more traditional orientation, or what we consider to be a more traditional organization or orientation of these particular proteins.

Vigoreaux: I was just wondering how you get chickens to volunteer for this kind of work? My question is whether you think the cellular titin that you are looking at, which is really a cytoplasmic titin, will be the same thing as the nuclear titin that we have seen.

Keller: I don't know. We have yet to pay much attention to nuclei, and actually some of those experiments were done with an old antibody that we basically ran out of trying to screen libraries trying to get this thing cloned. Now that we have some newer monoclonals we'll pay some attention to nuclei.

MECHANICAL PROPERTIES
OF TITIN ISOFORMS

Henk Granzier[1], Michiel Helmes[1], Olivier Cazorla[1],
Mark McNabb[1], Dietmar Labeit[2], Yiming Wu[1],
Rob Yamasaki[1], Alka Redkar[1], Miklós Kellermayer[3],
Siegfried Labeit[2], and Karoly Trombitás[1]

[1]Department of Veterinary and Comparative Anatomy, Pharmacology
and Physiology, Washington State University, Pullman, WA
[2]Department of Anesthesiology and Intensive Operative Medicine,
University Hospital Mannheim, Mannheim, Germany and European
Molecular Biology Laboratory, Heidelberg, Germany
[3]Department of Biophysics, Pécs University Medical School,
Pécs , Hungary

Abstract: Titin is a giant filamentous polypeptide of multi-domain construction spanning between the Z- and M-lines of the sarcomere. As a result of differential splicing, length variants of titin are expressed in different skeletal and
cardiac muscles. Here we first briefly review some of our previous work
that has revealed that titin develops force in sarcomeres either stretched
beyond their slack length (passive force) or shortened to below the slack
length (restoring force) and that titin's force underlies a large fraction of
the diastolic force of cardiac muscle. Next we present our mechanical and
immunoelectron microscopical (IEM) studies of skeletal and cardiac
muscles that express titin isoforms. The previously deduced molecular properties of titin were used to model titin's extensible region in the sarcomere
as serially linked WLCs: rigid segments (containing folded Ig/Fn domains)
and more flexible segments (PEVK segment). The model was tested on
skeletal muscle fibers that express titin isoforms with tandem Ig and PEVK
length variants. The model adequately predicts titin's behavior along a
wide sarcomere length range in skeletal muscle, but at long sarcome lengths
(SLs), predicted forces are much higher than those determined experimentally. IEM reveals that this may result from Ig domain unfolding. Experi-

Elastic Filaments of the Cell, edited by Granzier and Pollack
Kluwer Academic/Plenum Publishers, 2000

ments were also performed on cardiac myocytes from mouse and cow that express predominantly a small cardiac titin isoform (N2B titin) or a large isoform (N2BA titin), respectively. The passive tension–SL relation of myocytes was found to increase more steeply with SL in mouse than in cow. IEM revealed an additional source of extensibility within both of these cardiac titins: the unique N2B sequence (absent in skeletal muscle). Furthermore, the PEVK segment of the N2BA isoform extended to a maximal length of ~ 200 nm, as opposed to ~60 nm for the N2B isoform. We propose that, along the physiological SL range, the long PEVK segment found in N2BA titins results in a low PEVK fractional extension and that this underlies the lower passive tensions of N2BA-expressing cow myocytes.

Introduction

When the striated-muscle sarcomere is stretched, a passive force is generated that tends to restore the sarcomere back to its original resting length. Many observations have indicated that in the generation of passive force a significant role is played by a unique protein called titin (also known as connectin). In addition titin also helps to maintain sarcomeric integrity during contraction, it has been implicated in myofibrillogenesis as a thick filament scaffold, and it may function as a cell-signaling molecule. (For reviews and original citations see Gregorio *et al.*, 1999; Labeit *et al.*, 1997; Maruyama, 1997; Trinick, 1996; Wang, 1996).

   Titin's force arises from its extensible I-band region, which consists of two main segment types: 1) a segment type rich in proline (P), glutamate (E), valine (V) and lysine (K) residues (the so-called PEVK segment) and 2) serially linked immunoglobulin-like domains (tandem Ig segments) flanking this PEVK segment (Labeit and Kolmerer, 1995). The extensible titin region of skeletal muscles also contains the N2A element (4 Ig domains and a 106- residue unique sequence). Skeletal muscles express isoforms of titin that contain PEVK and tandem Ig segments that differ in length (Labeit and Kolmerer, 1995). In this chapter we present structural and mechanical results on fibers of soleus and psoas muscles that express a large and small titin isoform, respectively.

   N2A titin transcripts have not only been found in skeletal muscles but also in cardiac muscles (Labeit and Kolmerer, 1995). On the other hand, a splice element known as the N2B element (3 Ig domains and a 572-residue unique sequence) is found exclusively in cardiac muscles. The N2B element is found together with a 163-residue PEVK segment and tandem Ig segments with 37 Ig domains (Labeit and Kolmerer, 1995). Recently, it was reported that a class of cardiac titin isoforms contains not only the N2B element but also the N2A element (hence the name used for these isoforms: N2BA titin), a ~600 residue PEVK and 12-25 additional Ig domains (depending on the splice pathway). Herein we present mechanical and ultrastructural studies on mouse/rat and bovine myocardium that express high levels of N2B and N2BA titin, respectively. The rat PEVK extends to a maximal value of ~60 nm and the bovine PEVK to ~200. The functional implications of these findings will be discussed.

Material and Methods

Passive tension measurements in cardiac myocytes and skeletal muscle fibers

Single fibers were dissected from rabbit (adult male New Zealand White rabbits weight ~3 kg) psoas and soleus skeletal muscle. Fibers were skinned with 1.0% Triton- X100 in relaxing solution, followed by mechanical skinning (for details see Granzier and Irving, 1995). Cardiac myocytes were isolated from rat left ventricle (LV) and bovine left atrium (LA). Atria and ventricles were rapidly cut in small strips that were added for 50 min to relaxing solution containing 1% Triton-X100 at 4°C (for solution compositions see Granzier and Irving, 1995). The preparations were then placed in relaxing solution and homogenized in a blender (Ultra-Turrax TP18 /S25N-10G, Ika Works Inc.) for 10 seconds at 12,500 rpm. This protocol, adapted from Hofmann *et al.*, (1991), resulted in a suspension of small clumps of myocytes, single myocytes and cell fragments. The preparations were skinned a second time by another 50-min treatment with 1% Triton-X100 in relaxing solution. Extensive rinsing with relaxing solution was used to remove the detergent. To limit titin degradation, all solutions contained protease inhibitors (leupeptin: 0.04 mmol/L and E64: 0.01 mmol/L).

The attachment protocol and experimental apparatus for passive tension measurements of cardiac myocytes and skeletal muscle fibers were as described earlier (Granzier and Irving, 1995). The preparations were stretched to a predetermined amplitude at a constant velocity (0.1 length/sec), immediately followed by the release at the same constant speed. To obtain reproducible results, a 10-min period of rest was used between stretch-release protocols. Force was normalized to the cross-sectional area, determined as explained in Granzier and Irving, (1995). Sarcomere length was measured on-line at a frequency of 60 Hz by using a discrete Fourier transformation of the striation image (Ion Wizard version 4.4, Ionoptix, Milton, MA). In addition to titin, intermediate filaments (IFs) also develop force in stretched myocytes (Granzier and Irving, 1995). To determine the force contributed by IFs, myocytes were incubated for 45-min with relaxing solution that contained 0.6 mol/L KCl solution followed by a 45 min incubation in relaxing solution with 1.0 mol/L KI. These solutions depolymerize the thick and thin filaments and thereby remove titin's anchors in the sarcomere. The F-SL length relation following extraction was assumed to represent the IF component in myocytes. The force decrease that resulted from extraction was assumed to be titin based. Force was converted to tension by dividing measured forces by the cross-sectional area of the myocyte.

Titin-based force was converted to force per titin molecule assuming that the fractional myofibrillar area is 85% for psoas and 70% for soleus fibers and that there are 3300 titin molecules per μm^2 myofibril (Granzier and Irving, 1995). To model passive force the extensible region of skeletal muscle titin was assumed to behave as two wormlike chains in series: the tandem Ig and PEVK segments. For poas and soleus fibers we assumed that the contour lengths of the tandem Ig segments were 390 nm and 480 nm and the contour length of the PEVK 350 and 820 nm, respectively. The persistence length of tandem Igs was assumed to be 15 nm and that of the PEVK was 1.5 nm. (For further details see Granzier *et al.*, 1997; Kellermayer *et al.*, 1997; Kellermayer *et al.*, 1998.)

SDS-PAGE based analysis of titin

Myocardial and skeletal muscle samples were quick-frozen in liquid nitrogen, pulverized to a fine powder and then rapidly solubilized and analyzed by SDS-PAGE using 2-7 % acrylamide gradient gels. (For details see Granzier and Irving, 1995; Granzier and Wang, 1993a.) Gels were stained with 0.1% Coomassie Blue G250 (Neuhoff *et al.*, 1988).

Immunoelectron microscopy (IEM)

In skeletal muscle, the antibodies T12 (anti-I12/I3) and N2A (anti-I72/I73) were used to mark the proximal tandem Ig segment, I20 (anti-I76/I77/I78) and MIR (anti-I101/I102/I103) to mark the distal tandem Ig segment (as an alternative to MIR, Ti102 was used in some experiments), and N2A and I20 to mark the PEVK. (For detailed information about these antibodies see Jin, 1995; Trombitás *et al.*, 1995; Linke *et al.*, 1996; Granzier *et al.*, 1996; Granzier *et al.*, 1997; Trombitás *et al.*, 1998a, 1998b; Trombitás *et al.*, 1999.) In cardiac muscle, the N2B heart-specific unique sequence was also studied using antibodies raised against Ig repeats N2B-I16/I17 (antibody referred to as Un) and N2B-I18/I19 (Uc). Thus, the Un and Uc epitopes demarcate the unique sequence contained within the N2B element at its N-terminal end and C-terminal end, respectively. For studying the N2B cardiac titin isoform we labeled rat cardiac myocytes with Un and Uc (unique sequence) and Uc and I20 (demarcates the PEVK segment), and for studying the N2BA cardiac titin isoform we labeled bovine cardiac myocyes with Un and Uc (demarcates unique N2B sequence) and N2A and I20 (demarcates the PEVK segment).

Skeletal muscle fibers and cardiac muscles were kept in relaxing solution and were stretched (at ~ 0.25 μm/sarcomere/s) to different lengths, held in position for several min (for technical reasons) and then fixed, immunolabeled, embedded, and processed for electron microscopy (EM) as explained in Granzier *et al.*, (1996). Briefly, the muscle samples were fixed in 3% formaldehyde/PBS solution for 20 min, then washed and blocked in 1% BSA/PBS solution for one hour. After dilution of antibodies to the appropriate concentrations (typically ~50 μg/ml), the samples were labeled with first the primary and then secondary antibodies for 24 hours each. The samples were fixed in glutaraldehyde and osmium-tetroxide solutions, and embedded in araldite. Sections were cut using a Leica microtome, the sections were stained with potassium permanganate and lead citrate. EM negatives were taken using a JEOL 1200 type electron microscope.

The Z-line to epitope distances were measured from EM negatives following high-resolution scanning (UMAX, UC-1260) and digital image processing using custom-written macros for the image analysis program NIH Image (v. 1.6, Wayne Rasband, National Institute of Health). For spatial calibration, the electron microscope's magnification was used.

Results

Titin's I-band region as a bidirectional molecular spring

It is now well established that the extensible region of titin is from near the T12 epitope (~100 nm from the middle of the Z-line) to near the edge of the A-band where the epitopes for MIR and Ti102 are located (Trombitás *et al.*, 1995, 1998b; Linke *et al.*, 1996). As indicated in Figure 1, the extensible region has a very short end-to-end length in slack sarcomeres (~1.9 μm) and greatly extends when sarcomeres are stretched. Titin degradation and extraction experiments (Granzier and Irving, 1995) have revealed that this extension of titin results in a passive force that underlies a large fraction of the passive force of single cardiac myocytes (Fig. 2). Furthermore, the titin-based force dominates the force developed in rat cardiac trabeculae up to a sarcomere length (SL) of ~ 2.2 μm, while at longer SLs, collagen becomes more dominant (Fig. 3).

In sarcomeres that shorten to below the slack length, the C-terminal end of the extensible region (Ti102 epitope) is closer to the Z-line than the N-terminal end (T12 epitope; Fig. 1). This is the result of a near Z-line segment of titin that is anchored to the thin filament (Trombitás and Granzier, 1997; Granzier *et al.*, 1997; Linke *et al.*, 1997) and that holds the N-terminal end of titin's extensible region a constant ~100 nm distance away from the Z-line (Fig. 1). Thus, in short sarcomeres the C-terminal end of the extensible region moves past its

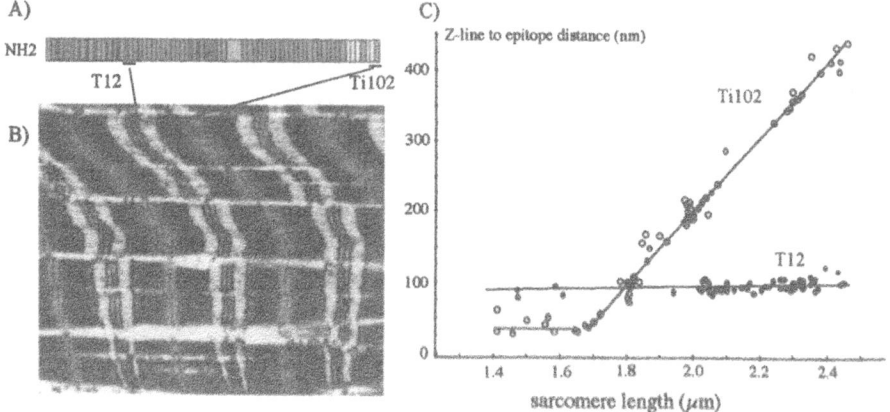

Figure 1. A) *Domain structure of the I-band region of N2B cardiac titin (based on Labeit and Kolmerer., 1995). Indicated are the binding sites of the anti-titin antibodies T12 and Ti102 which demarcate the N-terminal and C-terminal ends of the extensible region of titin, respectively (Trombitás et al., 1995).* **B)** *Rat cardiac myocyte labeled with T12 and Ti102 antibodies.* **C)** *Distance between epitopes and middle of Z-line vs. SL. The T12 epitope maintains a constant ~100 nm distance from the Z-line, while the Ti102 epitope is less than ~100 nm from the Z-line at SLs <1.8 and more than ~100 nm at SLs >1.8 μm.*

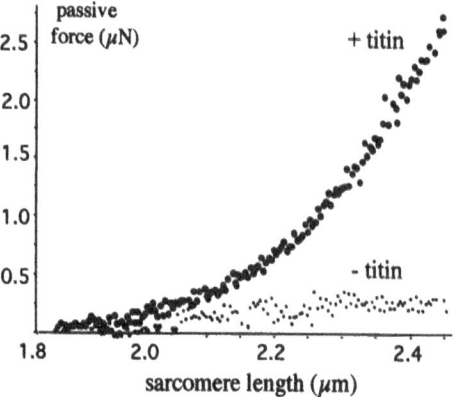

Figure 2. *Passive force–SL relation of single cardiac myocyte (rat) before and after titin degradation (for details see Granzier and Irving, 1995). Titin is the main source of passive force in the cardiac myocyte.*

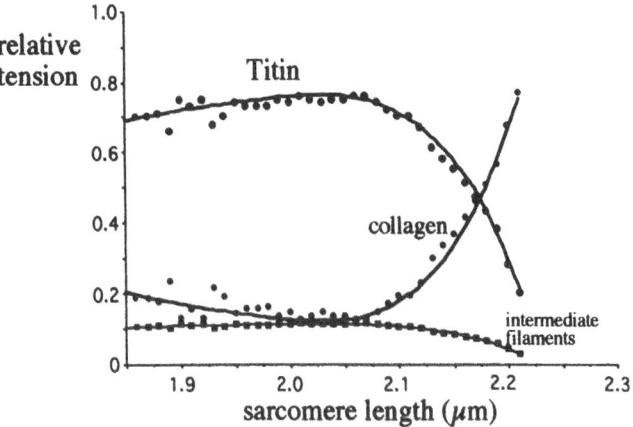

Figure 3. *Relative contribution of titin, collagen and intermediate filaments to the total tension develop by rat trabeculae (from Granzier and Irving, 1995). At SLs below 2.2 μm titin is the main source of passive tension.*

N-terminal end into the stiff titin region near the Z-line, extending titin in a direction opposite from that seen during sarcomere stretch. Our previous work on cardiac myocytes has revealed that titin extension in sarcomeres shortened to below the slack length results in a titin-based force (restoring force) that pushes the Z-lines away from the A-band, and that restores the slack length upon relaxation (Helmes *et al.*, 1996). Thus, the extensible region of titin is a molecular spring that can extend in two directions and that underlies both passive force in sarcomeres stretched above the slack length and restoring force in sarcomeres shortened below the slack length.

Structure-function relations of titin isoforms in skeletal muscle

It has been shown previously that different skeletal muscles express isoforms of titin (Hu *et al.*, 1986; Wang *et al.*, 1991; Granzier and Wang, 1993b; Granzier *et al.*, 1996). Using SDS-PAGE we found (Fig. 4) that the T1 band of psoas muscle (rabbit) has a greater mobility than T1 of soleus muscle (rabbit and human). These mobility differences are consistent with recent sequencing results (Labeit and Kolmerer, 1995; Freiburg *et al.*, submitted) that revealed much shorter proximal tandem Ig and PEVK segments in psoas than in soleus muscle.

To explore the passive properties of psoas and soleus titins, single muscle fibers were mechanically characterized. The fibers were stretched (velocity 100 nm / SL sec) to a predetermined SL, and their length was then held con-

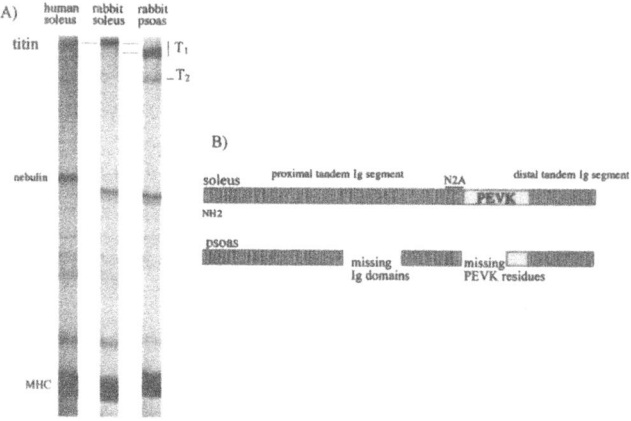

Figure 4. *A) SDS-PAGE of soleus and psoas muscles. T1 of psoas muscle has a greater mobility than T1 of soleus. B) Domain structure of the I-band segment of soleus and psoas titins (based on Labeit and Kolmerer, 1995 and Freiburg et al., submitted).*

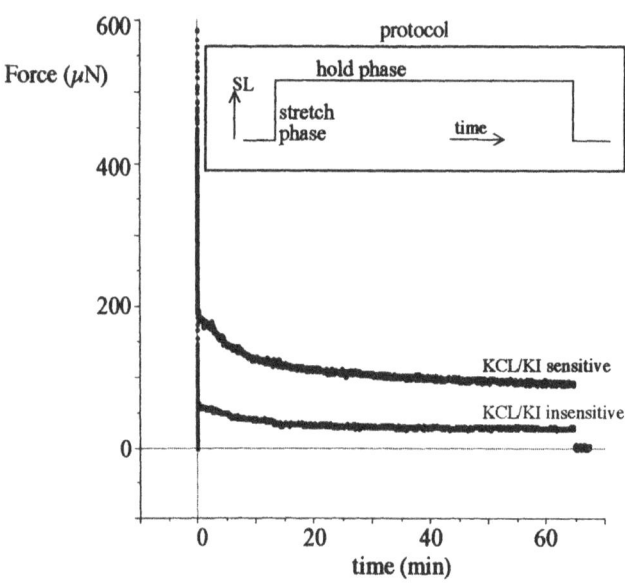

Figure 5. *Passive force response of a mechanically skinned psoas fiber when stretched from a SL of 1.95 μm to 4.05 μm (protocol explained in inset). The same protocol was imposed before and after extracting the fiber with 0.6 M KCl and 1 M KI (see Methods). To obtain the KCL/KI-sensitive force (assumed to be titin based), the KCL/KI-insensitive force (intermediate filaments) was subtracted from the total force before extraction.*

stant (hold phase) followed by a release back to the slack length (protocol explained in inset of Fig. 5). Passive force increased during stretch and decayed during the hold phase (a process known as force/stress relaxation). To determine the steady-state force, we first experimentally determined the minimal duration of the hold phase for force to reach a plateau value (defined as absence of a measurable force decrease during a 5-min period). This duration varied with SL and ranged from ~15 min (SL< 3.0 μm) to ~60 min (SL>4.0 min). Considering that at long SLs both titin and intermediate filaments (IFs) develop passive force (Granzier and Wang, 1993), we dissected their individual contributions to passive force. Fibers were incubated for 45 min each with 0.6 M KCl or 1 M KI added to the relaxing solution to remove thick and thin filaments, respectively (for details see Wang and Granzier, 1993b; Granzier and Irving, 1995). The fibers were characterized before and after extraction. Since extraction removes titin's anchors in the sarcomere, the extraction-sensitive force was assumed to be titin-based and the extraction insensitive force IF-based. An example of a result obtained with a psoas fiber that was stretched from a SL of 1.95 μm (slack) to a SL of 4.05 μm is shown in Figure 5. The results reveal that a ~60-min hold phase is required for force to attain a steady value. Furthermore, both titin (KCL/KI sensitive force) and IFs (KCL/KI insensitive force) contribute to passive force. In order to be able to compare results from different fibers, the measured titin-based force was divided by the fiber's cross-sectional area to obtain passive tension.

The titin-based peak tensions (tension at the end of the stretch) and steady-state tensions of psoas and soleus fibers are shown in Figure 6A. In both fiber types, peak tension is several-fold higher than the steady-state tension, thus stress relaxation is a prominent feature of both titin isoforms. Figure 6 also reveals that the peak and steady-state tensions increase much more steeply with SL in psoas than in soleus fibers. Steady-state tensions were converted to force per titin molecule (for calculations see Methods). This force also increased much more steeply in psoas fibers than in soleus fibers (Fig. 6B). We also compared the force per molecule with that predicted by the serially linked WLCs model (see Methods). This revealed that at forces less than ~10 pN per titin molecule, the model predicts values close to those found experimentally, while at higher forces the predicted values far exceed the measured values (Fig. 6B).

In order to understand the molecular basis of the mechanical differences between psoas and soleus titins, we used immunoelectron microscopy (IEM) to study the extensible behavior of tandem Ig and PEVK segments. Examples of labeled sarcomeres are shown in Figure 7A and the end-to-end lengths of the proximal tandem Ig and PEVK segment are shown in Figures 7B and C (soleus: solid lines; psoas: broken lines). At SLs less than 3.0 μm soleus and psoas results are similar. However at longer SLs the PEVK data deviate. The rate of

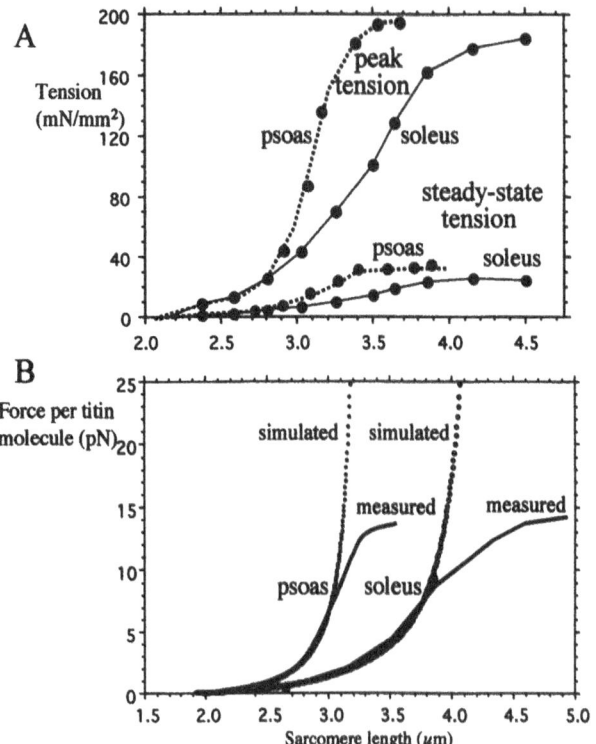

Figure 6. A) *Titin-based passive tension vs. SL of psoas and soleus fibers. Peak and steady-state tensions increase much more steeply with SL in psoas than in soleus fibers.* **B)** *Force per titin molecule vs. SL. Obtained values increase much more steeply with SL in psoas than in soleus fibers. The simulated force–SL relation was obtained by modeling titin's extensible region with the serially linked WLCs model (see Methods). At forces less than ~ 10 pN, the simulation predicts forces close to those measured while at higher forces the simulation far exceeds the measured values. (Steady-state tension of muscle fibers was converted to force per titin molecule, as explained in Methods.)*

PEVK extension is greatly reduced in psoas fibers at a SL of ~3.0 μm and the PEVK reaches a maximum length of ~ 350 nm at a SL of ~4.0 μm; in soleus fibers the rate of PEVK extension slows at ~3.5 μm and the PEVK reaches a maximum length of ~ 650 nm at a SL of ~ 4.0 μm (Fig. 7C). Extension of the proximal tandem Ig segment of soleus fibers revealed a plateau-like phase at SLs from ~3.0-3.5 μm while at longer SLs the segment extended again (Fig.

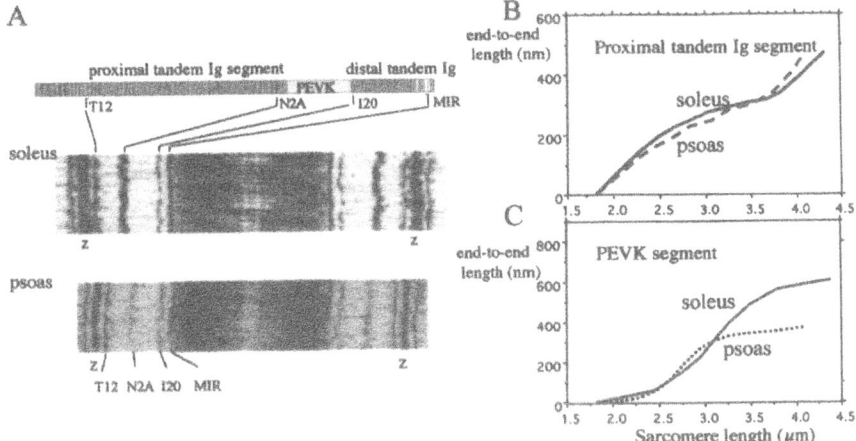

Figure 7. A) Examples of sarcomeres labeled simultaneously with the T12, N2A, I20 and MIR antibodies. Top indicates the binding sites of these antibodies in the I-band sequence of soleus titin. B) End-to-end length of proximal tandem Ig segment vs. SL. C) End-to-end length of PEVK segment vs. SL. See text for details.

7B). We speculate that this extension at long SLs results from Ig domain unfolding and that unfolding underlies the low measured forces relative to those predicted by the serially linked WLCs model (Fig. 6B; See Discussion below.)

Structure function relations of titin isoforms in cardiac muscle

It is now well established that cardiac muscle expresses a relatively small titin isoform (named N2B titin) that contains tandem Ig segments with only 37 Ig domains, a 163-residue PEVK segment, and the N2B element (4 Ig domains and a 572 residue unique sequence). In addition it was recently found that large mammals also express a larger titin isoform that contains both the N2B and the N2A element (N2BA titin) as well as a longer PEVK segment and additional Ig domains (Freiburg *et al.*, submitted; Cazorla *et al.*, 2000). Here we show that mouse ventricular cardiac myocytes and bovine atrial myocytes express predominately the N2B isoform and the N2BA isoform, respectively (see inset of Fig. 8).

The effect of expressing different titin isoforms on the passive tension–SL relation of single cardiac myocytes was studied. It was found that mouse myocytes have a much steeper passive tension–SL relation than bovine myocytes (Fig. 8). To explore the molecular basis for these differences we immunolabeled cardiac myocytes and measured the extensible properties of

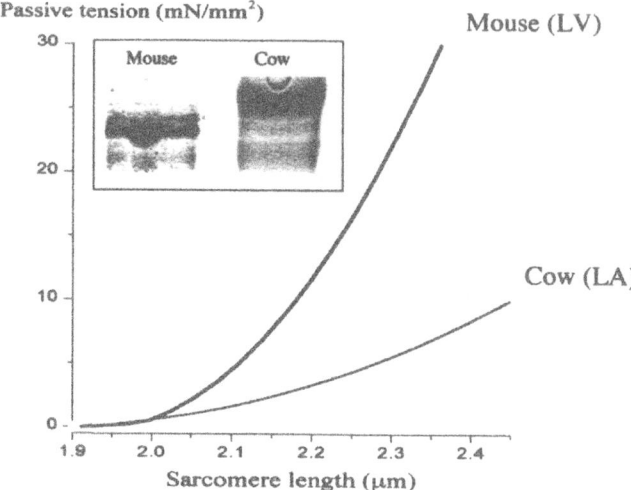

Figure 8. Passive tension–SL relation of single cardiac myocytes isolated from mouse left ventricle (LV) or bovine left atrium (LA). Tension increases much steeper in mouse than in cow. The inset shows a SDS-PAGE result of mouse LV and bovine LA. T1 of mouse migrates much further than T1 of cow, consistent with our previous finding of high expression levels of N2B cardiac titin in small rodents and high levels of N2BA titin in large mammals (Cazorla et al., 2000).

the PEVK segment and the N2B unique sequence (an example of a labeled sarcomere is shown in the inset of Fig. 9). Results show that the PEVK segment of rat cardiac myocytes (which express predominantly N2B titin; Helmes *et al.*, 1999) reaches a maximal length of ~60 nm at a SL of 2.4 μm, while in the cow, the PEVK segment continues to extend over the full SL range studied and reaches a length of ~200 nm at 3.2 μm SL (Fig. 9). Recently the unique sequence contained within the N2B element was identified as an extensible segment within N2B titin (Helmes *et al.*, 1999; Linke *et al.*, 1999; Trombitás *et al.*, 1999). In this study we found that the unique sequence of both rat and bovine cardiac titins extended, but that the length of the unique sequence increased more steeply with SL in rat than in bovine titin (Fig. 9).

Discussion

The I-band region of titin functions as a bidirectional molecular spring that develops passive force when sarcomeres are stretched beyond their slack length as well as restoring force when sarcomeres shorten to below this length. Due to

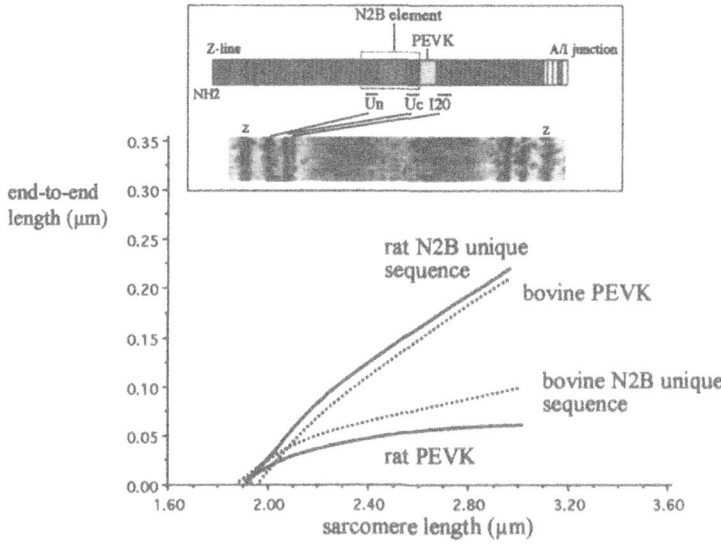

Figure 9. *End-to-end length of PEVK segment and unique N2B sequence vs. SL. The inset shows a rat sarcomere labeled simultaneously with Un, Uc and I20 antibodies. The distance between Un and Uc was taken as the end-to-end length of the unique sequence and the distance between Uc and I20 as the PEVK segment length. (The length of the bovine PEVK was determined from the distance between N2A and I20; Trombitás et al. unpublished.) Note that the bovine PEVK extends much further than the rat PEVK, but that the extension of the rat unique sequence greatly exceeds that of the bovine unique sequence.*

differential splicing the molecular spring segment of titin isoforms found in different muscles varies in length. We investigated isoforms of titin that are expressed in skeletal muscles (psoas and soleus muscles from rabbit) and myocardum (mouse, rat and bovine), using mechanics and immunoelectron microscopy (IEM) with antibodies that demarcate the tandem Igs, the PEVK and the unique N2B sequence. Results reveal that the passive properties of fibers/cells that express distinct titin isoforms vary greatly. IEM revealed the extension of the various subsegments of titin's extensible region and this allows us to evaluate the mechanisms that may underlie the observed passive tension variation.

Passive tension increases more steeply with SL in psoas than in soleus fibers. This finding is likely to be related to the different lengths of the tandem Ig and PEVK segments expressed in psoas and soleus muscles. Differences between titin's extensible properties in poas and soleus fibers are most obvious for the PEVK segment (Fig. 7C). In both psoas and soleus fibers the PEVK lengths approach a maximal value, but this value is much less in psoas than in soleus (~350 vs. ~650 nm). These maximal length differences are consistent with sequence studies (Labeit and Kolmerer, 1995; Freiburg *et al.*, submitted) that revealed a much shorter PEVK segment in psoas than in soleus muscle. The shorter maximal PEVK length of psoas fibers results in a higher fractional extension (end-to-end length divided by the maximal length) in sarcomeres shorter than 3.0 μm (where the PEVK end-to-end lengths are similar in psoas and soleus; Fig. 7C). A similar analysis can be applied to the proximal tandem Ig segment. The maximal length of this segment is also much shorter in psoas than in soleus fibers (Labeit and Kolmerer, 1995; Freiburg *et al.*, submitted) and our finding that tandem Ig extensions are similar in the two fiber types (Fig. 7B) results in a higher fractional extension of the proximal tandem Ig segment in psoas than in soleus fibers. Considering the entropic spring properties of tandem Ig and PEVK segments (Kellermayer *et al.*, 1997; Trombitás *et al.*, 1998a, 1998b; Linke *et al.*, 1998a, 1998b) higher fractional extensions are likely to result in higher forces. Thus, the much higher passive tensions measured in psoas fibers (Fig. 6) may be explained by the shorter PEVK and tandem Ig segments and the ensuing higher fractional extension of these segments when sarcomeres are stretched.

Extension of the proximal tandem Ig segment of rabbit soleus fibers (Fig. 7B) revealed a plateau-like phase at SLs from ~3.0-3.5 μm, similar to what we have reported earlier for human soleus fibers ((Trombitás *et al.*, 1998a; Trombitás *et al.*, 1998b)). However, unlike our previous results, we found here for the rabbit soleus a second phase of tandem Ig extension that starts at a SL of ~ 3.75 μm (Fig. 7B). This elongation may be explained by Ig domain unfolding. Ig domain unfolding is consistent with the low measured forces at long SLs, relative to those predicted by the serially linked WLCs model (Fig. 6B). Considering the reduced number of Ig domains that is contained within the proximal tandem Ig segment of psoas fibers (Freiburg *et al.*, submitted), and the end-to-end length of the proximal tandem Ig of psoas fibers that is similar to that of soleus (Fig. 7B), it seems likely that unfolding also takes place in psoas fibers when stretched to long SLs.

Currently it remains to be explained why our previous experiments with human soleus fibers failed to provided evidence for Ig domain unfolding along the full SL range that was studied (2.0 - 4.2 μm), while Ig unfolding may take place in rabbit soleus at SLs as short as 3.75 μm. Perhaps species differences in Ig domain sequences result in less stable domains in rabbit. Alternatively, differences in I-band composition (isoforms of regulatory proteins? nebulin?) may stabilize Ig domains in human soleus fibers but not in rabbit soleus fibers. To resolve this issue, further research is required. Research into the physiological

SL ranges of rabbit muscles is necessary in order to establish whether Ig unfolding may take place in psoas and soleus muscle under physiological SL conditions. Considering the relatively long SLs at which unfolding takes place in rabbit (>3.0 μm for psoas and >3.5 μm for soleus) we speculate that unfolding is unlikely to take place under physiological conditions, and that one of the functions of length variation of titin's extensible region is to accommodate varying physiological SL ranges in different muscles without requiring Ig unfolding.

Based on our findings we propose a model of passive force development in skeletal muscle in which stretching of slack sarcomeres results in (1) a phase in which tandem Ig extension dominates, (2) a subsequent phase in which PEVK extension dominates, and (3) a phase in which Ig domain unfolding reduces the steepness of the force rise (see Fig. 10). In this model, force differences be-

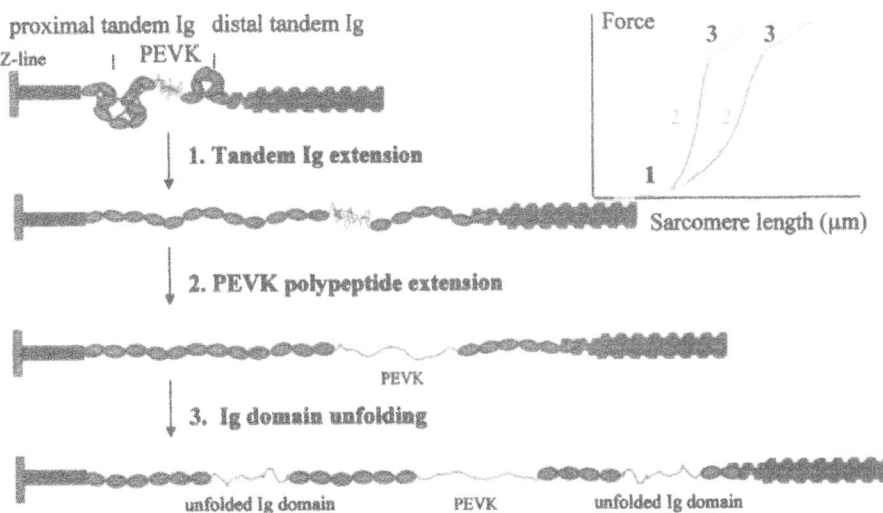

Figure 10. *Model of passive force development in skeletal muscle fibers. Top left schematically displays titin's I-band region in slack sarcomeres. When sarcomeres are stretched tandem Ig segment extension initially dominates, followed by PEVK segment extension, and finally by Ig domain unfolding. The inset schematically depicts the force-SL curve of two titin isoforms. The left curve is from an isoform with shorter tandem Ig and PEVK segments than the isoform represented by the right curve (as in psoas vs. soleus titins). The left curve therefore has a more limited SL range along which tandem Ig and PEVK extension takes place. Ig domain unfolding (3) of the two isoforms is assumed to take place at similar force levels. (Note that nonspecific binding takes place between different parts of the tandem and PEVK segments, indicated by thin black lines in the top left schematic. These nonspecific bonds are assumed to break upon stretch and to give rise to stress relaxation.)*

tween fibers that express titin isoforms can be explained by variation of the contour lengths of tandem Ig and PEVK segments that give rise to variation in the segments' fractional extensions and, thus, variation of force.

Passive properties of cardiac myocytes isolated from myocardium that expresses high levels of N2B titin (mouse/rat) are very different from those that express high levels of N2BA titin (Fig. 8). We explain these findings as follows. IEM reveals that the PEVK segment of N2BA titin is much longer than that of N2B titin (Fig. 9), a finding that is consistent with sequence studies (Freiburg *et al.*, submitted). Due to the long PEVK segment, a given SL will result in a lower fractional PEVK extension in N2BA titin than in N2B titin. The additional extensibility derived from the longer N2BA PEVK segment is also expected to result in a lower fractional extension of the unique N2B sequence, explaining the much shorter end-to-end length of the unique sequence of the bovine (Fig. 9). These lower fractional extensions of bovine's N2BA titin are likely to result in lower entropic forces and, thus, are likely to underlie the much more shallow passive tension–SL relation of bovine myocytes, relative to that of mouse myocytes.

In summary, differential splicing gives rise to length variation of titin's extensible region. This length variation results in variation of the fractional extension of tandem Igs, PEVK segments and the unique N2B sequence (cardiac muscle), and this provides skeletal and cardiac muscles with a means to modulate their passive mechanical properties.

Acknowledgments

We thank Danielle Higgins for technical support. This work was supported by the Deutsche Forschungsgemeinschaft La 668/5-1 (to S. Labeit), a postdoctoral fellowship awarded to OC from the American Heart Association (Washington State Affiliate, #98-WA-115) and the National Institute of Health National Heart, Lung, and Blood Institute (HL61497 and HL62881) to HG. HG is an Established Investigator of the American Heart Association.

References

Cazorla O, Freiburg A, Helmes M, Centner T, McNabb M, Trombitás K, Labeit S, Granzier H. Differential expression of cardiac titin isoforms and modulation of cellular stiffness. *Circ Res* 2000;26:59-67.

Granzier H, Helmes M, Trombitás K. Nonuniform elasticity of titin in cardiac myocytes: a study using immunoelectron microscopy and cellular mechanics. *Biophys J* 1996;70(1):430-442.

Granzier H, Kellermayer M, Helmes M, Trombitás K. Titin elasticity and mechanism of passive force development in rat cardiac myocytes probed by thin-filament extraction. *Biophys J* 1997;73(4):2043-2053.

Granzier HLM, Wang KW. Interplay between passive pension and strong and weak binding cross-bridges in insect indirect flight muscle. *J Gen Physiol* 1993;101:235-270.

Granzier HL, Irving TC. Passive tension in cardiac muscle: contribution of collagen, titin, microtubules, and intermediate filaments. *Biophys J* 1995;68(3):1027-1044.

Granzier HL, Wang K. Gel electrophoresis of giant proteins: solubilization and silver- staining of titin and nebulin from single muscle fiber segments. *Electrophoresis* 1993a;14(1-2):56-64.

Granzier HL, Wang K. Passive tension and stiffness of vertebrate skeletal and insect flight muscles: the contribution of weak cross-bridges and elastic filaments. *Biophys J* 1993b;65(5):2141-59.

Gregorio C, Granzier H, Sorimachi H, Labeit S. Muscle assembly: a titanic achievement? *Current Opinion in Cell Biology* 1999;11(11):18-25.

Helmes M, Trombitás K, Centner T, Kellermayer M, Labeit S, Linke A, Granzier H. Mechanically driven contour-length adjustment in rat cardiac titin's unique N2B sequence: titin is an adjustable spring. *Cir Res* 1999;84:1139-1352.

Helmes M, Trombitás K, Granzier H. Titin develops restoring force in rat cardiac myocytes. *Circ Res* 1996;79(3):619-626.

Hofmann PA, Hartzell HC, Moss RL. Alterations in Ca2+ sensitive tension due to partial extraction of C-protein from rat skinned cardiac myocytes and rabbit skeletal muscle fibers. *J Gen Physiol* 1991;97:1141-1163.

Hu DH, Kimura SK, Maruyama M. Sodium dodecyl sulfate gel electrophoresis studies of connectin-like high molecular weight proteins of various types of vertebrate and invertebrate muscles. *J Biochem* 1986;99:1485-1492.

Jin J-P. Cloned rat cardiac titin class I and class II motifs. *J Biol Chem* 1995;270(12):6908-6916.

Kellermayer MS, Smith SB, Granzier HL, Bustamante C. (1997). Folding-unfolding transitions in single titin molecules characterized with laser tweezers [see comments] [published erratum appears in *Science* 1997 Aug 22;277(5329):1117]. *Science* 1997;276(5315):1112-6.

Kellermayer MSZ, Smith SB, Bustamante C, Granzier HL. Complete unfolding of the titin molecule under external force. *J Struct Biol* 1998;122:197-205.

Labeit S, Kolmerer B. Titins: giant proteins in charge of muscle ultrastructure and elasticity. *Science* 1995;270(5234):293-296.

Labeit S, Kolmerer B, Linke WA. The giant protein titin. Emerging roles in physiology and pathophysiology. *Circ Res* 1997;80(2):290-294.

Labeit S, Kolmerer B. The complete primary structure of human nebulin and its correlation to muscle structure. *J Mol Biol* 1995;248(2):308-15.

Linke WA, Ivemeyer M, Olivieri N, Kolmerer B, Ruegg JC, Labeit S. Towards a molecular understanding of the elasticity of titin. *J Mol Biol* 1996;261(1):62-71.

Linke WA, Ivemeyer M, Labeit S, Hinssen H, Ruegg JC, Gautel M. Actin-titin interaction in cardiac myofibrils: probing a physiological role. *Biophys J* 1997;73(2):905-19.

Linke WA, Ivemeyer M, Mundel P, Stockmeier MR, Kolmerer B. Nature of PEVK-titin elasticity in skeletal muscle. *Proc Natl Acad Sci USA* 1998a;95(14):8052-8057.

Linke WA, Stockmeier MR, Ivemeyer M, Hosser H, Mundel P. Characterizing titin's I-band IG domain region as an entropic spring. *J Cell Science* 1998b;111(11):1567-1574.

Linke W, Rudy D, Centner T, Gautel M, Witt C, Labeit S, Gregorio C. I-band titin in cardiac muscle is a three-element molecular spring and is critical for maintaining thin filament structure. *J Cell Biol* 1999;146:631-644.

Maruyama K. Connectin/titin, giant elastic protein of muscle. *FASEB J* 1997;11(5):341-345.

Neuhoff V, Arold N, Taube D, Ehrhardt W. Improved staining of proteins in polyacrylamide gels including isoelectric focusing gels with clear background at nanogram sensitivity using Coomassie Brilliant Blue G-250 and R-250. *Electrophoresis* 1988;9:255-262.

Trinick J. Titin as a scaffold and spring. Cytoskeleton. *Curr Biol* 1996;6(3):258-260.

Trombitás K, Granzier H. Actin removal from cardiac myocytes shows that near Z line titin attaches to actin while under tension. *Am J Physiol* 1997;273(2 Pt 1):C662-70.

Trombitás K, Greaser M, French G, Granzier H. PEVK extension of human soleus muscle titin revealed by immunolabeling with the anti-titin antibody 9D10. *J Struct Biol* 1998a;122:188-196.

Trombitás K, Greaser M, Labeit S, Jin JP, Kellermayer M, Helmes M, Granzier H. Titin extensibility in situ: entropic elasticity of permanently folded and permanently unfolded molecular segments. *J Cell Biol* 1998b;140(4):853-859.

Trombitás K, Jin JP, Granzier H. The mechanically active domain of titin in cardiac muscle. *Circ Res* 1995;77(4):856-61.

Trombitás K, Freiburg A, Centner T, Labeit S, Granzier H. Molecular dissection of N2B cardiac titin's extensibility. *Biophysical J.* 1999;77:3189-3196.

Wang K. Titin/connectin and nebulin: giant protein rulers of muscle structure and function. *Adv Biophys J* 1996;33:123-134.

Wang K, McCarter R, Wright J, Beverly J, Ramirez-Mitchell R. Regulation of skeletal muscle stiffness and elasticity by titin isoforms: a test of the segmental extension model of resting tension. *Proc Natl Acad Sci USA* 1991;88(16):7101-5.

Discussion

TerKeurs: That was a beautiful presentation, Henk. I have a question about hysteresis. Did you study the hysteresis while the cells or muscles were stimulated?

Granzier: No, we have never done that.

TerKeurs: So was that in the skinned fibers or in skinned cells?

Granzier: Everything I presented was in skinned fibers and skinned myocytes. The obtained mechanical properties could indeed be modulated during contraction.

TerKeurs: Then the basis of my question is: would variation in calcium modify hysteresis as you have measured it? You could test that, of course, by working with intact muscle or cells. If I compare data that we have from intact muscle while twitching, then it is interesting to see that the hysteresis takes approximately 5-6 beats to settle down instead of persisting for a very long time, and it recurs quickly, also.

Granzier: It may depend on the sarcomere length to which you stretch. I think that many things can modulate these passive force-length curves that I showed. Activation is one of them.

TerKeurs: May I ask one more question? If you take myocardium of the pig and compare the epicardium and endocardium, or from the free wall and the septum, do you find differences in the distribution of the heavy titin verses the lighter titin?

Granzier: You are asking about different regions across the wall? Yes, we do. Epicardium has more of the large isoform.

Linke: Beautiful presentation, Henk. The first question is a very brief one, because I wonder, if the N2BA isoform is so large, will it ever be important when you stretch the muscle?

Granzier: I don't know. The experiments that I showed, where we dissected the contribution of various elements to passive forces of rat myocardium have to be preformed on myocardium of different species. Then we also need to know more about the physiological sarcomere lengths range in these species. These are very important questions.

Linke: I would like to come back to the skeletal muscle because you say the stress relaxation may be due to the breakage of bonds at low stretch. But when we use, for example, myofibrils, single myofibrils, and stretch them out, stress relaxation becomes increasingly higher. Like when you stretch from 2.2 to 2.8 or 3 microns sarcomere lengths, where Ig domain unfolding is still not prominent, or still very unlikely, but nevertheless stress relaxation becomes more important. And at that point you will have already broken your proposed bonds. So how do you explain that?

Granzier: At those lengths the mechanical properties are mainly determined by the PEVK, which still has non-covalent bonds. The only way to get rid of stress relaxation is to reversibly stretch release the fibers, as is the case *in vivo*. So I think that *in vivo* situations, stress relaxation may not be that prominent. The condition that we typically impose in our experiments, starting out with a very well rested cell at slack length, gives opportunity for bond formation.

Linke: Yes, it is hard to predict these bond formations. In our hands, we have stretched out the (rat psoas) PEVK to 50-60% at 2.9 microns sarcomere length. Bond formation is maybe less likely at that stage.

Granzier: Actually you could see in our data that stress relaxation becomes less prominent at those long lengths.

Linke: I just wanted to add that maybe yesterday there was some confusion in the audience about what's going on now. Is Ig domain unfolding important for titin extensibility or not? But what our results now clearly show is that wormlike chain behavior is likely in skeletal muscle at low stretch. Only when it comes to high stretch, like in our experiments to about 3 microns sarcomere length, then it all deviates apparently. And it was nice to see that also in your psoas titin data there is a deviation at 3 microns sarcomere length from the predicted curve, which fits exactly with our prediction, except that your explanation is a little bit different. But who knows whether the muscle ever reaches that length *in vivo*.

Granzier: I don't know.

Labeit: Beautiful presentation, Henk. Just one comment, not a question. I think the variation of isoforms you see in the mammalian heart also explains why in our 1995 science paper we have some significant corrections to make in the predicted isoform patterns. Because in the experiments we did at that time we used N2A skeletal probes, human, and hybridized them to slot blotted rabbit RNA. And since we know now from your beautiful data that the larger isoform is hardly present in rabbit, it is obvious that you miss some Ig domains that are expressed.

Trinick: Could you clarify what you think is going on with the proximal Ig segment? Do you think, if I understand you correctly then, the bit close to the Z-line would be attached all the time? But when you get further away from the Z-line, that would be transiently attached to the same filaments?

Granzier: Well, close to the Z-line the situation is pretty clear. Charles showed convincingly that there is a near Z-line segment that is functionally stiff, probably as a result of interaction with the thin filament. If you remove the thin filament, that segment collapses and the slack sarcomere is now shorter. Away from the Z-line the proximal tandem Ig segment, I think, in steady-state follows wormlike chain behavior, as the segment length and also the force measurements suggest. This behavior may be modulated transiently by perhaps a weak interaction with the thin filament that you only see under dynamic conditions

Trinick: OK, sorry to ask you a supplementary. I think the evidence that the titin molecules are self-associating in the distal Ig segment to form end filaments is not bad. So do you think that your modeling would detect that, because the segment length in the chain models would be quite different then; it would be a much stiffer structure.

Granzier: Yes, we spoke about that earlier. I was planning to point that out on my second slide, it would take me too long to scroll back and pull it up, but it actually shows that the distal terminal Ig segment collapses in short sarcomeres. That is why I was struck by the end filaments always being so straight and stiff. I don't really understand that.

Trinick: Your work uses antibodies.

Granzier: Yes, we label the ends. And we know that in short sarcomeres the ends are close together. But I would expect the ends to be apart from each other by at least 100 nm if the distal tandem Ig segment is stiff, as suggested by your images of the end filaments. I don't know what the explanation is.

Trinick: Well, I don't think you can rule out that antibody perturbs the structure.

Granzier: As Charles pointed out earlier, the tissue was first fixed and then labeled. Maybe that does not fully rule out that the antibody disturbs the structure, but it makes it less likely.

Andrew: Can you measure the mechanical properties of sarcomeres from much older animals, such as the heart in older animals?

Granzier: That is very interesting. Preliminary data suggest that there are changes in isoform expression.

Linke: I would like to comment on what John said. If you think in terms of wormlike chain modeling and if you look at the distal Ig domain region, the contour length is relatively short compared to the predicted persistence length. So you have a ratio which would predict that you probably get a semi-flexible chain. Then, if the end filaments stick out freely, they may not entirely collapse. However, they may be pushed to collapse in epitope-stained short sarcomeres because of additional constraints imposed by the filament arrangement. That is one possible explanation. So in your experiments the end filament is intrinsically semi-flexible but it is basically pushed towards the A-band.

TerKeurs: In one of your first slides you compared the contribution of elastic forces of titin to collagen and other elements. Did you also do that for the magnitude of the opposing forces?

Granzier: Those experiments are very difficult.

TerKeurs: We measured compressive forces of cells that have been isolated with collagenase versus compressive forces of trabeculae. The difference between the two is about a factor of 3; that is, the opposing force in the collagen containing trabeculae 3 times larger than that of the isolated collagenase-treated cells.

Granzier: We have done measurements at the myocyte level and compared them with your measurements. I don't remember exactly the fraction of the total that we ascribe to titin, maybe about half. I think our value is a little higher than yours, but what we have to do is go back and repeat the experiments for different species and see if the opposing force, let's say in a pig, is different than that in the rat.

INTACT CONNECTING FILAMENTS CHANGE LENGTH IN 2.3-nm QUANTA

Felix Blyakhman*, Anna Tourovskaya, and Gerald H. Pollack

Dept. of Bioengineering, University of Washington, Seattle, WA
*Ural State University, Ekaterinburg, Russia

Abstract: In isolated titin molecules, length changes may occur in discrete steps (Tskhovrebova *et al.*, 1997; Rief *et al.*, 1997). The extent to which such steps are preserved in the intact muscle-filament lattice has remained unclear. We carried out experiments on single isolated insect-flight-muscle myofibrils in which thin filaments had been functionally removed either by stretch beyond overlap or by a "rigor-stretch" protocol, leaving the connecting (titin) filaments as the sole length-absorbing agent. The myofibril was released or stretched by a motor in ramp-like fashion. The time course of length change in the single sarcomere was stepwise. The same was true for half-sarcomere lengths. The presence of steps at the sarcomere level implies that parallel filaments step synchronously, with high cooperativity. Step sizes showed a consistent distribution: The smallest size was ~2.3 nm, and others were integer multiples of that value. Similar results were found for stretch and release. To our knowledge, the ~2.3-nm step quantum is the smallest consistent biomechanical event ever demonstrated. This quantum is an order of magnitude smaller than anticipated from the folding/unfolding of a complete Ig- or fibronectin-like domain, and may imply that folding occurs in sub-domain increments. The 2.3-nm incremental length change corresponds to a single turn of the domains' beta sheet.

Elastic Filaments of the Cell, edited by Granzier and Pollack
Kluwer Academic/Plenum Publishers, 2000

Introduction

Although length changes in the connecting filament were initially presumed to occur in smooth, linear fashion, a hint of discrete change was initially revealed by morphological studies. These studies demonstrated the presence of discrete knob-like elements running along the length of the filaments (Trinick *et al.*, 1984; Maruyama *et al.*, 1984; Wang and Greaser, 1985).

The implied discreteness of length change was confirmed in two important studies published simultaneously. In highly loaded single titin molecules and titin constructs consisting of a string of Ig-like domains, Rief *et al.* (1997) found that linear stretch ramps produced sawtooth-tension waveforms. This was interpreted to imply discrete Ig-domain unfoldings. Similarly in single isolated titin molecules, Tskhovrebova *et al.* (1997) induced rapid stretch and observed that the subsequent tension decay was stepwise. In the former study step size was 25 – 28 nm, and in the latter, most steps were in the range 5 – 20 nm. The source of the difference is unclear; nor is it clear whether steps smaller than 5 nm might have been observable with higher resolution. Nevertheless, at least under the conditions studied, titin has the capacity to manifest stepwise length changes.

The intact sarcomere is different from the isolated titin molecule in that many connecting filaments lie in parallel. The cross-section of a typical myofibril may contain several hundred filaments. Even if such filaments changed length in steps, the steps would show up in the half-sarcomere only if all parallel filaments stepped synchronously. A hint of this unexpected phenomenon appeared in earlier studies of relaxed single frog-muscle fibers (Granzier and Pollack, 1985). When whole muscle fibers were stretched or released in ramp-like fashion, sarcomere-length changes occurred in steps. These "passive" steps were observed by three independent methods.

In the current study, we developed a newer method to study dynamics of considerably smaller preparations. This method has allowed us to investigate single sarcomere dynamics in single isolated myofibrils. The results confirm earlier measurements and show with new precision that length changes occur in steps that are integer multiples of a ~2.3-nm quantum.

Methods

Single myofibrils of honeybee-flight muscle were isolated as described earlier and mounted in an apparatus built around a Zeiss Axiovert-35 microscope (Yang *et al.*, 1998). Specimens were either stretched to sarcomere lengths with little or no overlap, or subjected to the "rigor-stretch" protocol, which dislodges thin filaments from their anchor points on the Z-line, leaving the connecting filaments as the sole force-bearing element in the I-band (White and Thorson,

1973; Trombitás and Tigyi-Sebes, 1977). One end of the myofibril was attached to the tip of a fixed glass needle, the other to the glass tip of a piezoelectric motor, which could impose desired length changes on the specimen. The phase-contrast image of the striations was projected onto the face of a 1024-element photodiode array, which was scanned every 50 ms. to produce a trace of intensity vs. position along the myofibril.

Single sarcomere length was computed as the span between medians of contiguous A-band intensity peaks. Successive computations gave the time course of sarcomere length. In earlier experiments medians were computed independently for each scan (Bartoo *et al.*, 1993). That algorithm gave satisfactory traces of sarcomere length vs. time, with rms noise typically on the order of 3 - 5 nm (Yang *et al.*, 1998; Blyakhman *et al.*, in press). Recently, a fourfold increase of signal-to-noise was achieved by using a differential approach that compared each scan to the one previous, based on minimum average risk (Sokolov *et al.*, submitted). With this approach, rms noise could be reduced to ~1 nm or less. Preliminary results obtained with this newer approach are included here.

Results

A representative specimen is shown in Figure 1. Generally, honeybee myofibrils were used because of their relatively large (~2 μm) diameter and clear striation pattern.

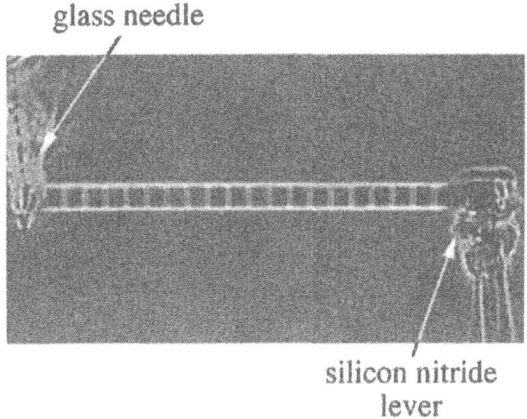

Figure 1. Phase-contrast image of honeybee myofibril mounted in the experimental apparatus. The specimen is attached between a silicon-nitride lever and the glass needle of a piezoelectric motor, which can impose a length change.

Figure 2 shows successive scans along a myofibril. Large upward deflections correspond to A-bands, while small upward peaks correspond to Z-lines. Sarcomere length was computed as the span between successive A-band medians.

Representative traces of sarcomere-length vs. time are shown in Figure 3. Shortening periods are punctuated by pauses, during which time shortening ceases. The pauses and steps are evident in the half-sarcomere as well. In some records the stepwise pattern was evident over the entirety of the shortening trace while in others, such features were fractionally obscured by noise.

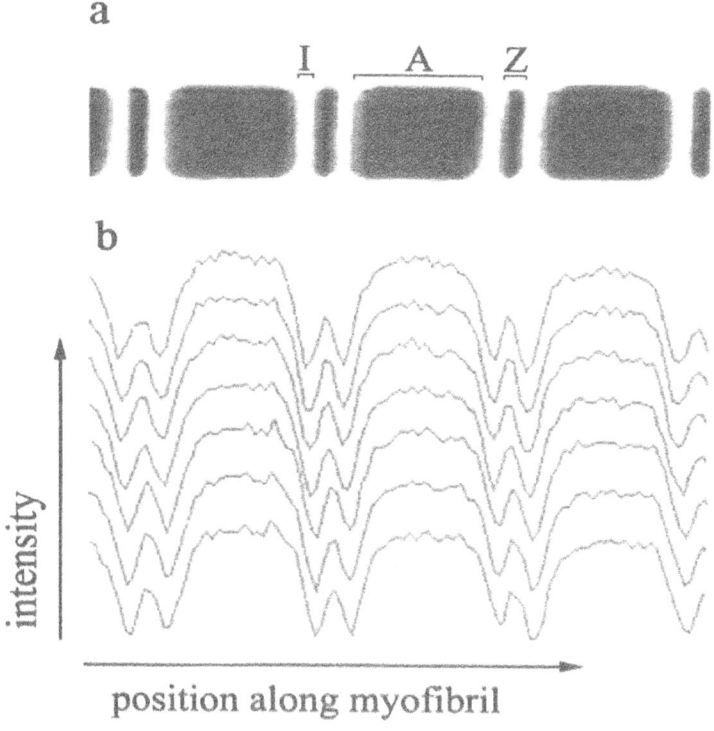

Figure 2. Sarcomere-length analysis. The myofibril image is shown in a. Successive scans made at 50 ms. time intervals during imposed length change are shown in b.

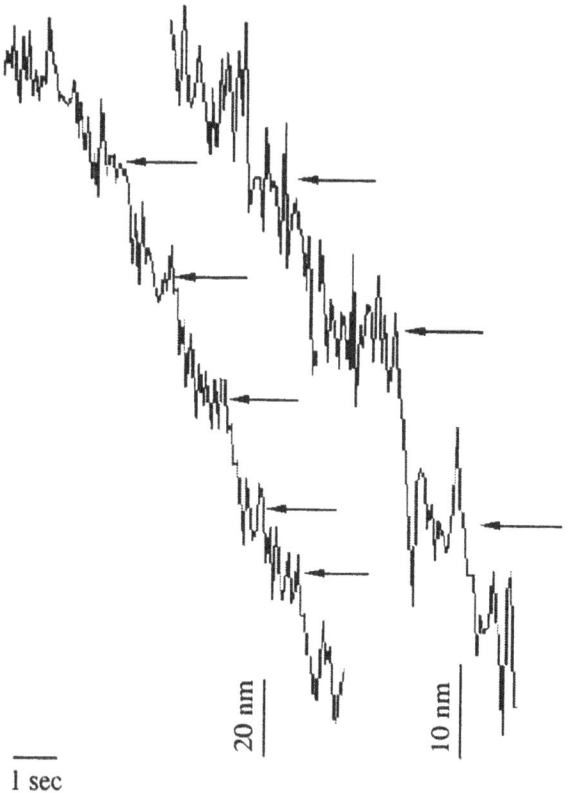

20 nm

10 nm

1 sec

Figure 3. *Representative sarcomere-shortening traces. Arrows denote pauses. Left trace obtained by computing the span between medians of successive A-bands. Right trace corresponds to half-sarcomere, obtained by computing the span between A-band and Z-line.*

Figure 4 shows representative traces obtained with the new differential method (Sokolov *et al.*, submitted). Because of the higher signal-to-noise ratio, pauses otherwise obscured by noise were more clearly delineated. Note scale difference relative to Figure 3. Steps on the order of 2 – 3 nm showed up consistently.

The steps did not arise out of jerkiness of the motor. The attachment point between motor and myofibril was imaged on the photodiode array and its median position was computed using the same algorithm used regularly to compute A-band median. Figure 5 shows that apart from noise, translational movement during the imposed length change was smooth and contained no evidence of steps or jerkiness.

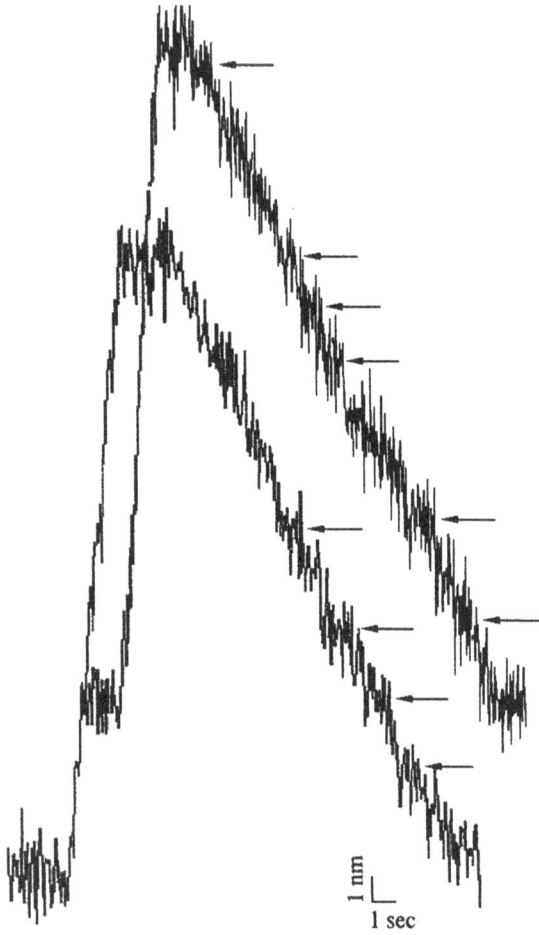

Figure 4. *Representative traces of sarcomere length change obtained using the newer differential method. Note difference of scale compared to Figure 3. Arrows denote pauses.*

To check whether steps might arise out of specimen translation over the photodiode array, we followed the displacement of a single A-band as it translated along the array during an imposed motor ramp. We concentrated on the A-band closest to the motor, which should translate smoothly because it was close to the smoothly translating motor. A-band median was computed as above. The bottom trace of Figure 5 shows that no step-like behavior is present. A similar conclusion was drawn earlier in experiments in which the discrete sensor was replaced by a continuous one (Yang *et al.*, 1998). The steps remained in evidence with this alternative sensor.

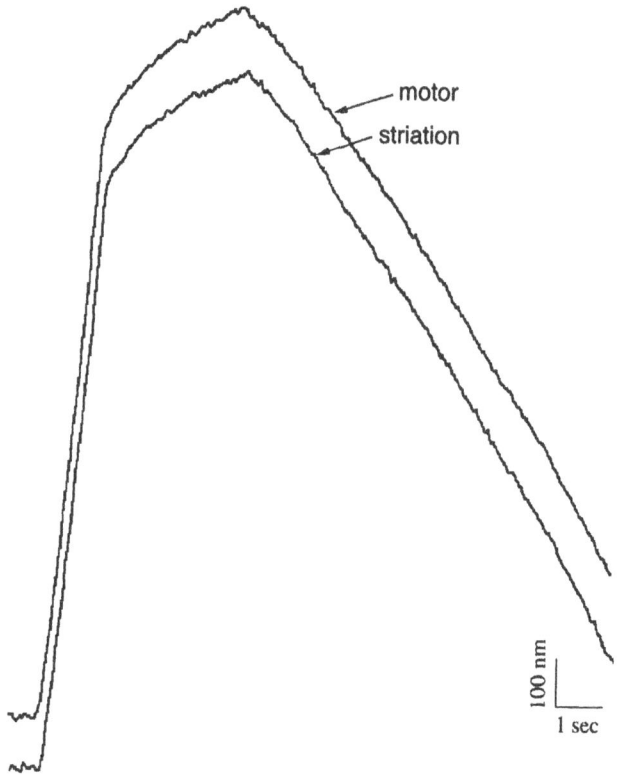

Figure 5. *Time course of motor and striation position during
imposed ramp motor movement.*

Figure 6 shows that steps similar to those seen during slow release are
seen during slow stretch. The character of the steps and pauses was qualitatively similar to that obtained during release, although the discrete features
tended to be slightly less clear. In the absence of thick-thin filament interaction, the steps obtained during stretch and release presumably arise out of length
changes in the connecting filament.

Step size was reckoned as the sarcomere-length difference between successive pauses (please see Blyakhman *et al.*, 1999 for details of the analytical
procedure). The resulting step-size histograms are shown in Figure 7. The
main panel shows the results obtained with the standard algorithm (gray) and
the high-resolution algorithm (black). Both histograms show regularly spaced
peaks. The best-fit value for peak separation is 2.3 nm in both cases, although
the high-resolution algorithm is able to resolve steps considerably smaller than
the standard algorithm. It appears that step size is an integer multiple of 2.3
nm.

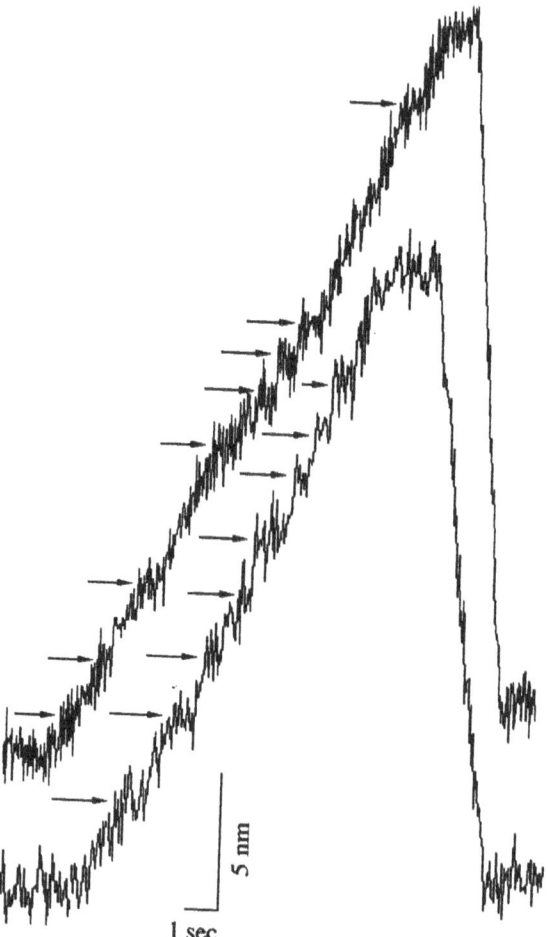

Figure 6. *Time course of sarcomere-length change during slow stretch. Records obtained with differential algorithm. Arrows denote pauses. Steps of a few nanometers are apparent.*

The inset of Figure 7 shows a similar analysis carried out for the stretch experiments. The results were largely similar to those obtained during release. The histogram contained two, perhaps three peaks. Peak positions were nominally 2.5 nm and 5 nm, close to those measured during shortening, although the reduced distinctiveness of the stretch steps resulted in broader histogram peaks, which did not allow us to conclude whether the step size was indeed identical. To some extent at least, the dynamics of stretch and release appear to be quantitatively reversible.

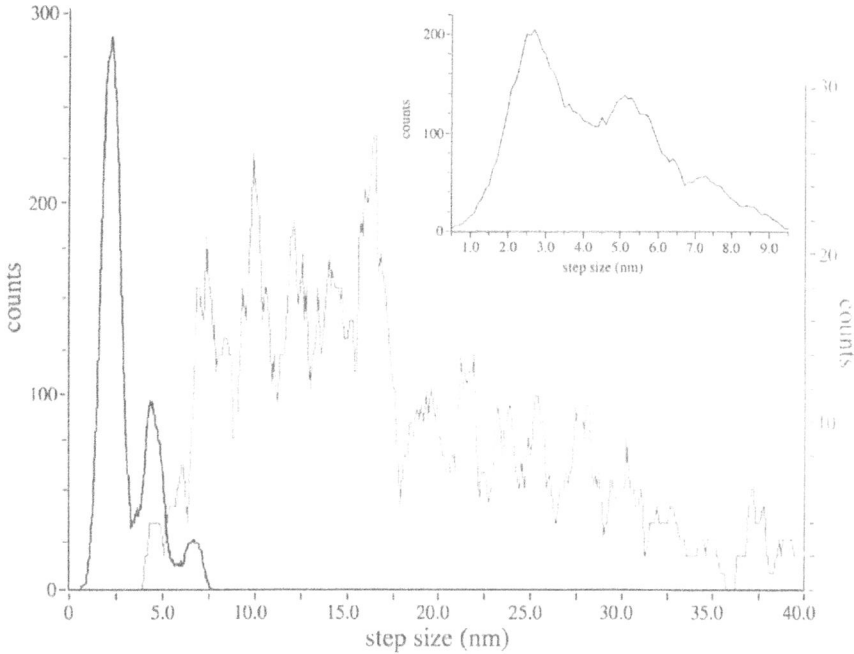

Figure 7. *Histograms of step size. Main panel: Step size measured during slow release, using standard algorithm (gray) and high-resolution algorithm (black). Inset: Step size measured during slow stretch. High-resolution algorithm used.*

Discussion

The observed connecting filament steps very likely correspond to the steps seen in studies of isolated titin molecules (Tskhovrebova *et al.*, 1997; Rief *et al.*, 1997). In the isolated titin studies the step size was mainly in the range 10 – 25 nm, which overlapped the range seen here with standard resolution methods. We found however that size is not necessarily a good criterion for similarity because size is resolution-dependent (Fig. 7). The higher the resolution the smaller the size of the step that can be discerned. Thus, many of the larger steps detected with low signal-to-noise may well be multiples of smaller steps.

The main finding is that step size is 2.3 nm or an integer multiple thereof. Thus, the step can be 2.3 nm, 2 x 2.3 nm, etc. These figures apply during release and as best we can tell during stretch as well, implying that extension and retraction are quantitatively reversible. Although these size values were measured for the full sarcomere, they actually apply to the half-sarcomere because the complementary half-sarcomeres do not generally step synchronously in the myofibril (Blyakhman *et al.*, in press). Thus, the 2.3 nm step value corresponds to the single connecting filament bundle of half of the sarcomere.

One question is why the connecting filaments of the bundle step synchronously. Although a general mechanism for synchrony has been put forth (Pollack, 1990), in the case of the myofibrillar half-sarcomere, individual filaments are cross-linked by radial struts (Trombitás *et al.*, 1988). Thus, steps in parallel filaments are constrained to occur at the same time, and the half-sarcomere is therefore a macroscopic reflection of sub-molecular events.

Another question is which of the several connecting-filament domains gives rise to the step. The connecting filament contains Ig, PEVK and Fn-III domains. However, only the Ig domain repeats many times along the filament. It is therefore a probable candidate. The Ig domain contains ~90 residues and has a sevenfold antiparallel beta-barrel structure (Erickson, 1994; Improta *et al.*, 1996). Each turn comprising the barrel surface will then be several nanometers in length. A step of several nanometers could therefore reflect the folding of a single turn, whereas folding of the entire domain would correspond to a step of 7 x 2.3 nm, or ~16 nm.

In this context, the sevenfold nature of the step-size histogram of Figure 7 is of interest. The Figure shows a sharp drop of amplitude between the seventh and eighth peak. This finding emphasizes the sevenfold nature of the response, implying the likely relevance of the sevenfold beta-barrel structure. It appears, then, that the Ig domain is involved in the stepping mechanism, although the possibility that this domain modulates the response of another, non-titin connecting-filament element (Pollack, 1990) is not ruled out.

In sum, the single myofibrillar sarcomere offers a convenient window of observation of molecular events. It is the smallest functional preparation that retains natural architecture. And because of its relatively large size, it offers higher signal-to-noise than molecular scale preparations. Thus, we have been able to deduce the fundamental event of connecting filament length change— the 2.3 nm step.

References

Bartoo ML, Popov VI, Fearn L, Pollack GH. Active tension generation in isolated skeletal myofibrils. *J Muscle Res Cell Motil* 1993;14: 498-510.

Blyakhman F, Shklyar T, Pollack GH. Quantal length changes in single contracting sarcomeres. *J Mus Res Cell Motil* 1999; in press.

Erickson HP. Reversible unfolding of fibronectin type III and immunoglobulin domains provides the structural basis for stretch and elasticity of titin and fibronectin. *Proc Nat'l Acad Sci USA* 1994;91:10114-10118.

Granzier H, Pollack GH. Stepwise shortening in unstimulated frog skeletal muscle fibers. *J Physiol* 1985;362:173-188.

Improta S, Politou AS, Pastore A. Immunoglobulin-like modules from titin I-band: extensible components of muscle elasticity. *Structure* 1996;4:323-337.

Maruyama K, Toshitada Y, Yoshidomi H, Sawada H, Kikuchi M. Molecular size and shape of ß-connectin, an elastic protein of striated muscle. *J Biochem* 1984;95:1423-1493.

Pollack GH. *Muscles and Molecules: Uncovering the Principles of Biological Motion.* Seattle: Ebner & Sons, 1990.

Rief M, Gautel M, Oesterhelt F, Fernandez JM, Gaub HE. Reversible unfolding of individual titin immunoglobulin domains by AFM. *Science* 1997;276:1109-1112.

Sokolov Y, Grinko A, Tourovskaia A, Reitz F, Pollack GH, Blyakhman F. "Minimum average risk" as a new peak detection algorithm applied to myofibrillar dynamics. 1999:Submitted.

Trinick J, Knight P, Whiting A. Purification and properties of native titin. *J Mol Biol* 1984;180:331-356.

Trombitás K, Tigyi-Sebes A. *In Insect Flight Muscle*, RT Tregear, ed. Amsterdam: N. Holland Publication Co. 1977;79-90.

Trombitás K, Baaatsen PHWW, Pollack GH. I-bands of striated muscle contain lateral struts. *J Ultras and Mol Str Res* 1988;100:13-30.

Tskhovrebova L, Trinick J, Sleep JA, Simmons RM. Elasticity and unfolding of single molecules of the giant muscle protein titin. *Nature* 1997;387:308-312.

Wang SM, Greaser ML. Immunocytochemical studies using a monoclonal antibody to bovine cardiac titin on intact and extracted myofibrils. *J Cell Biol* 1985;107:1075-1083.

White DCS, Thorson J. The kinetics of muscle contraction. *Prog Biophys Mol Biol* 1973;27:173-255.

Yang P, Tameyasu T, Pollack GH. Stepwise dynamics of connecting filaments measured in single myofibrillar sarcomeres. *Biophys J* 1998;74:1473-1483.

Discussion

Jin: If we use single molecule information to fit this model, how can we imagine that in an intact sarcomere all those molecules with all structural and functional domains synchronize to give you this detectable step?

Pollack: Of course that is the $64 question. In the case of the single myofibril, I think there is no problem. I showed early on that there are interconnections among parallel filaments. The existence of these interconnections is not widely accepted, but papers that we published some years ago and also in my book (Pollack, 1990) demonstrate quite clearly that there are interconnections between filaments in the I-band. Whether the interconnections are between actin filaments—which you see all over biology so why shouldn't they occur in muscle?—or between actin filaments and connecting filaments or between connecting filaments and connecting filaments, is not exactly clear. But there do seem to be cross connections in the I band, and if they exist they probably act to synchronize adjacent filaments. So if you get steps in one connecting filament, perhaps you would get them in another. That is a simple answer.

Jin: That is what I wanted to hear.

TerKeurs: They are also interconnected by force, of course. So synchrony in motion could be induced by the structure and by simultaneous support of force.

Pollack: Yes, what you are suggesting is that as soon as the first connecting filament begins its shortening step, it relieves the tension on the next one, and that predisposes the next one to undergo the same shortening step, so it is a catastrophic process.

TerKeurs: I have a question that relates to, for example, Henk's talk. If you would be stretching different domains of a titin filament or a connectin filament, you would expect that you explored the properties of an IgG or N2B or N2A or a PEVK domain. Would you expect different step sizes? Did you explore that hypothesis?

Pollack: No, we haven't really done any serious exploration. That is why we are beginning to get into vertebrate muscle, where the PEVK becomes much more important. In the next few months, we should have a lot to say on the step distribution in these species and then we could make an intelligent comment. I would suppose that if you have 2 or 3 different domains, that the same kind of model could occur in each one of those domains. And indeed, if the molecular structure is different in the PEVK, which it is, I would expect either no steps or steps of perhaps different size. The problem is that in all the experiments that we have done in the past with larger preparations, it has been difficult to find any condition where you don't find steps. So if PEVK shortening or stretching is not giving you steps, I would be surprised. The step size would be crucial, yes.

Vigoreaux: First of all, pardon my ignorance here. Can you explain to me what role calcium activation plays in this model and where this tension comes from?

Pollack: In the proposed model, changes in length induced by a change in tension will occur in some sort of coordinated sequence along the filament, one domain at a time. At a given length, tension arises from the retractive force of each unfolded domain. The second part of the question refers to calcium. That is a very interesting question. I don't know if Henk ter Keurs is going to be talking about the effect of calcium, but there is an old study by Endo (CSH XXXVII, 505-510, 1972), who did careful studies of resting tension as a function of calcium concentration. He found that at very low calcium levels, as you increase the calcium from the resting level to not quite the activated level, the resting length-tension curve shifts up progressively; it has the same shape, but just shifts upward. So it is possible that the equilibrium of the Ig, fibronectin, or PEVK domain, is shifted by calcium. Andre Krammer showed in his presentation that there is a calcium binding site on titin. So it is possible that the equilibrium between folded and unfolded states is affected by calcium binding. That could explain Endo's results. It could have some impact on calcium-activated contraction.

Vigoreaux: But you did mention you see the same steps in the active muscle.

Pollack: I am sorry if I confused the issue. We do find steps in activated muscle as well as passive muscle. In the activated case the histogram of step size was also multi-peaked, but the spacing between peaks was different—it was 2.7 nm. The mechanism may be somewhat different. We have tried to interpret the result using a model in which actin filaments crawl, or reptate, over the myosin filament. That would give the 2.7 nm quantum, which is equal to the linear advance of actin molecules along the thin filament.

Sugi: I wanted to make some comments. First, your step phenomenon in active muscle is very puzzling. So I don't want to make any comment. Next I would like to make a suggestion, to make a new type of experiment. This is the role of titin for connecting active relaxation. Probably you know the famous work on relaxation in which just single muscle fibers are contracted to very short lengths, and then when stimulation is stopped, the muscle jumps back to its original length. My suggestion is to study whether this rapid lengthening occurs in steps.

Pollack: Thank you for the suggestion. It reminds me of the so-called diastolic suction phenomenon that exists in the heart. The heart will contract, and then refill by sucking blood from the veins by some sort of active restoring force. The question is, whether this has something to do with titin-length restoration or something else. My own bias would be something else—maybe the thick filament. As you know—but many others are maybe less aware of—evidence that the thick filaments can shorten is appreciable. There are 30-35 published reports showing that during active contraction thick filaments shorten appreciably (Pollack, *Physiol. Rev.*, 1983). If they really do shorten appreciably, then they must also re-lengthen. If they re-lengthen, this would be another way that the fiber could restore its initial length. So whether it is titin, the thick filament, or something else, remains to be seen.

Granzier: Do you think titin is involved?

Pollack: It may be involved, yes, sure.

Granzier: The thick filament could never return the sarcomere to lengths above 1.6 or 1.7 microns, true? But the slack length of the sarcomere is 1.8 or 1.9 μm.

Pollack: I agree with you. Titin may be involved.

Linke: Jerry, you have certainly beautiful tools to also study the forces and I am sure you have already done it. Since we know so much about the single molecule mechanics and unfolding-refolding of titin, it would be easy to compare, for example, the forces, and see whether they are in the same range and maybe distinguish better whether it is indeed Ig domain unfolding or whether it could be something else.

Pollack: Well, we've done some force measurements, but not many. For technical reasons it is somewhat more demanding to measure force at the same time, as you know. That is something for the future.

Bullard: I wanted to ask you a similar question. Have you compared the force needed to get these steps from bee muscle with that needed for rabbit psoas? Is the bee muscle noticeably stiffer?

Pollack: No, I am sorry, we haven't done that yet. But I want to make a comment about force levels in general. When you look at a single molecule in isolation, you have to make some assumptions about the conditions. One question is whether the milieu that surrounds the molecule is the same as the milieu that surrounds the molecule when it is in the intact lattice. I gave one example showing the radial interconnecting links that exist in the intact lattice, but they are not present in the single molecule. They could have some effect. Also one is a bit uncertain about the local ionic strength and microenvironment. So, when one measures absolute levels of force in single molecules I think that has to be taken with a grain of salt. It is certainly a valid measure, but whether it applies in the intact lattice quantitatively is uncertain. So I think a direct comparison between the two is not so simple to make without assumptions.

TITIN-THIN FILAMENT
INTERACTION AND POTENTIAL
ROLE IN MUSCLE FUNCTION

Jian-Ping Jin

Department of Physiology and Biophysics, Case Western Reserve University School of Medicine, Cleveland, OH

Abstract: Titin (connectin) is a giant polypeptide that forms a single-molecule filamental structure extending from the M-line to the Z-line in the sarcomere of striated muscle. The primary structure of titin consists mainly of repeats of two types of ~100-amino acid modules (fibronectin type III and immunoglobulin-like motifs, respectively) and a Pro rich segment named the PEVK domain. The I-band region of titin shows an elasticity important to the passive properties of the myofibril. To investigate the biological function of titin, we cloned cDNA segments encoding single or linked structural modules of titin into expression vectors to produce non-fusion titin fragments in *E. coli*. High level expression of titin fragments was achieved and effective purification procedures were developed. We also developed specific monoclonal antibodies against the titin fragments and solid-phase protein-binding assays to investigate the interaction of the titin structural modules and other sarcomeric proteins. The results show that the immunoglobin-like module that are enriched in the I-band titin binds to F-actin. In contrast to the rigid association of A-band titin with the thick filament, the relatively weak titin-actin binding suggests that the I-band titin may reversibly interact with the thin filament during muscle contraction. This hypothesis is supported by an epitope similarity between the actin-binding site of caldesmon and the immunoglobulin-like module of titin, which suggests analogous functions of caldesmon and titin in organizing the contractile proteins. Together with data from mechanical studies demonstrating that the titin-actin interaction may contribute to the passive property of cardiac muscle in a Ca^{2+}-dependent manner, we speculate that the Ca^{2+}-mediated thin filament regulation may coordinate the function of titin during muscle contraction and relaxation.

Elastic Filaments of the Cell, edited by Granzier and Pollack
Kluwer Academic/Plenum Publishers, 2000

Introduction

Titin, also known as connectin, is a giant polypeptide that forms a single-molecule filament extending from the M-line to the Z-line in the sarcomere of striated (cardiac and skeletal) muscles (see Wang, 1985; Trinick, 1991; Maruyama, 1997 for reviews). The titin filaments show an elastic behavior in the I-band portion in contrast to the rigid A-band titin associated with the thick filaments. The large titin polypeptide consists mainly of multiple repetitive sequence modules (Labeit and Kolmerer, 1995) that may have evolved through an adaptation in which the single-molecule titin filament performs a role together with the polymer myosin and actin filaments in the sarcomere. The complete primary structure of titin determined through cDNA cloning and sequencing revealed that titin contains two types of ~100-residue repeating molecules (fibronectin type III and immunoglobulin domains, respectively) and a Pro rich segment named the PEVK domain (Labeit and Kolmerer, 1995). The fibronectin type III (FnIII) and immunoglobulin (Ig)-like sequence modules have also been found in several other proteins, such as twitchin (Benian *et al.*, 1989), projectin (Ayme-Southgate *et al.*, 1991), C-protein (Einheber and Fischman, 1990), myosin light chain kinase (Olson *et al.*, 1990), skelemin (Price and Gomer 1993), and the cellular titin in nonmuscle cells (Eilertsen *et al.*, 1994). On the contrary, the PEVK segment is a novel structure unique to titin with highly variable lengths in different types of muscle.

The FnIII and Ig modules are proposed to function as the basic units for titin's myosin filament association as well as its elastic behavior in the I-band region (Labeit *et al.*, 1992; Linke *et al.*, 1997; 1998). In order to understand the biological function of titin and titin-like proteins, it is necessary to characterize the function of the three classes of titin primary structures. Direct testing of the function of titin by protein interaction assays is difficult due to titin's large size and limited solubility. The fact that each class of the repeating modules in titin are homologous structures with similar sequences makes it possible to investigate their structure-function relationship by analysis of titin fragments containing defined single or combinations of the sequence modules. The elastic feature of titin is an interesting focus of research for its contribution to the passive property of muscle (Granzier and Irving, 1995; Granzier *et al.*, 1997). For the integrated organization of contractile and cytoskeletal proteins in the sarcomere, it is important to consider the I-band titin elasticity together with the Ca^{2+}-regulated structure-function transition of the thin filament during the activation and relaxation of muscle. Results from our studies on the structure and function of molecular-engineered titin modules have led to a hypothesis that the interaction between I-band titin and the thin filament may contribute to the passive property of muscle.

Material and Methods

Details of the recombinant DNA and biochemical techniques applied can be found in our previous publications (Jin and Wang, 1991; Jin, 1995; Ogut and Jin, 1996, 1998; Wang and Jin, 1997, 1998).

Cloning of cDNA fragments encoding rat cardiac titin sequence motifs into expression vector

To obtain cDNA templates encoding single or combinations of the titin Fn III and Ig modules (Fig. 1), polymerase chain reaction (PCR) was applied to amplify titin cDNAs from a λgt11 rat cardiac cDNA library as previously described (Jin, 1995). The PCR products of anticipated sizes were identified by agarose gel electrophoresis and purified.

To express nonfusion titin structural modules in *E.coli* using the T7 polymerase-based system (Studier *et al.*, 1990), pAED4 expression plasmids (Jin, 1995) were constructed. The recombinant plasmids were screened by directly examining the transformed bacterial colonies by PCR (Fig. 2). To verify the orientation, translation reading frame, and full sequence of the coding template, the recombinant plasmids were sequenced.

Gel electrophoresis

To monitor the expression and purification of the small titin fragments, we applied a sodium dodecyl sulphate-polyacrylamide gel electrophoresis (SDS-PAGE) using the Tris-Tricine buffer system modified from that described by

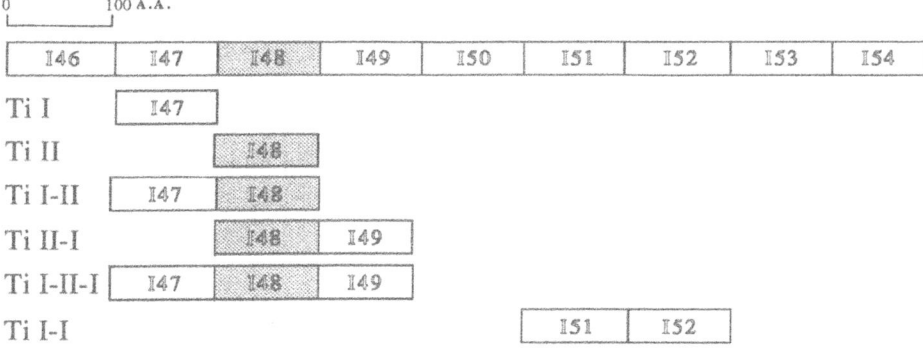

Figure 1. Molecular engineering of six rat cardiac titin fragments containing single or various combinations of the FnIII and Ig modules. Six cDNAs encoding single or linked rat cardiac titin sequence modules were cloned into expression vectors. Their locations in the A-I junction region of the titin polypeptide chain (the top map) were determined and numbered according to Labeit and Kolmerer (1995).

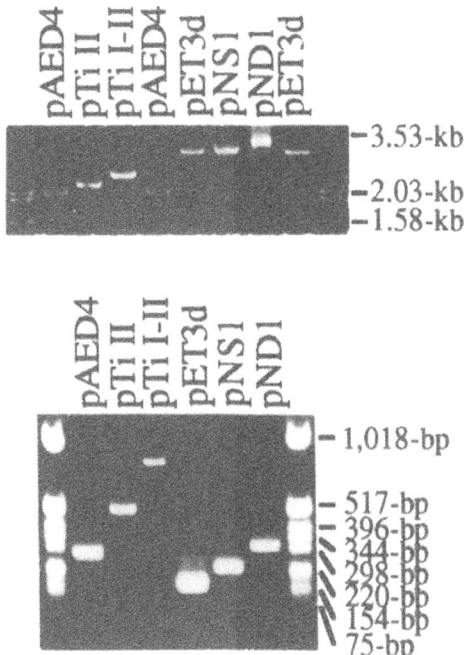

Figure 2. *PCR screening of recombinant expression plasmids bearing a short coding template. The effective PCR identification of several recombinant expression plasmids bearing short cDNA fragments was demonstrated by 0.9% agarose gel electrophoresis. The expression plasmids pTi II and pTi I-II containing cDNA coding inserts of 314-bp and 629-bp, respectively, were analyzed together with two other expression constructs (pNS1 and pND1) bearing cDNA coding inserts of 116-bp or 209-bp. Since construction of the expression plasmids involved a deletion of the polylinker sequences in the vector, the insignificant size-shift of the recombinant plasmids vs. the vector controls was further reduced (the upper panel), making the screening difficult. In contrast, the sizes of DNA inserts amplified by PCR from the four recombinant expression plasmids can be accurately identified by the gel electrophoresis (515-bp for pTi II, 830-bp for pTi I-II, 248-bp for pNS1 and 341-bp for pND1, shown in the lower panel), which can be distinguished from the poly-linker/primer regions amplified from the pAED4 and pET3d vectors without an insert (305-bp and 172-bp, respectively).*

Schagger and von Jagow (1987) with the BioRad mini-gel apparatus (Jin, 1995). Using 14% resolving and 4% stacking gels with an acrylamide:bisacrylamide ratio of 20:1 (SDS was omitted from the resolving gel), a good resolution of the small proteins has been achieved.

Expression and purification of the titin fragments

BL21(DE3)pLysS *E. coli* cells (Studier *et al.*, 1990) were transformed with the recombinant expression plasmids and ampicillin-chloramphenicol dual-resistant colonies were selected. The transformed host cells were expanded and induced by isopropylthiogalactoside (IPTG). The induced bacterial cells were harvested and analyzed by SDS-PAGE to verify the expression of titin fragments. Although intact titin is known to have a limited solubility under physiological conditions, the engineered titin fragments containing 1 to 3 of the ~100 amino acid modules were present in the supernatant of the bacterial lysates. Large scale purification procedures have been developed for the titin fragments according to their physical properties.

The titin fragments with an alkaline theoretical isoelectric point (pI) as calculated from sequence data were purified by the following procedure: The bacterial cells were disrupted by French press. The lysate was subjected to ammonium sulfate fractionation and the fraction of 50-80% saturation was collected and dialyzed against 0.1 mM EDTA. After adjusting the pH to 4.5 by sodium acetate, the supernatant was loaded onto a CM-52 cation-exchange column equilibrated in 10 mM sodium acetate, pH 4.5, containing 0.1 mM EDTA. After washing with the equilibration buffer, the column was eluted with a linear gradient of KCl. Fractions containing the titin fragment, as revealed by SDS-PAGE on the A_{280nm} peak, were collected.

A different strategy was designed for the purification of the highly soluble, relatively acidic titin fragments. The French pressed bacterial lysate was diluted with low ionic strength Tris-HCl buffer to reduce conductivity to <2 mS/cm for being fractionated on a DE-52 anion-exchange column equilibrated at pH 9.0. The column was eluted with a linear gradient of KCl and the 280_{nm} absorbency peak was analyzed by SDS-PAGE to identify fractions containing the titin fragment.

The titin fragments were further purified by gel filtration chromatography to remove the small amount of high or low molecular weight contaminants. The CM-52 or DE52 column fractions were concentrated by lyophilization and redissolved in 5 -10 ml of 10 mM imidazole-HCl, pH 7.0, 15 mM β-mercaptoethanol containing 6 M urea to load on a G75 sizing column. The column was developed with the same buffer and the A_{280nm} peak was examined by SDS-PAGE for fractions containing highly purified titin fragment. After dialysis against 0.5% formic acid, the proteins were lyophilized for long-term storage. All of the titin fragments were subjected to amino acid analysis for their authenticity and purity.

Preparation of monoclonal antibodies against the titin structural modules

Female Balb/c mice were immunized by an intraperitoneal injection of 50 μg of the purified titin fragment antigen in Freund's complete adjuvant followed by boosts of the same amount of antigen at three-week intervals. The mouse sera were tested for the specific anti-titin antibody titer by indirect enzyme-linked immunosorbant assay (ELISA). Development of hybridoma cell lines and monoclonal antibody (mAb) production were carried out as previously described (Jin *et al.*, 1996).

To characterize the mAb specificity against the cloned titin structural modules by Western blotting, the titin fragments were resolved by 14% SDS-PAGE as described above and transferred to nitrocellulose membranes as previously described (Jin, 1995). The subsequent blocking and incubations with the anti-titin mAbs and alkaline phosphatase-labeled anti-mouse IgG second antibody were done as previously described (Wang and Jin, 1998).

ELISA-mediated solid-phase protein-binding assay

As previously described in troponin T-tropomyosin binding studies (Jin, 1989; Wang and Jin, 1998; Ogut and Jin, 1998) and the investigation of nebulin-actin interaction (Jin and Wang 1991), we have developed an ELISA-mediated solid-phase protein-binding assay for characterization of the titin structural modules (Jin, 1995). As illustrated in Figure 3, microtiter plates were coated with purified rabbit skeletal muscle F-actin (30 µg/ml) or myosin (50 µg/ml) in 100 mM KCl, 3 mM $MgCl_2$, 10 mM Tris-HCl, pH 8.0 (Buffer A) at 4 °C overnight. After washing with Buffer A plus 0.05% Tween-20 (Buffer T), the plates were blocked with 1% BSA in Buffer T at room temperature for 2 hr. After washes with Buffer T, the plates were incubated with serial dilutions of purified titin fragments in Buffer A at room temperature for 2 hr. The plates were then washed with Buffer T and incubated with anti-titin polyclonal or monoclonal antibody in Buffer T containing 0.1% BSA (Buffer B) at room temperature for 1 hr. Following washes with Buffer T, horseradish peroxidase-conjugated rabbit anti-mouse immunoglobulin second antibody (HRP-RAM) in Buffer B was added to the plate and incubated at room temperature for 45 min. The plates were finally washed with Buffer T and H_2O_2-ABTS [2, 2'-azinobis-(3-ethybenzthiazoline sulfonic acid)] substrate was added for color development at room temperature. The A_{405nm} of each assay well was monitored at a series of time points by a BioRad Lab automated microplate reader and the data from a time point within the linear range of the color development were used for plotting each set of the binding curves.

Results

Cloning of cDNAs encoding titin structural modules and construction of expression vectors

Six overlapping or adjacent rat cardiac titin cDNA coding templates have been cloned and constructed into pAED4 expression plasmids for expressing titin structural modules in various configurations (Fig. 1). The position of the cloned modules in the whole titin polypeptide chain was determined according to the published human titin sequence and motif repeating pattern (Laibeit and Kolmerer, 1995). The unique module organization in the cloned rat cardiac

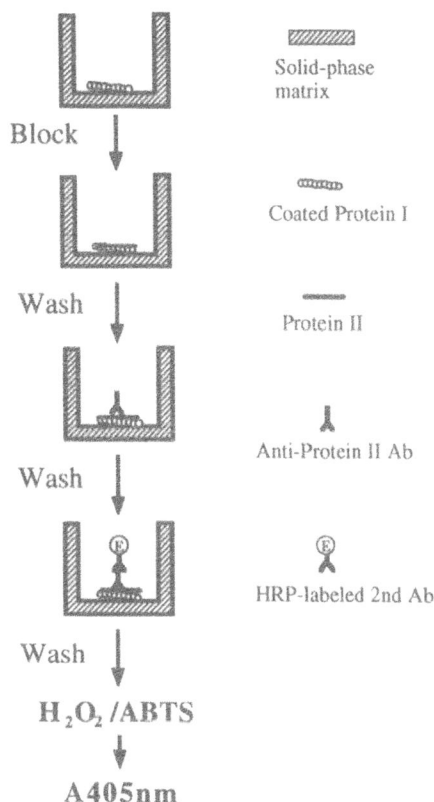

Figure 3. *An ELISA-mediated protein-binding assay. One of the two proteins in the binding reaction is coated on microtitering plates. After blocking the remaining plastic surface with BSA in the presence of detergent (Tween-20), the plate is incubated with serial dilutions of the other protein, followed by an antibody specifically recognizing the second protein. HRP-conjugated second antibody specific to the first antibody is used to detect the binding of protein II to protein I via the H_2O_2-ABTS substrate color reaction.*

titin segment indicates its location at the A-I junction (Labeit and Kolmerer, 1995).

Purification of the titin fragments

The six titin fragments were all successfully expressed in the transformed *E. coli* culture. Although toxicity of titin fragments to the host bacterial cell has been observed, we have achieved a good expression of these single or linked titin structural modules. The purification methods developed for the two classes of titin fragments (Figs. 4 and 5, respectively) have produced large quantities of highly purified titin fragments for functional characterization and specific antibody production. The experimentally determined amino acid compositions of the purified titin fragments closely resembled the residue molar ratio of that

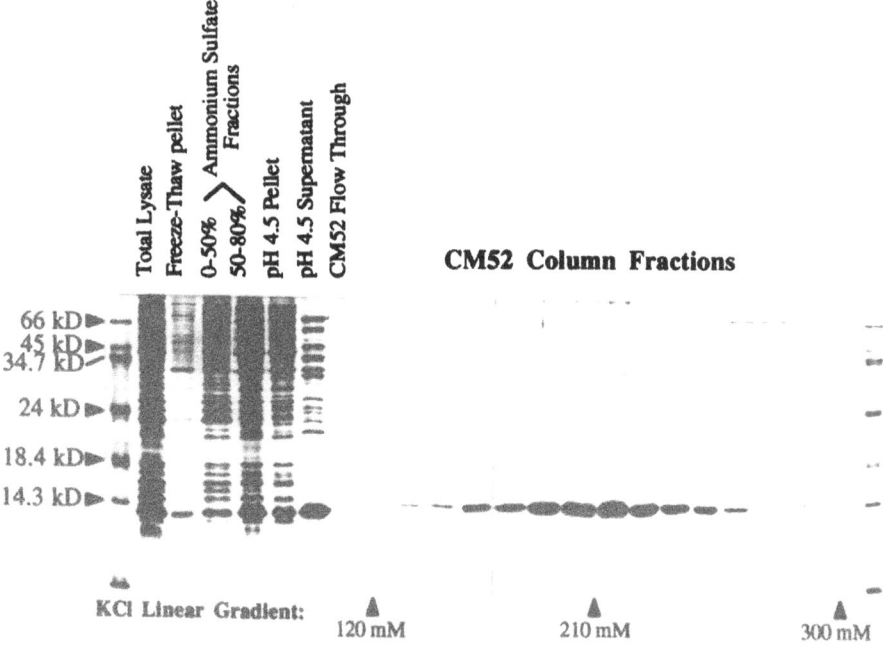

Figure 4. *Expression and purification of the cloned alkaline Ti II module. The transformed BL21(DE3)pLysS E. coli cells collected from the IPTG-induced culture were lyzed using a French press. The sample was stored at -20 °C overnight and a small amount of precipitate was removed. The ammonium sulfate fraction of 50-80% saturation was dialyzed to remove the salt. After adjusting the pH to 4.5, a small amount of precipitated proteins was removed by centrifugation. The pH 4.5 supernatant enriched with the titin fragment was loaded on a CM-52 ion-exchange column and eluted with a linear KCl gradient. The fractions were analyzed by 14% Tris-Tricine SDS-PAGE and the Ti II protein peak was around 210 mM KCl.*

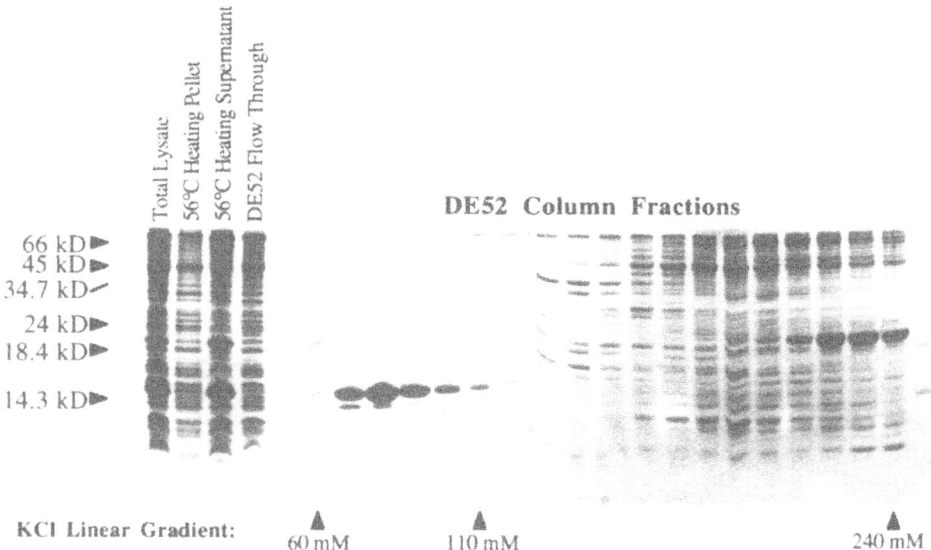

Figure 5. *Expression and Purification of the cloned acidic Ti I module. Bacterial cells collected from the IPTG-induced culture were lyzed by French press. The total lysate was heated to 56 °C to denature the heat-sensitive bacterial proteins. After adjusting the pH to 9.0, the precipitated proteins were removed by centrifugation. The supernatant enriched with the Ti I fragment was loaded on a DE-52 ion-exchange column and eluted with a linear KCl gradient. The fractions were analyzed by 14% Tris-Tricine SDS-PAGE and the Ti I protein peak was between 60 mM and 110 mM KCl.*

calculated from the cDNA-derived protein sequences, indicating the accurate cloning and expression of the nonfusion titin fragments in *E. coli* and the effectiveness of the purification methods. Three examples of the results are shown in Table 1. A mouse polyclonal antiserum raised using the Ti I-II fragment (Fig. 1) as immunogen recognized intact titin in Western blots verifying the authenticity of the cloned titin fragments (Jin, 1995). Indirect immunofluorescence microscopy further confirmed the presence of epitopes recognized by the mouse anti-Ti fragments antibodies in both rat cardiac and skeletal muscle myofibrils (Jin, 1995).

Monoclonal antibodies against cloned titin structural modules

Six mAbs have been developed by immunization using the cloned rat cardiac titin fragments. As summarized in Table 2, mAbs Ti105 and Ti107 recognize the Ti I-I fragment containing two FnIII modules (I51-I52) whereas mAbs Ti102, Ti104, Ti106, and Ti108 recognize the I48 Ig-like module. It is worth noting that the reactivities of the 4 mAbs to the I48 module are different.

Table 1. Amino acid analysis of three cloned rat cardiac titin fragments

Amino Acid	Ti I		Ti II		Ti I-II	
	C [a]	E [b]	C [a]	E [b]	C [a]	E [b]
Ala	9	9.0	8	9.1	16	17.3
Arg	6	5.6	4	4.2	10	9.6
Asx	11	11.4	9	9.5	20	21.2
Glx	12	12.2	7	8.1	19	21.1
Gly	8	8.6	6	8.6	14	16.1
His	0	0.2	0	0.6	0	0.8
Ile	4	4.2	8	5.3	12	11.5
Leu	3	4.0	8	7.7	11	12.4
Lys	10	10.0	10	9.9	20	19.9
Met	2	0.9	1	1.0	2	1.3
Phe	1	1.4	1	1.0	2	3.0
Pro	12	12.3	9	10.2	20	20.5
Ser	6	5.3	3	4.2	9	8.1
Thr	7	7.0	14	12.2	21	17.5
Tyr	3	3.0	1	1.6	4	4.5
Val	8	8.1	11	8.1	19	18.6
Cys	2	N.D.	1	N.D.	3	N.D.
Trp	4	2.4	1	N.D.	4	0.5

[a] C, amino acid composition calculated from sequence data; [b] E, amino acid composition experimentally determined from the bacterially expressed protein; N.D.: not determined.

For example, Ti102 recognizes the isolated I48 module (TiII) significantly weaker compared to its reactivity to I48 residing within the linked fragment (Ti I-II, Ti II-I and Ti I-II-I). In contrast, mAb Ti104 recognizes the isolated I48 module similarly to that within the longer fragments.

Interaction of the titin structural modules with myosin and F-actin

ELISA-mediated solid-phase protein binding experiments demonstrated interactions of the titin fragments Ti I-II and Ti I (Fig. 1) with immobilized rabbit muscle myosin or F-actin. The results showed that both the two-module Ti I-II fragment and the single FnIII module (Ti I) had saturable binding curves to both myosin and F-actin with Ti I-II being ~10-fold higher in binding affinity than Ti I to both myosin and F-actin (Jin, 1995). In contrast, the single Ig module Ti II fragment bound to F-actin but not myosin (Fig. 6A). The titin-F-actin interaction detected by the solid-phase binding assays has been confirmed in solution by the classical F-actin co-sedimentation assay (Jin, 1995). Figure

Table 2. **Monoclonal antibodies against cloned titin fragments**

mAb	Titin Fragment					
	Ti I	Ti II	Ti I-II	Ti II-I	Ti I-I	Ti I-II-I
Ti102	–	+	+	+	–	+
Ti104	–	+	+	+	–	+
Ti105	–	–	–	–	+	–
Ti106	–	+	+	+	–	+
Ti107	–	–	–	–	+	–
Ti108	–	+	+	+	–	+

The specificity between six mAbs and different titin fragments was determined by Western blotting. The positive recognitions are indicated and by "+", and the negative interactions are represented by "–".

6B further shows that the Ti I-II binding to F-acting was not Ca^{2+}-dependent and there was no binding between the Ti I-I fragment (containing two FnIII modules, I51-I52; Fig. 1) and F-actin.

Discussion

Because of the presence of multiple repeating structural modules and the insolubility of intact titin, molecular cloning and expression of titin fragments have become useful for biochemical characterization of the structure-function relationship of this giant sarcomeric protein. The expression of titin structural modules without any fusion peptide will provide authentic materials for functional studies but requires an optimized purification method for each fragment expressed. Good yield, purity, and biological activity of the cloned titin fragments expressed and isolated from bacterial culture are essential for their use in functional characterization. Using conventional biochemical procedures, we have developed two effective methods for the purification of acidic and basic titin fragments from *E. coli* lysate. Accurate cloning and authenticity of these titin fragments was verified by amino acid analysis, immunological reactivity and sarcomere localization (Jin, 1995). The titin fragments purified have shown the predicted secondary structure feature and binding activity to muscle myosin and F-actin (Fig. 6). Using similar approaches, we have also successfully expressed and purified a PEVK segment of titin for functional characterization.

Using the molecular-engineered titin fragments as immunogen, we have developed 6 mAbs specifically recognizing the different types of structural modules of titin (Table 2). These mAbs provide useful tools for investigating the structure and function of titin and its sequence modules. In addition to applications in sarcomere localization of titin epitopes in electron microscopy (Trombitás *et al.*, 1995; 1998), these module-specific mAbs have been useful in the solid-phase ELISA binding assays for titin's interaction with various myofilamental proteins (Jin, 1995).

The ELISA-mediated solid-phase protein-binding analysis (Fig. 3) is a high performance method to investigate the interaction of titin fragments with other sarcomeric proteins. This method has several major advantages: a) a large number of samples and experimental conditions can be tested simultaneously; b) the buffer conditions can be optimized for different steps of the reaction; c) quantitative data are produced; and d) as a result of its high sensitivity, relatively low concentrations of protein are required, allowing the examination of myofibrillar proteins with limited solubility.

The biological significance of the titin-F-actin interaction remains to be established. Titin-actin interactions have also been observed using *in vitro* motility assay (Li *et al.*, 1995) and in actin polymerization experiments (Astier *et al.*, 1998). Projectin-F-actin interaction has been found to affect actomyosin ATPase activation (Weikamp *et al.*, 1998). The different module organizations (Labeit and Kolmerer, 1995) may confer the functional difference between the A- and I-band titins. The observation of titin-actin interaction suggests that the I-band portion of titin is not a "naked" filament or a "free spring." The interaction between titin and the thin filament may contribute to the resting tension of myofibril and the elasticity of I-band titin. Instead of the proposed stretch-regulated titin motif secondary structure unfolding hypothesis which is challenged by the nonphysiologically high force required (Rief *et al.*, 1998), a reversible "uneven zipping" between the I-band titin and the thin filament during contraction or stretching of the muscle may be a mechanism responsible for the titin filament's elastic behavior. This hypothesis deserves further experimental investigation towards a better understanding of muscle contraction and relaxation. We have demonstrated that the actin-binding Ti I-II, but not the nonactin-binding Ti I-I, fragment (Fig. 6B) can compete with sarcomeric titin filaments to reduce the Ca^{2+}-dependent resting stiffness of rat ventricular muscle (Stuyvers *et al.*, 1998a; 1998b). Therefore, titin-actin interaction contributes to the passive property of cardiac muscle which is critical to myocardial function. Since the titin-actin interaction is not directly dependent on Ca^{2+} (Fig. 6B), the results suggest that Ca^{2+} may regulate the titin-thin filament interaction through the troponin-tropomyosin system.

The mAb Ti104 generated against the I48 Ig module of rat cardiac titin (Ti II, Fig. 1) was found to cross-react with smooth muscle caldesmon. Peptide mapping of the Ti104 epitope by Western blotting and amino acid sequencing of the chymotryptic fragments of the Ti II motif revealed that the Ti104 epitope is located in a 37-amino acid peptide. Ti104 and an anti-caldesmon mAb C21 (Lin *et al.*, 1991) showed very similar reaction patterns to various caldesmon isoforms. Using several deletion mutants of caldesmon, we have identified that

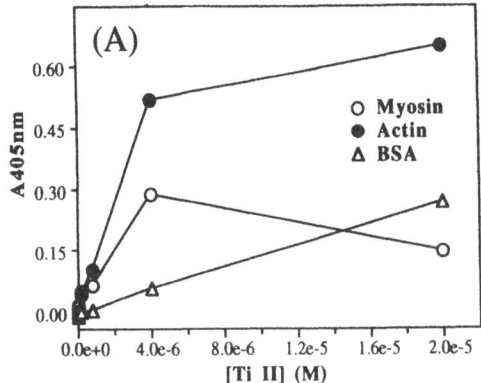

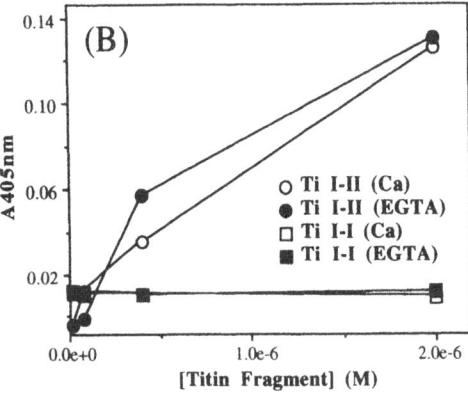

Figure 6. *Interaction of the cloned titin modules with F-actin.*

(A) Interaction of Ti II with F-actin and myosin: Serial dilutions of purified Ti II were incubated with myosin, F-actin or BSA coated on microtitering plates. The amount of Ti II bound to the immobilized proteins was determined by the ELISA procedure (Fig. 3) via a mouse anti-Ti I-II antiserum and HRP-GAM second antibody. The results show that Ti II binds to F-actin but not myosin.

(B) Interaction of the Ti I-II and Ti I-I fragments with F-actin: Serial dilutions of the two titin fragments were incubated with F-actin coated on microtitering plates in the presence or absence of 0.1 mM CaCl₂. The ELISA results show that the Ti I-II but not Ti I-I bound to F-actin in a Ca²⁺-independent manner.

the C21/Ti104 epitope locates in the actin-binding domain (corresponding to Leu693-Leu721) of the smooth muscle caldesmon (Raharjo *et al.*, 1996). The relationship between the Ti104/C21 epitope and the actin-binding site is supported by the fact that C21 competitively inhibits the binding of caldesmon to F-actin-tropomyosin filaments (Lin *et al.*, 1991). The structure/function similarity between Ti II and the actin-binding domain of caldesmon suggests that these two proteins perform analogous roles in organizing the contractile elements and/or in modulating the contraction/motility within muscle and nonmuscle cells (Raharjo *et al.*, 1996).

Since our titin fragments are originated from the A-I junction region of the titin filament in the sarcomere, their biological activity may reflect the specific feature of the A-I junction titin. Structure-function characteristics of the A-I junction titin may be important for titin's role in restoring the sarcomere to its slack length (Helmes *et al.*, 1996). The potential role of the A-I junction titin as a reversible anchoring structure against the thin filament can be further investigated using the cloned titin structural modules (Fig. 1) and the specific mAbs (Table 2). With a combination of molecular cloning, ultrastructural imaging, and biochemical and biophysical characterizations, the interaction between titin and actin thin filament will be revealed to further understand the Ca^{2+}-regulated muscle contraction and relaxation.

Acknowledgments

I would like to thank Mary Resek and Gail McMartin for their technical assistance. This work was supported in part by Grants-in-Aid from the Heart and Stroke Foundation of Canada and the American Heart Association Northeast Ohio Affiliate.

References

Astier C, Raynaud F, Lebbart MC, Roustan C, Benyamin Y. Binding of native titin fragment to actin is regulated by PIP_2. *FEBS Lett* 1998;429:95-98.

Ayme-Southgate A, Vigoreaux J, Benian G, Pardue ML. *Drosophila* has a twitching/titin-related gene that appears to encode projectin. *Proc Natl Acad Sci USA* 1991;88:7973-7977.

Benian GM, Kiff JE, Neckelmann N, Moermann DG, Waterston RH. Sequence of an unusually large protein implicated in regulation of myosin activity in C. elegans. *Nature* 1989;342:45-50.

Eilertsen KJ, Kazmierski ST, Keller TC III. Cellular titin localization in stress fibers and interaction with myosin II filaments in vitro. *J Cell Biol* 1994;74:361-364.

Einheber S, Fischman DA. Isolation and characterization of a cDNA clone encoding avian skeletal muscle C-protein: an intracellular member of the immunoglobulin superfamily. *Proc Natl Acad Sci USA* 1990;87:2157-2161.

Granzier H, Irving TC. Passive tension in cardiac muscle: contribution of collagen, titin, microtubules, and intermediate filaments. *Biophys J* 1995;68:1027-1044.

Granzier H, Kellermayer M, Helmes M, Trombitás K. Titin elasticity and mechanism of passive force development in rat cardiac myocyte probed by thin-filament extraction. *Biophys J* 1997;73:2043-2053.

Helmes M, Trombitás K, Granzier H. Titin develops restoring force in rat cardiac myocytes. *Circ Res* 1996;79:619-626.

Jin J-P. *Molecular studies of cardiac troponin T isoforms*. Ph.D. thesis. The University of Iowa, Iowa City, Iowa, USA 1989.

Jin J-P. Cloned rat cardiac titin class I and class II motifs: Expression, purification, characterization and interaction with F-actin. *J Biol Chem* 1995;270:6908-6916.

Jin J-P, Wang K. Cloning, expression and protein interaction of human nebulin fragments composed of varying number of sequence modules. *J Biol Chem* 1991;266:21215-21223.

Jin J-P, Walsh MP, Resek ME, McMartin GA. Epitope structure and expression of calponin in different smooth muscles and during development. *Biochem Cell Biol* 1996;74:187-196.

Labeit S, Gautel M, Lakey A, Trinick J. Towards a molecular understanding of titin. *EMBO J* 1992;11:1711-1716.

Labeit S, Kolmerer B. Titin: giant protein in charge of muscle ultrastructure and elasticity. *Science* 1995;270:293-296.

Li Q, Jin J-P, Granzier H. The effect of genetically expressed cardiac titin fragments on in vitro actin motility. *Biophys J* 1995;69:1508-1518.

Lin JJ-C, Davis-Nanthakumar EJ, Jin J-P, Lourim D, Novy RE, Lin JL-C. Epitope mapping of monoclonal antibodies against caldesmon and their effects on the binding of caldesmon to Ca^{2+}/calmodulin and to actin or actin-tropomyosin filaments. *Cell Motil Cytoskel* 1991;20:95-108.

Linke WA, Ivemeyer M, Labeit S, Hinssen H, Ruegg JC, Gautel M. Actin-titin interaction in cardiac myofibrils: Probing a physiological role. *Biophys J* 1997;73:905-919.

Linke WA, Stockmeier NR, Ivemeyer M, Hosser H, Mundel P. Characterizing titin's I-band Ig domain region as an entropic spring. *J Cell Sci* 1998;111:1567-1574.

Maruyama K. Connectin/titin, giant elastic protein of muscle. *FASEB J* 1997;11:341-345.

Ogut O, Jin J-P. Expression, zinc-affinity purification and characterization of a novel metal-binding cluster in troponin T: Metal-stabilized α-helical structure and effects of the NH_2-terminal variable region on the conformation of intact troponin T and its association with tropomyosin. *Biochemistry* 1996;35:16581-16590.

Ogut O, Jin J-P. Developmentally regulated, alternative RNA splicing-generated pectoral muscle-specific troponin T isoforms and role of the NH_2-terminal hypervariable region in the tolerance to acidosis. *J Biol Chem.* 1998;273:27858-27866.

Olson NJ, Pearson RB, Needleman D, Hurwitz MY, Kemp BE, Means AR. Regulatory and structural motifs of chicken gizzard myosin light chain kinase. *Proc Natl Acad Sci USA* 1990;87:2284-2288.

Price MG, Gomer RH. Skelemin, A cytoskeletal M-disc periphery protein, contains motifs of adhesion/recognition and intermediate filament proteins. *J Biol Chem* 1993;268:21800-21810.

Raharjo WH, Lin JJ-C, Mak AS, Jin J-P. An epitope structure shared by titin and caldesmon. *Biophys J* 1996;70:A379.

Rief M, Gautel M, Schemmel A, Gaub HE. The mechanical stability of immunoglobulin and fibronectin III domains in the muscle protein titin measured by atomic force microscopy. *Biophys J* 1998;75:3008-3014.

Schagger H, von Jagow G. Tricine-sodium dodecyl sulfate-polyacrylamide gel electrophoresis for the separation of proteins in the range of 1 ~ 100-kDa. *Anal Biochem* 1987;166:368-379.

Studier FW, Rosenberg AH, Dunn JJ, Dubendorff JW. Use of T7 RNA polymerase to direct expression of cloned genes. *Method Enzymol* 1990;185:60-89.

Stuyvers BD, Miura M, Jin J-P, ter Keurs HEDJ. Ca^{2+}-dependence of diastolic properties of cardiac sarcomeres: involvement of titin. *Prog Biophysics Mol Biol* 1998a;69:425-443.

Stuyvers BD, Jin J-P, ter Keurs HEDJ. Involvement of titin in stiffness-$[Ca^{2+}]$ relationship of cardiac sarcomeres. *Biophys J* 1998b;74:A351.

Trinick J. Elastic filaments and giant proteins in muscle. *Curr Opin Cell Biol* 1991;3:112-118.

Trombitás K, Jin J-P, Granzier, H. The passive-tension domain of titin in cardiac muscle. *Circ Res* 1995;77:856-861.

Trombitás K, Greaser M, Labeit S, Jin J-P, Kellermayer M, Helmes M, Granzier H. Titin extensibility *in situ*: Entropic elasticity of both permanently folded and permanently unfolded molecular segments. *J Cell Biol* 1998;40:853-859.

Wang J, Jin J-P. Primary structure and developmental acidic to basic transition of 13 alternatively spliced mouse fast skeletal muscle troponin T isoforms. *Gene* 1997;193:105-114.

Wang J, Jin J-P. Conformational modulation of troponin T by configuration of the NH_2-terminal variable region and functional effects. *Biochemistry* 1998;37:14519-14528.

Wang K. "Sarcomere associated cytoskeletal lattice in striated muscle." In *Cell and Muscle Motility* Vol. 6, JW Shay, ed. New York: Plenum Press, 1985;315-369.

Weikamp B, Jurk K, Beinbrech G. Projectin-thin filament interaction and modulation of the sensitivity of the actomyosin ATPase to calcium by projectin kinase. *J Biol Chem* 1998;273:19802-19808.

Discussion

Pollack: My question actually links your talk with part of my talk and has to do with TnT and the various mutants that you've been talking about. On one of my slides I showed evidence that we had published earlier for the lateral interconnections between the filaments in the I bands. I didn't discuss them to any great extent. However, the hypothesis that we had come up with is that these interconnections are TnT. The reason is that we found that the interconnections repeated every 40 nm in axial direction. The length of the TnT molecule is 20-30 nm, something like that. It was about the right size, had the right periodicity, and was the only sarcomeric protein present in enough abundance. So my question to you is what do you think of that hypothesis and second, whether you have done any EM studies on any of the TnT mutants, to see what kind of structural changes you might see in cases like that.

Jin: For the first part of your question, we believe that titin and the thin filament interact in the sarcomere. In *in vitro* protein binding assays, we have demonstrated this interaction. Since the thin filament conformation is regulated by troponin-tropomyosin during the contractile cycles, the position and functional states of troponin may be important for the titin-actin interaction. However, our *in vitro* binding assays did not show any direct interaction between titin fragments and troponin or tropomyosin. As to the second part of the question, we only did a rather preliminary EM on those transgenic mouse hearts, and the picture showed very short sarcomeres. We don't understand this result and need to do more experiments.

Gregorio: I noticed in one of your figures, I think you used one from Kuan Wang's review, you had nebulin binding along the actin filament in a groove, independent from the tropomyosin-binding groove. Is this the current model in the field?

Jin: That is the only comprehensive figure I found in literature. Many points in this model are hypothetical because we really don't know how nebulin functions in the sarcomere. I remember during my training in Texas, when I demonstrated nebulin-actin interaction for the first time, Kuan was so excited that he offered me a chance to rename the protein. However, we didn't do so because of the lack of a functional model.

Sanger: A question about your early gel; which smooth muscle was it?

Jin: That was chicken gizzard.

Sanger: And two, did you do any "guesstimate" about that high titin-like band with respect to the myosin? It looked as a much lower ratio than you got for the other two muscles.

Jin: Yes, that's why we have ignored it for years. But Tom Keller demonstrated that it also exists in non-muscle cells. We should really pay some attention to them. But, you are right, the stoichiomery is very different from that of striated muscle.

Ter Keurs: When you made a transgenic TnT animal, Frank Brozovich measured the force calcium relations. Did he try at all to see whether there is an effect or stretch on these muscles? You would expect an effect in the control muscle and you could test whether it is changed as a result of a TnT change in the transgenic muscle.

Jin: This is an important point. We didn't do any stretch experiments, just Ca^{2+} activation. Actually, our first speaker of this session (Le Guennec) presented rather convincing data showing the stretch of titin causes a change in sensitivity to Ca^{2+} activation, and this could be via the troponin regulatory system. Then, there is a question that I didn't have time to ask earlier, if a stretch increases calcium sensitivity, which was removed by mild protease digestion, this effect probably will alter the cooperativity of the activation. The hypothesis is that when the sarcomere shortens, the proposed stretch-dependent sensitivity will decrease, reducing the overall cooperativity of the activation curve. When the stretch-induced sensitization is removed by trypsin digestion, the sarcomere length-dependent activation would show a steeper curve due to the more decrease in sensitivity at the longer sarcomere length. Since the changes in troponin function alter the Ca^{2+} sensitivity in a tension-dependent manner, reflected by a change in cooperativity, I think we can address these questions using our TnT transgenic cardiac muscle model.

IS TITIN THE LENGTH SENSOR IN CARDIAC MUSCLE?
PHYSIOLOGICAL AND PHYSIOPATHOLOGICAL PERSPECTIVES

Jean-Yves Le Guennec, Olivier Cazorla,
Alain Lacampagne[1], and Guy Vassort[1]

*Laboratoire de Physiologie des Cellules
Cardiaques et Vasculaires, Tours, France[1]
Laboratoire de Physiopathologie Cardiovasculaire,
Montpellier, France*

Abstract: One of the most salient physiological characteristics of cardiac muscle is that a dilated heart pumps more vigorously, a phenomenon known as the Frank-Starling relationship (see Allen and Kentish, 1985). At least two cellular mechanisms participate in this phenomenon: the reduction of the interfilament lattice spacing which favors the formation of cross-bridges (Wang and Fuchs, 1995) and the increased affinity of troponin C (TnC) for calcium (Ca^{2+}) (Babu *et al.*, 1988). In the latter case, it has been established that TnC itself is not the length sensor (Moss *et al.*, 1991). The intracellular structure(s) able to sense changes in cell length has always been challenged and is still not known. We previously observed on intact isolated cardiac cells that active tension is more closely related to passive tension than to sarcomere length *per se* (Cazorla *et al.*, 1997). This might have some physiological implications in the working heart since we found that sub-epicardial cells are more supple than sub-endocardial cells. In the present work on skinned cells, we studied the relationship between different levels of passive tension (modulated by a mild trypsin digestion) and the shift in pCa_{50} of tension-pCa relations induced by a stretch of cells from 1.9 to 2.3 μm sarcomere length. A significant correlation was obtained between passive tension and the stretch-induced shift in pCa_{50}, or stretch-sensitivity of the active force. These observations led us to assume that titin might play a role in sensing cell length to modulate the contractile activity. Besides, it is known that myocardial infarcted cells are less sensitive to stretch. We propose that, in such a rat model, alterations of titin might participate in heart failure.

Introduction

Striated muscle cells, including cardiac cells, are mainly constituted by three kinds of filaments: thin filaments mainly composed of actin, thick filaments mainly composed of myosin, and titin. Titin (also known as connectin) was discovered about two decades ago (Maruyama *et al.*, 1976) and is the largest polypeptide ever described (around three megadaltons). A single molecule spans half a sarcomere, connecting I- and A-bands with its amino terminus associated to the Z-line and its carboxy terminus to the M-line (see Labeit *et al.*, 1997). Along the sarcomeric structure, several proteins have been shown to interact with titin. They include α-actinin (Sorimachi *et al.*, 1997), actin near the Z line (Trombitás *et al.*, 1997; Linke *et al.*, 1997) and probably outside it (Trombitás and Granzier, 1997; Stuyvers *et al.*, 1998), myosin (Houmeida *et al.*, 1995), calpain p94 (Sorimachi *et al.*, 1995), C-protein (Freiburg and Gautel, 1996) and myomesin (Soteriou *et al.*, 1993; Obermann *et al.*, 1997). All these interactions strongly suggest that titin is a ruler during myofibrillogenesis (Freiburg and Gautel, 1996; Mayans *et al.*, 1998). Also, titin helps maintain the structural integrity of the contracting sarcomere (Horowits *et al.*, 1986).

Intrinsically, titin exhibits elastic properties in its I-band domain. This mechanical property is mainly due to a serially linked immunoglobulin (Ig) - like domain and a PEVK segment (constituted of 70% proline (P), glutamate (E), valine (V) and lysine (K)) (Labeit and Kolmerer, 1995; Linke *et al.*, 1996; 1998). PEVK and the tandem-Ig domains compose a two-spring system acting in series. Cardiac cells are much stiffer than are skeletal muscle cells. Granzier and Irving (1995) showed that this difference is related to the shorter length of cardiac titin. Furthermore, the PEVK segment is shorter in cardiac cells than in skeletal cells (163 *vs.* 2200 residues) (Labeit and Kolmerer, 1995).

The elastic properties of titin are involved in another typical cardiac characteristic, relaxation, known as lusitropism in the whole heart. Contraction is a consequence of an increase in cytoplasmic Ca^{2+} concentration. Ca^{2+} binding to troponin C unmasks the interaction site of actin for myosin, leading to force generation by cross-bridges and sliding of thin and thick filaments. After Ca^{2+} ions have been extruded from the cytoplasm, actin and myosin dissociate. To fully relax, cells need an internal system of restoring forces which brings back myofilaments to their resting position. This is mainly attributed to titin (Helmes *et al.*, 1996).

It cannot be excluded that titin might have other roles since it interacts with so many proteins. Some of these interactions can be regulated by molecules like PIP_2 (Astier *et al.*, 1998) and Ca^{2+} (Kellermayer and Granzier, 1996; Stuyvers *et al.*, 1998). In this chapter, we report that titin participates in the cardiac length sensing mechanism with deleterious consequences during physiopathological situations.

Discussion

Effects of stretch on intact isolated cells

A technique has been developed that allows us to attach intact isolated cardiac mammalian cells using thin carbon fibers (Le Guennec *et al.*, 1990). Using this technique, we showed that stretching isolated cardiomyocytes positively modulates the inotropic state. This allowed the establishment of length-tension relationships, which are representative at the cellular level of the pressure-volume relationship known as the Frank-Starling law.

We previously reported that some guinea-pig cardiac cells could be stretched several times while others developed a calcium overload and membrane depolarization after just one or two stretches. These observations were initially attributed to a lesion of the membrane produced by the attachment to carbon fibers. It was later observed that the increase in resting Ca^{2+} ($[Ca^{2+}]_i$) can be prevented by a stretch-activated channel blocker, streptomycin, at a concentration of 40 µM. We concluded that this phenomenon was physiological: some cells, but not all, responded to a stretch by an arrhythmogenic behavior, probably due to the presence of a particular class of stretch-activated channels (Gannier *et al.*, 1994; 1996).

To avoid these stretch-induced arrhythmia, 10 µM streptomycin was introduced in all physiological solutions. Then the large increase in $[Ca^{2+}]_i$ associated with cell depolarization was rarely seen. However, in such conditions, cells could be divided in two groups on the basis of their mechanical resting properties. Some cells are lengthened upon stretching and developed weak passive force, we named them supple or compliant cells. The others, the stiff cells lengthened much less with stretch and developed more passive force (Gannier *et al.*, 1994). In a more detailed study, we reported that these differences in stiffness depended on the cellular origin within the layer of the ventricular wall: stiff cells are mainly localized in the sub-endocardium while supple cells are mainly in the sub-epicardium (Cazorla *et al.*, 1997).

When the active properties of both cell types were compared, we got surprising results. Supple cells exhibited a well-known behavior upon applied stretch: stretch induces sarcomere lengthening and increases both passive and active tensions. This is the basis of the Frank-Starling law where sarcomere lengthening is associated to positive inotropic effects. However, in stiff cells, the behavior was quite different. Stretching these cells increased both passive and active tensions despite a small or even negligible sarcomere lengthening. After pooling the data, the two cell populations were clearly distinguishable when we plotted active tension *vs.* resting sarcomere length. However, both populations were mixed when we plotted active tension *vs.* passive tension. These results were interpreted as follows: The passive tension rather than the length *per se* is very important in modulating active force and contributes to the Frank-Starling law. We also postulated that titin might participate in the length-sensing mechanism, since this protein is responsible for about 90% of cardiac cell passive tension.

Is titin the length sensor in cardiac cells?

Modulation of the cardiac contraction by stretch involves mainly a modification of the Ca^{2+} sensitivity of the contractile machinery. The hypothesis that titin could be the length-sensor was tested by quantifying the changes in Ca^{2+}-sensitivity of the contractile machinery with stretch (stretch sensitivity) and by reducing the amount of "active" titin.

The sensitivity of the contractile machinery to Ca^{2+} ions was assessed in chemically skinned or "permeabilized" cells. In brief, cardiac cells were enzymatically isolated as in the above experiments and then submitted to a detergent (Triton X-100) that permeabilized all cell membranes (sarcolemma, mitochondria and sarcoplasmic reticulum), allowing the medium around the myofibrils to be easily controlled. Cells were attached to glass rods using optical glue (Pucéat *et al.*, 1990). One of the rods was linked to a micromanipulator, allowing for cell length changes and the other rod was attached to a strain gauge (Fig. 1) which measured the force developed by the cell during a stretch (passive force) and during application of sub-threshold pCa solutions (active force).

Using this technique, we showed that passive properties of sub-endocardial and sub-epicardial cells are different (Cazorla *et al.*, unpublished). Sub-endocardial cells are stiffer than are sub-epicardial cells (Fig. 2).

These results confirm that the differences in stiffness observed in the previous experiments comparing sub-epicardial and sub-endocardial cells were due to the intrinsic passive properties of the cells and not to special features in metabolism leading to different $[Ca^{2+}]_i$ or pH_i.

Since we found, in intact isolated cells, that sub-endocardial cells are more sensitive to stretch than are sub-epicardial cells, we investigated the sensitivity to Ca^{2+} of the contractile machinery with stretch. For this purpose, the active force developed at various pCas was normalized to the maximal active force developed at pCa 4.5 (see inset, Fig. 1). This was done at 2 SLs in both types of cells.

As shown in Figure 3, stretch sensitivity is higher in sub-endocardial cells than in sub-epicardial cells. This finding is in agreement with our previous results on intact cardiac cells and strengthens the hypothesis that passive tension plays an important role in the modulation of active tension. In the following, to test whether this modulation of active tension by stretch is under the control of resting tension, we checked variations in the stretch sensitivity on altering resting tension. These experiments were performed on rat ventricular cardiac cells since we observed that under our experimental conditions, these cells are more sensitive to stretch than guinea-pig cardiac cells (0.4 *vs.* 0.2 shift in pCa_{50} when increasing the SL from 1.9 to 2.3 µm).

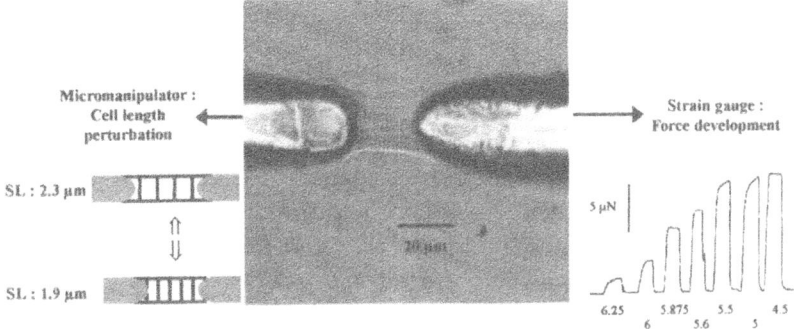

Figure 1. Microphotograph of a permeabilized cell attached to glass rods. Mean sarcomere length (SL), was determined using a fast Fourier transform algorithm of the acquired video image (Gannier et al., 1993). Inset at right shows the active force developed by a cardiac cell when applying the different pCa solutions as indicated below the recording.

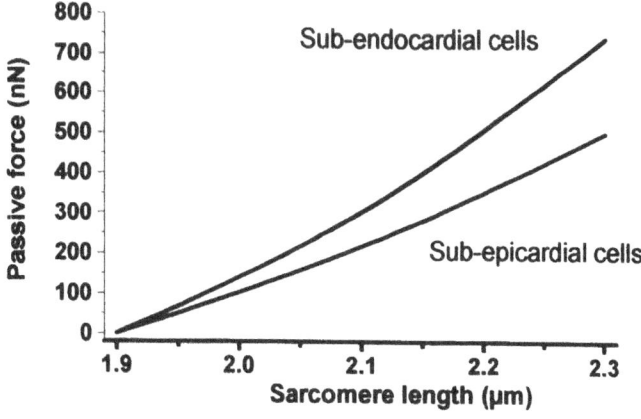

Figure 2. Passive force-SL relationships established in permeabilized guinea-pig ventricular cells isolated from the sub-epicardium and the sub-endocardium. The curves correspond to the fits obtained from experimental data.

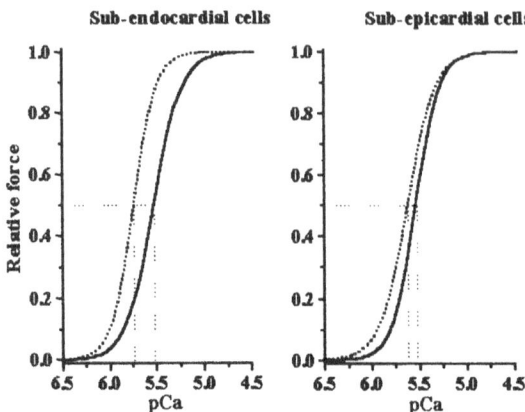

Figure 3. *Comparison of length-dependent effects on force-pCa curves of permeabilized cells isolated from the sub-epicardium and the sub-endocardium. These curves were established at 2 sarcomere lengths : 1.9 μm (solid line) and 2.3 μm (dotted line) and normalized to maximal tension value. The vertical dotted lines indicate the pCa necessary to evoke 50 % of the maximal force (pCa$_{50}$). The difference in pCa$_{50}$ at the 2 SL is an index of the stretch sensitivity of the contractile machinery. The curves are fits*

Titin content, and thus passive tension, was modulated in the cells by submitting them to a mild trypsin digestion protocol. This protocol has been used first by Yoshioka *et al.* (1986) in skeletal muscle and later by Granzier and Irving (1995) in cardiac muscle cells. This protocol consists of superfusing isolated permeabilized cells with a solution containing a low concentration of trypsin (0.25 μg/ml) for a period of time varying between 5 to 30 min. In such conditions, trypsin exclusively degrades titin and not the contractile proteins nor the regulatory proteins (Yoshioka *et al.*, 1986; Granzier and Irving, 1995; Cazorla *et al.*, 1999). The degradation of titin was estimated from the reduction in passive force when stretching cells from 1.9 to 2.3 μm SL. The time needed to reduce passive tension by half was around 12 min. Eighty to 85% of the initial resting force was lost after a 30-min trypsin digestion. The digestion was stopped at different times, corresponding, thus, to different levels of passive tension, by substituting trypsin with a protease inhibitor, leupeptine (80 μM). After confirming that titin was the only protein affected by this protocol (see Cazorla *et al.*, 1999), we found a negative correlation between the stretch-induced shift in pCa$_{50}$ (stretch sensitivity) and the remaining level of passive tension in the range 20 to 70% of the initial passive tension (Fig. 4).

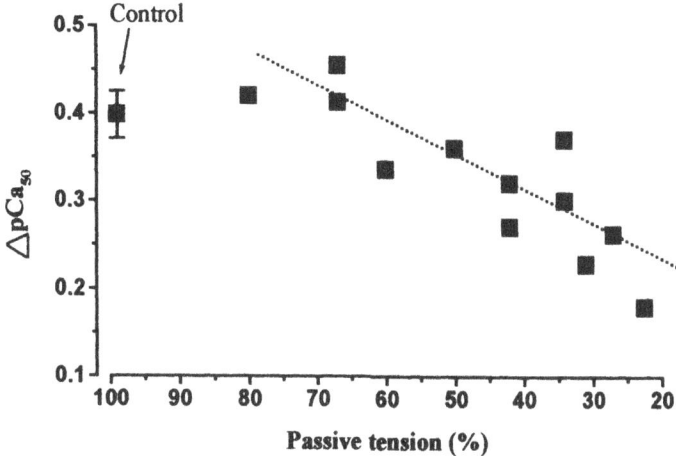

Figure 4. *Correlation between stretch sensitivity of the contractile machinery ($DpCa_{50}$) and passive tension (normalized to the passive tension before starting trypsin digestion).*

Above 70%, a more detailed study is needed to determine how stretch sensitivity is changed. Also, even after almost all the titin had been degraded (below 20% of passive tension) there was always some stretch-sensitivity (about $0.15\ pCa_{50}$ shift). This might represent the component of stretch sensitivity that is attributable to other regulatory mechanisms such as modifications of the interfilament lattice spacing (Wang and Fuchs, 1995).

Physiopathological consequences

These experiments demonstrate that the passive tension is able to modulate the active tension. Moreover, this newly described regulation seems to involve titin with the titin contributing to the length-sensing mechanism that accounts for the Frank-Starling law. This is of physiological and physiopathological importance. Physiologically, we showed that sub-epicardial and sub-endocardial cells have different stiffnesses. The sub-endocardial stiff cells are more sensitive to stretch than are sub-epicardial supple cells. This should have some implication during ventricular contractions. Epicardial cells are expected to lengthen more during ventricular blood filling, due to their peripheral position. However their lower sensitivity to stretch favors a force development in the same range as sub-endocardial cells. Such a behavior might participate in the efficiency of cardiac work.

Furthermore, one might anticipate that perturbations in titin content and/ or isoforms have consequences in the functioning of the heart. It has been reported that titin content is reduced in the human failing heart (Morano *et al.*, 1994) and that volume-pressure relationship is altered (Schwinger *et al.*, 1994).

Both observations can be linked in a simplified model in agreement with the length-sensing role of titin. A reduced titin content might make the heart less sensitive to changes in ventricular wall distention, contributing, thus, to the failure.

To test this hypothesis, we performed experiments on rats that underwent left anterior coronary artery ligature. After 24 weeks post-myocardial infarction, the hearts were diagnosed failing as indicated by the presence of pleural infusions and ascites, together with a roughly doubled end-diastolic volume. In preliminary experiments, we checked for the decreased resting tension, and if this was present, we checked to see whether this was associated with a decreased Ca^{2+}-sensitivity of the contractile machinery.

As shown in Figure 5, passive tension is significantly reduced in cells from failing rats compared to sham rats (rats that were submitted to all the surgical procedures without the coronary ligature).

We tested indirectly if a reduced titin content can explain the lower passive force observed in failing rat hearts. To do that, we applied the mild trypsin digestion protocol to permeabilized cells as was done previously in control rat cells. A typical result is shown in Figure 6.

In failing rat cells, the mild trypsin digestion protocol reduced passive tension to about 50% of the initial passive tension instead of 15-20% in control rat cells. The reduced trypsin-sensitive component of passive tension in failing heart cells suggests a reduced titin content, as was already reported in human heart cells (Morano *et al.*, 1994). More studies are necessary to explain why the titin-based force is reduced in heart failure (both qualitative and quantitative modifications are expected).

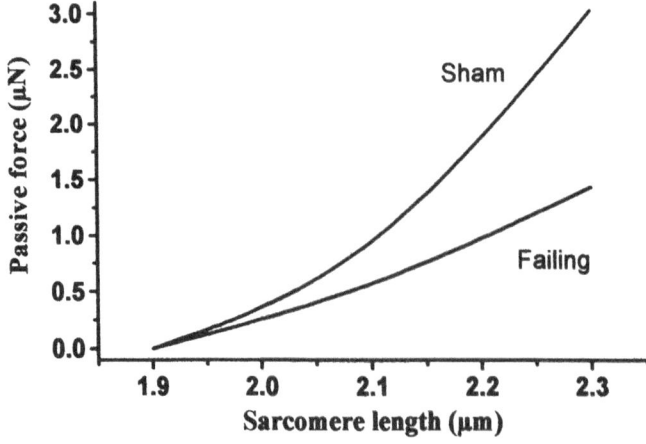

***Figure* 5**. *Passive force-SL relationships established in sham and failing heart permeabilized cells. The curves are fits obtained from experimental data.*

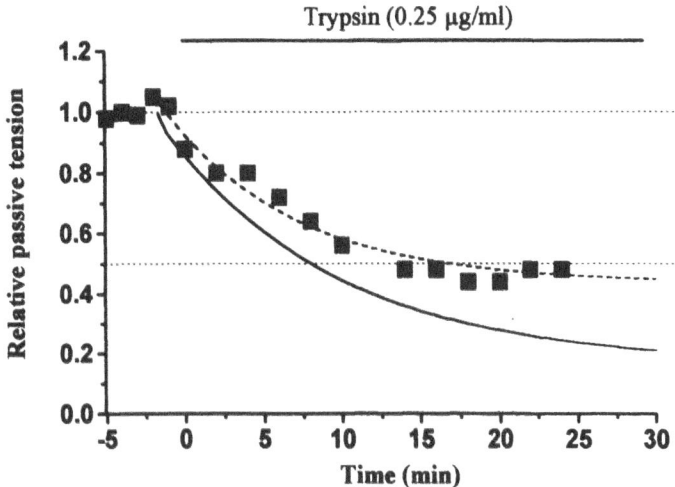

Figure 6. *Effects of a mild trypsin digestion on relative passive tension of a cell isolated from a failing rat. For comparison, the continuous curve represents mean time-dependent decrease in passive tension of a cell from a control rat.*

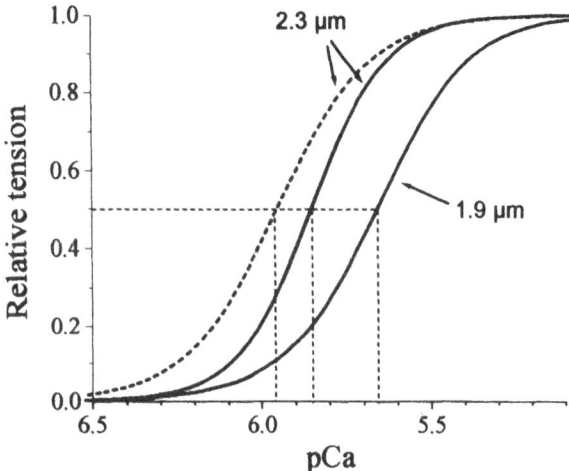

Figure 7. *Tension-pCa relationships in sham (dotted line) and in failing (solid line) heart cells. These curves were obtained at 1.9 and 2.3 μm SL. The two curves at 1.9 μm SL are superimposed. Curves are fits obtained from experimental data.*

The reduced titin content that accounts for the reduced passive tension in such failing hearts might as well induce alterations in the stretch sensitivity of the contractile machinery if titin participates in the length-sensing mechanism. Such is shown in Figure 7, which illustrates a lower sensitivity to stretch of failing heart cells when compared to sham. It thus seems thus that even if a direct link of titin is not clearly established and complementary experiments are needed, there is a strong relationship between the reduced passive tension and the reduced sensitivity to stretch of the active force in the failing rat heart cells.

Conclusions

In intact cells, we previously established that active tension is more closely related to passive tension than to sarcomere length *per se*. This close relationship between passive tension and stretch-sensitivity of the contractile machinery to generate active force was confirmed in permeabilized cells. In both, control cells after mild trypsin digestion and failing cells, the reduced passive tension was associated with a reduced sensitivity of active force to stretch. A common feature between these two cell models was their titin content: a mild trypsin digestion of control cells is known to degrade primarily titin while in failing cells titin content is reduced, at least in the human heart.

How can titin be the length-sensor in heart cells? The first hypothesis is that titin can modify the position of myosin heads since it interacts with myosin molecules (Houmeida *et al.*, 1995). Titin can also contribute to the reduction in interfilament lattice spacing with stretch, which is known to play a role in the Frank-Starling mechanism (Wang and Fuchs, 1995). Once titin properties can be fully understood, other hypotheses can be proposed.

References

Allen D, Kentish J. The cellular basis of the length-tension relation in cardiac muscle. *J Mol Cell Cardiol* 1985;17:821-840.

Astier C, Raynaud F, Lebart M-C, Roustan C, Benyamin, Y. Binding of a native titin fragment to actin is regulated by PIP_2. *FEBS Letters* 1998;429:95-98.

Babu A, Sonnenblick E, Gulati J. Molecular basis of the influence of muscle length on myocardial performance. *Science* 1988;240:74-76.

Cazorla O, Pascarel C, Garnier D, Le Guennec J-Y. Resting tension participates in the modulation of active tension in isolated guinea pig ventricular myocytes. *J Mol Cell Cardiol* 1997;29:1629-1637.

Cazorla O, White E, Le Guennec J-Y. Regional differences in the passive properties in single rat and ferret ventricular myocytes. *Biophys J*, 1998;74:A155.

Cazorla O, Vassort G, Garnier D, Le Guennec J-Y. Length modulation of active force in rat cardiac myocytes: Is titin the sensor? *J Mol Cell Cardiol*, 1999;31:1215-1227.

Freiburg A, Gautel M. A molecular map of the interactions between titin and myosin-binding protein C. *Eur J Biochem* 1996;235:317-323.

Gannier F, White E, Lacampagne A, Garnier D, Le Guennec J-Y. Streptomycin reverses a large stretch-induced increase in $[Ca^{2+}]_i$ in isolated guinea-pig ventricular myocytes. *Cardiovasc Res* 1994;28:1193-1198.

Gannier F, White E, Garnier D, Le Guennec J-Y. A possible mechanism for large stretch-induced increases in [Ca^{2+}]$_i$ in isolated guinea-pig ventricular myocytes. *Cardiovasc Res* 1996;32:158-167.

Granzier H, Irving T. Passive tension in cardiac muscle: contribution of collagen, titin, microtubules, and intermediate filaments. *Biophys J* 1995;68:1027-1044.

Helmes M, Trombitás K, Granzier H. Titin develops restoring force in rat cardiac myocytes. *Circ Res* 1996;79:619-626.

Horowits R, Kempner E, Bisher M, Podolsky R. A physiological role for titin and nebulin in skeletal muscle. *Nature* 1986;626:160-164.

Houmeida A, Tskhovrebova L, Trinick J. Studies of the interaction between titin and myosin. *J Cell Biol* 1995;131:1471-1481.

Kellermayer M, Granzier H. Calcium-dependent inhibition of in vitro thin-filament motility by native titin. *FEBS Letters* 1996;380:281-286.

Labeit S, Kolmerer B. Titins, giant proteins in charge of muscle ultrastructure and elasticity. *Science* 1995;270:293-296.

Labeit S, Kolmerer B, Linke W. The giant protein titin. Emerging roles in physiology and pathophysiology. *Circ Res* 1997;80:290-294.

Le Guennec J-Y, Peineau N, Argibay J, Mongo K, Garnier D. A new method of attachment of isolated mammalian ventricular myocytes for tension recording: length dependence of passive and active tension. *J Mol Cell Cardiol* 1990;22:1083-1093.

Linke W, Ivemeyer M, Olivieri N, Kolmerer B, Rüegg J, Labeit S. Towards a molecular understanding of the elasticity of titin. *J Mol Biol* 1996;261:62-71.

Linke W, Ivemeyer M, Labeit S, Hinssen H, Rüegg J, Gautel M. Actin-titin interaction in cardiac myofibrils: probing a physiological role. *Biophys J* 1997;73:905-919.

Linke W, Ivemeyer M, Mundel P, Stockmeier M, Kolmerer B. Nature of PEVK-titin elasticity in skeletal muscle. *P N A S USA* 1998;95:8052-8057.

Maruyama K, Natori R, Nonomura Y. New elastic protein from muscle. *Nature* 1976;262:58-60

Mayans O, Van Der Ven P, Wilm M, Mues A, Young P, Fürst D, Wilmanns M, Gautel M. Structural basis for activation of the titin kinase domain during myofibrillogenesis. *Nature* 1998;395:863-869.

Morano I, Hädicke K, Grom S, Koch A, Schwinger R, Böhm M, Bartel S, Erdmann E, Krause E. Titin, myosin light chains and C-protein in the developing and failing human heart. *J Mol Cell Cardiol* 1994;26:361-368.

Moss R, Nwoye L, Greaser M. Substitution of cardiac troponin C into rabbit muscle does not alter the length dependence of Ca^{2+} sensitivity of tension. *J Physiol (Lond)* 1991;440:273-289.

Obermann W, Gautel M, Weber K, Furst D. Molecular structure of the sarcomeric M band: mapping of titin and myosin binding domains in myomesin and the identification of a potential regulatory phosphorylation site in myomesin. *EMBO J* 1997;16:211-220.

Pucéat M, Clément O, Lechêne P, Pelosin J, Ventura-Clapier R, Vassort G. Neurohormonal control of calcium sensitivity of myofilaments in rat skinned heart cells. *Circ Res* 1990;67:517-524.

Schwinger R, Böhm M, Koch A, Schmidt U, Morano I, Eissner H, Überfuhr P, Reichart B, Erdmann E. The failing human heart is unable to use the Frank-Starling mechanism. *Circ Res* 1994;74:959-969.

Sorimachi H, Kinbara K, Kimura S, Takahashi M, Ishiura S, Sasagawa N, Sorimachi N, Shimada H, Tagawa K, Marayama K, Susuki K. Muscle-specific calpain, p94, responsible for limb girdlemuscular dystrophy type 2A, associates with connectin through IS2, a p94-specific sequence. *J Biol Chem* 1995;270:31158-31162.

Sorimachi H, Freiburg A, Kolmerer B, Ishiura S, Stier G, Gregoro C, Labeit D, Linke W, Susuki K, Labeit S. Tissue-specific expression and α-actinin binding properties of the Z-disc titin: implications for the nature of vertebrate Z-disks. *J Mol Biol* 1997;270:688-695.

Soteriou A, Gamage M, Trinick J. A survey of the interactions made by titin. *J Cell Science* 1993;14:119-123.

Stuyvers B, Miura M, Jin J-P, Ter Keurs H. Ca²⁺-dependence of diastolic properties of cardiac sarcomeres: involvement of titin. *Progress Biophys Mol Biol* 1998;69:425-443.

Trombitás K, Granzier H. Actin removal from cardiac myocytes shows that near Z line titin attaches to actin while under tension. *Am J Physiol* 1997;273:C662-C670.

Trombitás K, Greaser M, Pollack G. Interaction between titin and thin filaments in intact cardiac muscle. *J Muscle Res Cell Motil* 1997;18:345-351.

Wang Y-P, Fuchs F. Osmotic compression of skinned cardiac and skeletal muscle bundles : Effects on force generation, Ca²⁺ sensitivity and Ca²⁺ binding. *J Mol Cell Cardiol* 1995;27:1235-1244.

Yoshioka T, Higuchi H, Kimura S, Ohashi K, Umazume Y, Maruyama K. Effects of mild trypsin treatment on the passive tension generation and connectin splitting in stretched skinned fibers from frog skeletal muscle. *Biomed Res* 1986;7:181-186.

Discussion

Pollack: Could you flash up the last slide please? Could you elaborate on these proposed mechanisms? I don't really understand how they could actually work. For example, the titin would be stretched in the I-band not the A-band and so how would it or how could it affect myosin head position? And then how would titin participate in the reduction of interfilament lattice. Maybe you could elaborate? Because I don't understand.

LeGuennec: For the myosin head position, it is known that titin interacts with C protein, which is close to the head of myosin. So it will be force which will be transmitted to change the head of the myosin in a more efficient position, like it happens during phosphorylation of MLC2.

Pollack: But is that region of titin actually under tension—the A-band portion?

LeGuennec: It is not certain...

TerKeurs: Is there a problem with the units that you used for stress development by these cells? A muscle generates 75 mN per square milimeter; that would be 75 nN per square micrometer. You have one nanoNewton per square micrometer.

LeGuennec: It did not make sense, huh?

TerKeurs: In the slides that I saw, I don't know which cells they were any more, but it was a surprising 100-fold lower than in muscles. And the funny thing, you would expect a higher stress.

LeGuennec: I have not studied the literature. But what we found, that is what you saw on the slide.

TerKeurs: And it is not the resting force, it is the active force development of your intact cells.

LeGuennec: Ah, for intact cells, but it was auxotonic conditions, so you cannot know. It depends on the compliance of the fiber which attaches the cell, because the more stiff it is, the more you would realize isometric conditions and the force would be higher.

TerKeurs: So in that case, the question is, what did you plot on the length axis? Sarcomere length prior to contraction or at the peak of contraction?

LeGuennec: Yes, in this slide we plotted diastolic sarcomere length before contraction, but we also did sarcomere lengths at the peak of contraction. I think there was no relationship at all in that case. It was not clear.

TerKeurs: I am very puzzled by the fact of 100-fold difference in stress development. That's a bit worrisome. The other question I have regards your heart failure muscles. This was a rat infarct model? If one makes an infarct in the left ventricle of rat, by ligating the anterior descending artery, you can induce a large infarct of one square centimeter. And everybody knows that the heart of a rat is about 1 cm in diameter. An infarct of 1 square cm does not need to cause heart failure. An infarct of 1 1/2 or 2 square cm induces evidence of clear heart failure. So my question to you is what was the evidence for heart failure?

LeGuennec: ...And also there were measurements made by echocardiography.

TerKeurs: All those hearts will be bigger whether there's heart failure or not.

LeGuennec: You think?

TerKeurs: I'm sure.

LeGuennec: I thought that a doubling of the end-diastolic volume and a reduced ejection fraction are characteristic of heart failure.

Sugi: I have a question about one of your conclusions. You mentioned that isometric tension is more closely related to passive tension than absolute sarcomere lengths. Is that right?

LeGuennec: Yes.

In the field of skeletal muscle mechanics there is a lot of literature indicating that isometric tension is heavily dependent on the past history with which the muscle fiber is brought to the isometric length where force is measured. So in your case, when passive tension is different, than past history with which the fiber becomes isometric may be also different. So I think this point should be carefully examined. My point is that it may depend on past history.

LeGuennec:I think that you are right. These are mainly preliminary results. We are planning more complete experiments. But I think that in skeletal muscle, Henk Granzier showed that passive tension may be related to active tension. There was a paper about that.

Granzier: I did work on insect flight muscle many years ago in Kuan Wang's lab. We looked at the phenomenon of stretch activation. Actually we drew the same conclusion as you did: it is not sarcomere length that sets the level of active tension, but instead the level of passive tension that you have when you initiate contraction.

Linke: I have a question related to one of the slides you showed about the relationship between the delta pCa 50 and the passive tension when you use the trypsin treatment to decrease it. I think there is that one paper by Higuchi where he did exactly that experiment and he did not see any change in calcium sensitivity. He did exactly that experiment. In your experiment you need a change, as far as I could get it from the quick slide, of at least a 40% decrease in passive tension before you see the shift. So for me that would mean actually that passive tension is not that important for this shift in calcium sensitivity.

LeGuennec:Yes, there is a big gap between 100 and 60-70%, I agree. We don't know exactly what we do with our trypsin digestion. We can imagine some opposite effects of trypsin; first an increase in sensitivity and next a decrease in sensitivity. But it may be something else. We have to work more carefully on this part, I agree.

Vigoreaux: I was wondering whether either you or Henk can speculate on the significance of the relationship between passive tension and active tension on the large effect that apparently passive tension has on active tension.

LeGuennec:For cardiac cells, it is known that when you stretch a cardiac cell there is an increase of calcium affinity of Tropinin-C, and it has been shown by the Moss group, by two different techniques, by protein substitution and by using transgenic mice. But in fact it seems the vector of expression of the cardiac Troponin-C is very important. It seems that if you put cardiac Troponin-C in skeletal contractile proteins, there is no stretch effect like that. So what you need to have is the contractile machinery of cardiac cells. This means that cardiac Troponin-C needs something else to sense the length. What we think is that one of the mechanisms that participates in length sensing is titin.

Vigoreaux: I understand that, but why would it be different than skeletal muscle? I mean I am trying to understand the significance, the physiological significance.

LeGuennec: Skeletal muscle force is much lower, much lower. And physiological it is not at all the same. Skeletal muscles work at maximum tension, but the heart works on the ascending limb of the length-tension curve while passive tension is high.

Ca²⁺-DEPENDENCE OF PASSIVE PROPERTIES OF CARDIAC SARCOMERES

Bruno D. Stuyvers, Masahito Miura, and Henk E.D.J. ter Keurs

University of Calgary, Health Sciences Centre, Departments of Medicine, Physiology and Biophysics, Calgary, Canada

Abstract: Rat cardiac trabeculae constitute a well-known experimental model in studies of cardiac contraction at the sarcomere level. Continuous measurement of length of the sarcomeres (SL) by laser diffraction technique permits one to monitor the active shortening of the contractile units during generation of force. When the preparation is stimulated repetitively (0.5 Hz) by electrical pulses, active shortenings are separated by periods corresponding approximately to the diastolic interval in the heart and wherein normally no major contractile event would have been expected. In contrast to this expectation, studies conducted with high-resolution (2-4 nm) SL measurements technique revealed that sarcomeres continuously lengthened (by 10-60 nm) from the end of twitch relaxation to the next stimulation. Such lengthening resulted from an internal expansion of the sarcomere and not from stretch exerted by extra-sarcomeric sources. We further characterized diastolic changes by measuring sarcomere stiffness (**Sarc-Stiff**) estimated from the response to short bursts (30 ms) of sinusoidal perturbations (frequency: 500 Hz) at 5 moments of the resting interval separating twitches. **Sarc-Stiff** increased continuously by ~ 30 % during the diastolic interval (29°C, pH: 7.4, $[Ca^{2+}]_o$ = 1 mM). We then investigated during the same period the intracellular dynamics of Ca^{2+}, as a major determinant of sarcomere motions in muscle. Intracellular free-Ca^{2+} concentration ($[Ca^{2+}]_i$) was measured continuously in trabeculae microinjected with the fluorescent Ca^{2+}-probe Fura-2 and stimulated at 0.5 Hz. It appeared that the Ca^{2+}-transient, which drives the twitch, did not end with the apparent relaxation of the force. Instead, $[Ca^{2+}]_i$ kept decreasing in an exponential manner throughout the diastolic interval. At $[Ca^{2+}]_o$ = 1 mM, $[Ca^{2+}]_i$ decreased from 230 to 90 nM with a time constant of ~ 250 ms. The similarity in time courses of Ca^{2+}-decline and of **Sarc-Stiff** increase suggested that properties of resting sarcomeres were related to $[Ca^{2+}]_i$ in the sub-micromolar range. In order to examine this possibility, **Sarc-Stiff** was measured in chemically skinned trabeculae, i.e. in a preparation allowing control of $[Ca^{2+}]$ surrounding the sarcomeres. **Sarc-Stiff** was measured at different $[Ca^{2+}]$ from 1 to 450 nM. We found that 1) below 70 nM, **Sarc-Stiff** was independent on $[Ca^{2+}]$, 2) between 70 and 200 nM, i.e., approximately the range wherein $[Ca^{2+}]_i$ decreased during diastole in intact muscle, **Sarc-Stiff** decreased by $\sim 50\%$ with increase of $[Ca^{2+}]$ and 3) above 200 nM, **Sarc-Stiff** increased steeply with increase of $[Ca^{2+}]$ as was expected from Ca^{2+}-dependent attachment of cross-bridges between actin and myosin. The data fitted accurately to the sum of 2 sigmoid functions: 1) at $[Ca^{2+}] < 200$ nM, **Sarc-Stiff** decreased with increase of $[Ca^{2+}]$ with a Hill coefficient $(n_H) = -2.6$ and $[Ca^{2+}]$ at half maximal activation $(EC_{50}) = 0.16 \pm 0.013 \mu M$; 2) at $[Ca^{2+}] > 200$ nM, **Sarc-Stiff** increased with $[Ca^{2+}]$ $(n_H: 2.1; EC_{50}: 3.4 \pm 0.3 \mu M)$ consistent with Ca^{2+}-dependent attachment of cross-bridges. It was possible to reproduce the diastolic variation of **Sarc-Stiff** observed in intact muscle by using the time course of $[Ca^{2+}]_i$ in the **Sarc-Stiff** $- [Ca^{2+}]$ relationship determined from skinned trabeculae.

We conclude that physical properties of the sarcomeres are inversely related to Ca^{2+} below 200 nM, i.e., in a range of concentrations where the myocytes operate during diastole while the influence of cross-bridges is negligible.

Introduction

Diastole is commonly described as a period of the cardiac cycle where blood entering the ventricle stretches the myocardial tissue passively, while intracellular ionic concentrations are in a steady state and cellular processes of contraction are silent. During this period, myocardial fibers elongate and generate passive tension on the ventricular content through the classical tension-length relation.

The shape of the relation between passive tension and level of stretch is determined by the visco-elastic properties of cardiac tissue. It is commonly admitted that these properties are steady and, at rest, the myocardium behaves like a passive visco-elastic system. Based on this concept of passive system, the myocardial fibers should respond to the same stretch with identical visco-elastic resistance as long as the muscle stays unstimulated. Actually, recent studies on rat cardiac trabeculae revealed that visco-elastic resistance of cardiac muscle varies during the resting period separating two twitches (Stuyvers *et al.*, 1997). We found that such changes resulted from two phenomena: 1) intracellular free-Ca^{2+} concentration $([Ca^{2+}]_i)$ varies continuously during the resting period (Stuyvers *et al.*, 1997) and 2) the visco-elastic properties are affected by small variations of Ca^{2+} through a mechanism other than normal

Ca^{2+} activation of contractile filaments (Stuyvers *et al.*, 1998). We review in this chapter the major experimental features leading to the novel and challenging idea that an important factor of the "active" properties of cardiac muscle such as calcium ions may regulate the "passive" properties as well.

Results and Discussion

Re-examination of diastole from study of isolated cardiac muscle

Diastolic lengthening of cardiac sarcomeres

Background. Trabeculae constitute a well-known model of properties of cardiac muscle. Continuous measurement of length of the sarcomeres (SL) by laser diffraction technique permits one to monitor the active shortening of contractile units during contraction. When the preparation is stimulated repetitively, resting periods wherein normally no major contractile event should be detected separate active shortenings of the sarcomeres. Contrary to this expectation, we noticed systematically a slow and small increase of SL trace during the time interval between two twitches. We decided to examine whether this SL increase reflected a physiological event or was simply a technical artefact.

Preparation. Trabeculae (n = 17) were dissected from right ventricles of rat hearts and mounted in an experimental chamber as described previously (Stuyvers *et al.*, 1997; De Tombe and ter Keurs, 1992). The preparation was stimulated every 2 seconds and left to equilibrate for 1 hour at 25°C in the perfusion medium. The standard solution was a modified Krebs-Henseleit buffer solution composed of (mM) 112 NaCl, 1.2 $MgCl_2$, 5 KCl, 2.4 Na_2SO_4, 2 NaH_2PO_4, 1 $CaCl_2$, 10 Glucose, 19 $NaHCO_3$, and equilibrated with 95% O_2 and 5% CO_2; pH 7.4 (adjusted with KOH 1 M).

Force, muscle length and sarcomere length measurements. Methods used to measure force (F), muscle length (ML) and SL have previously been described in detail (Stuyvers *et al.*, 1997; DeTombe and ter Keurs, 1992; Daniels *et al.*, 1984). Variations of muscle length were determined from the displacement of a motor arm controlled by a dual servo-amplifier (model 300s, Cambridge Technology, Watertown, MA, USA). Force was measured by a silicon strain gauge (model X17625, SensoNor, Horten, Norway) with a resolution of 0.63 µN (Stuyvers *et al.*, 1997). SL was measured by laser diffraction techniques as previously described (Stuyvers *et al.*, 1997; Daniels *et al.*, 1984). The resolution of SL measurements was 2-3 nm.

Results. Figure 1 shows the typical SL changes following a twitch. Complete relaxation of twitch force was accompanied by a sarcomere shortening transient followed by a gradual lengthening (ΔSL) (Fig. 1A). The course of

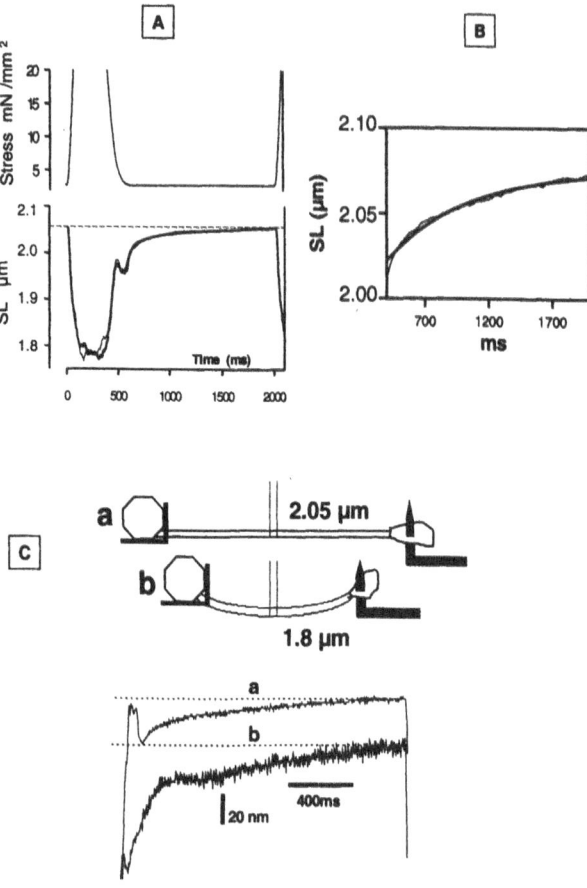

Figure 1. *SL change during diastolic interval in intact rat cardiac trabeculae. **A**: Typical time course of stress and sarcomere length (SL) during and in between twitches of a cardiac trabecula of rat. **B**: Diastolic sarcomere lengthening from Panel A represented on a larger scale; the superimposed continuous line represents the nonlinear regression fit to the data: SL= -0.02 . e $^{(-t/618)}$ + 2.1 (r=0.92); temperature: 25°C; pH: 7.4; [Ca^{2+}]$_o$ = 1mM). **C**: Lengthening of the sarcomere in the central segment of a trabecula at two different sarcomere lengths; note that diastolic lengthening observed at SL: 2.05μm (**a**) still occurred below slack length (**b**).*

ΔSL was individually fitted by a mono-exponential function (Fig. 1B). The variation of SL during the resting period was reproducible from muscle to muscle, but the amplitude varied under otherwise similar conditions (ΔSL = 25 ± 9 nm, n = 15).

Misalignment of the preparation with respect to the laser beam could not explain this observation since SL changes derived from both first-order lines of the laser diffraction were identical (see superimposed SL traces in Fig. 1A). A similar SL increase still occurred below slack length (Fig. 1C); we noticed that this lengthening was, then, accompanied by increased buckling of the muscle (not shown). Hence, this lengthening must have resulted from an internal expansion rather than from a stretch due to a slow shortening of the ends of the trabeculae. For this reason, the resting interval separating two twitches was referred herein as "diastolic interval" or "diastole" (from Greek, *Diastole*: expansion). Interestingly, during SL increase, force changes never exceeded 0.3 ‰ of the maximal twitch force (Fig. 1A).

Sarcomere Stiffness during diastole

In order to investigate further the modifications occurring during diastole, we measured the stiffness of the sarcomeres with the expectation that such a parameter measured at different time of the diastolic interval would provide information about changes occurring in the conformation of the system.

Measurements of stiffness of the sarcomeres. Stiffness of the sarcomeres (Stiff-Sarc) was determined from 30 ms bursts of 500 Hz sinusoidal perturbations imposed on muscle length (Fig. 2A). The corresponding oscillations of F and SL were analyzed by spectral analysis. The apparent stiffness modulus of the sarcomere (MOD) and the phase delay between F and SL oscillations (Φ) were calculated from Fast Fourier Transform. Stiff-Sarc was measured at 10, 30, 50, 70 and 90% of total duration of diastole. The 100% duration corresponded approximately to 1500 ms when trabeculae were stimulated at 0.5 Hz and $[Ca^{2+}]_0 = 1$ mM. MOD was expressed in mN /mm^2 /μm, i.e., in unit of stress per unit of SL variation.

Results. In standard conditions ($[Ca^{2+}]_0 = 1$ mM), a 30% increase of MOD occurred in the initial 450 ms of the diastole (MOD: 9.3 ± 0.6 at 10 % vs. MOD: 12.2 ± 0.5 mN /mm^2 /μm at 50%, n = 158; p <0.05, 18 muscles, Fig. 2B, left panel). MOD slightly decreased at the end of the diastole to 11.6 ± 0.65 mN /mm^2 /μm. On average, MOD increased during diastole following a monoexponential function (see Fig. 2B, left panel, dotted line) with a time constant of ~300 ms.

During the same period, Φ decreased significantly from 84 ± 3 degrees to 73 ± 4 degrees (n=158; p<0.05; 18 muscles; Fig. 2B, right panel). At $[Ca^{2+}]_0 = 1$ mM, 15 of 18 trabeculae studied showed discrete sarcomere spontaneous activity with a minimal effect on Force and restricted to a brief period at the end of diastole (see example of Fig. 2B, bottom traces). The reversal of the time course of stiffness during diastole always occurred simultaneously with the spontaneous motion of the sarcomeres (Fig. 2B).

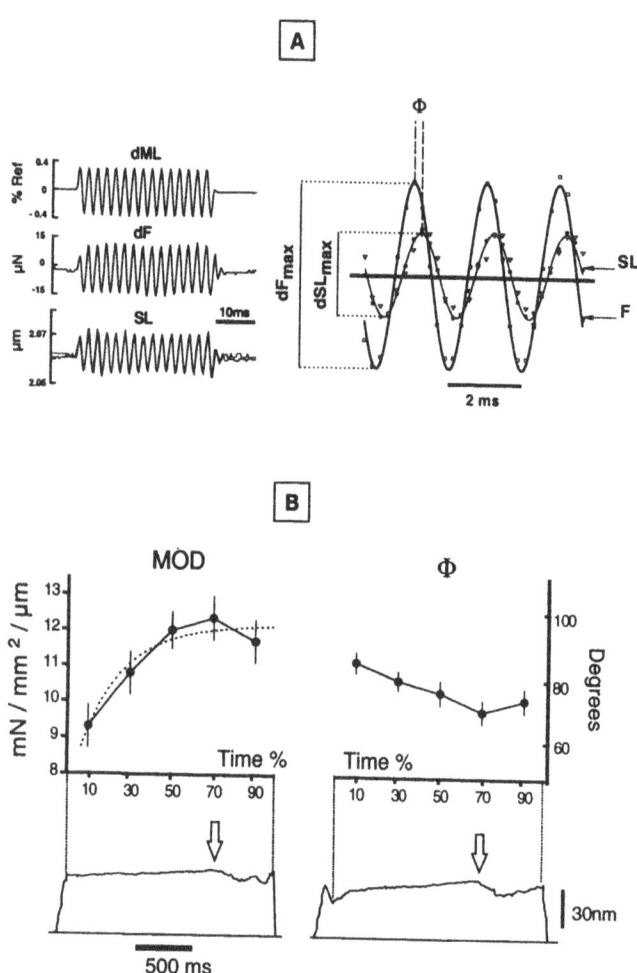

Figure 2. Sarcomere stiffness during diastole. A*, left panel:
Typical records of response of sarcomere length (SL) and force
(dF) during a 30 ms burst of 500 Hz sinusoidal perturbations of
muscle length (dML). SL was recorded from 2 detectors as
described in Daniels et al. (1984).* **A****,** *right panel: SL and F
sinusoids represented on slower time base.* ***B****: Time Course of
stiffness during diastole at [Ca²⁺]$_o$ = 1 mM. Data are expressed
as Mean ± SEM (n= 18). The time scale was expressed in % of
the diastolic interval (see text). The time course of MOD was
fitted with a monoexponential function (see dotted line): MOD =
-9 . e (-t/308) + 12 (r = 0.92). Lower traces show an example of
time course of sarcomere length obtained from a representative
trabeculae showing slight spontaneous fluctuations of SL starting
at 70% of diastole (see arrows).*

Diastolic variation of [Ca²⁺]ᵢ

Because Ca^{2+} is considered as a main regulator of the sarcomere activity, we investigated $[Ca^{2+}]_i$ during diastole in order to evaluate a possible relation with mechanical changes of the sarcomeres.

$[Ca^{2+}]_i$ measurements. Ca^{2+} probe Fura-2 was microinjected iontophoretically into the trabeculae as described previously (Stuyvers *et al.*, 1997; Backx and ter Keurs, 1993). The epifluorescence of Fura-2 from trabeculae was collected by a photomultiplier tube (PMT-R2693 with a C1053-01 socket, Hamamatsu) through a 500 nm bandpass filter (Melles Griot, Irvine, CA). $[Ca^{2+}]_i$ was given by the classical equation (after subtraction of the autofluorescence of the muscle):

$$[Ca^{2+}]_i = \beta \cdot K'_d \cdot (R - R_{min}) / (R_{max} - R)$$

where K'_d: effective dissociation constant, β: value of the fluorescence of the Ca^{2+}-free dye divided by the value of the fluorescence of the Ca^{2+}-bound dye at 380 nm, R: ratio of the fluorescence at 340 nm excitation to that at 380 nm excitation (340/380), R_{min}: R at zero $[Ca^{2+}]_i$ and R_{max}: R at a saturating $[Ca^{2+}]$. The values for β, K'_d, R_{min}, and R_{max} were determined by *in vitro* calibration. R_{min} and R_{max} were 0.129 and 3.71, respectively; K'_d was 4.57 mM; β was 12.1.

Results. As illustrated in Figure 3, relaxation of the Ca^{2+} transient did not end with the active force development. Instead, $[Ca^{2+}]_i$ continued to decrease during diastole. When $[Ca^{2+}]_o = 1$ mM, the amplitude of the latter decline was approximately 120 nM. Decay of $[Ca^{2+}]_i$ was completed approximately 200 ms before the end of the diastolic interval, i.e., at 90% of total duration. The time constant of the decay during the diastole was calculated from the fit of a monoexponential function to the decline of $[Ca^{2+}]_i$. The time constant (t_{Ca}) was 210-325 ms, i.e., similar to the time constant of increase of sarcomere stiffness during the same period (see Fig. 2B). Similarity between time courses of both parameters suggested that visco-elastic properties of cardiac sarcomeres were influenced by variations of $[Ca^{2+}]_i$ in the submicromolar range.

Evidence for Ca²⁺-dependence of passive properties of cardiac sarcomeres

Ca²⁺ and sarcomere stiffness in skinned trabeculae

Background. The above results suggested the existence of a relation between visco-elastic properties of the sarcomeres and Ca^{2+} in a range of concentrations wherein no contractile process was expected to occur. In order to investigate this possibility further, we used the same approach of sarcomere stiffness in trabeculae permeabilized with detergent. This preparation enabled us to

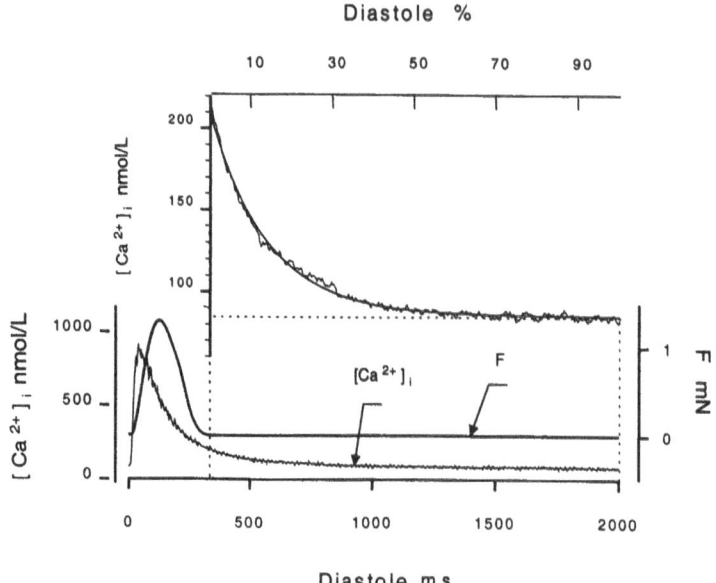

Figure 3. $[Ca^{2+}]_i$ *during the diastolic interval: assessment of* $[Ca^{2+}]_i$ *simultaneously with F showed an exponential decline of* $[Ca^{2+}]_i$ *from the peak systolic to the end-diastolic level. The diastolic part of* $[Ca^{2+}]_i$ *decay is represented on a larger scale on the upper part of the diagram; amplitude of the decay during diastole was 123 nM; data were fitted through a monoexponential function. See superimposed continuous line:* $[Ca^{2+}]_i$ = 738 $\cdot e^{(-t/224)}$ + 86; r = 0.98. *The time scale was expressed in ms (lower part) and in % of diastole (upper part).*

control $[Ca^{2+}]$ surrounding the myofibrils. We focused our study on Ca^{2+} -concentrations measured during diastolic decrease of $[Ca^{2+}]_i$ in intact trabeculae and we calculated the corresponding stiffness of the sarcomeres using the same method of sinusoidal perturbations.

Protocol. Trabeculae were incubated for 30-45 min with a Ca^{2+}-free solution (RS) containing Saponin (50 µg/mL) (Stuyvers *et al.*, 1998). Then, the preparation was washed four times with RS. Solutions with different free Ca^{2+}-concentrations were prepared by mixing RS with varied amounts of a solution containing 10 mM of Ca^{2+} ("*activating solution*": AS). Composition of RS was (total concentrations in mM): 100 BES^-, 10 $EGTA^{4-}$, 6.8 ATP^{4-}, 10 CP^{3-}, 6.30 Mg^{2+}, 55 $CH_3CH_2COO^-$ (propionate), 12.6 Cl^-, 33.6 Na^+, 1 Dithiothreitol; temperature: 25°C; pH 7.1 (adjusted with KOH 1 M). Total ionic concentrations ($[\ ]_{total}$) were calculated in order to obtain final solutions with appropriate free ion concentrations ($[\ ]_{free}$), with net charge = 0, ionic strength = 0.2 M and ionic equivalent = 0.17 M at 25°C and pH 7.1. The composition of AS was identical except that $[Mg^{2+}]_{total}$ and $[Ca^{2+}]_{total}$ were respectively 5.85 and 10 mM (equivalent $[Mg^{2+}]_{free}$ and $[Ca^{2+}]_{free}$: 3.1 mM and 400 µM respectively). All solutions contained protease inhibitors Leupeptin (40 µM) and P.M.S.F. (0.5 mM).

Sarcomere stiffness was first measured in RS. The solution was then rapidly removed and replaced by 1 mL of solution with a different $[Ca^{2+}]$. Free Ca^{2+}-concentrations between 0 and 430 nM were tested randomly. All measurements were performed at 25°C.

Results were expressed as mean ± S.E.M. Significance of relations between stiffness and $[Ca^{2+}]$ was judged on the basis of ANOVA and the correlation coefficients of the regressions. Student's t-test was used for comparisons of two means; the difference was considered significant when $p < 0.05$. Data were fitted using a Marquardt-Levenberg algorithm.

Results. MOD was first measured in intact muscle at 1350 ms after the stimulation, i.e., approximately at half of diastole, and was taken as the reference (MOD_{ref}): MOD_{ref} was 36.6 ± 4.8 mN / mm² /µm (n = 17). In Fura-2 microinjected trabeculae, $[Ca^{2+}]_i$ had, then, reached about 100 nM. We found that MOD of skinned fibers at $[Ca^{2+}]$ = 100 nM was 36.8 ± 6.5 mN /mm² /µm (n = 17), i.e. identical to MOD_{ref} measured under equivalent conditions in skinning experiments while the muscle was still intact. The phase delay between SL and F oscillations (Φ) measured at $[Ca^{2+}]$ = 100 nM, was similar to F measured in skinned fibers before skinning ($Φ_{ref}$): Φ = 160 ± 16 DEG and $Φ_{ref}$ = 170 ± 27 DEG (n=17). Below 450 nM, MOD varied with $[Ca^{2+}]$ following a triphasic relation (17 muscles; Fig. 4A). At $[Ca^{2+}]$ below 70 nM, MOD was independent of $[Ca^{2+}]$. Between 70 and 200 nM, MOD decreased with increase of $[Ca^{2+}]$. Above 200 nM, a steep increase in MOD was observed with increase of $[Ca^{2+}]$. Over the total range of $[Ca^{2+}]$ tested, the data could be fitted with the sum of two sigmoidal relationships. Figure 4A shows that values of MOD obtained below 200 nM of $[Ca^{2+}]$ fitted well with a sigmoidal relationship with a negative Hill coefficient. Polynomials fitted through the data between 90 and 200 nM (see insert Fig. 4A) permitted a model independent estimate of the maximal variation of MOD versus $[Ca^{2+}]$ and showed that MOD decreased by 50% over an increase in $[Ca^{2+}]$ of 100 nM. Under identical conditions, similar polynomials showed that Φ increased by 80 DEG (see insert Fig. 4B). Between 1 and 200 nM, changes of Φ with $[Ca^{2+}]$ were opposite to variations of MOD (Fig. 4A and Fig. 4B). Above 200 nM, Φ decreased again with further increase of $[Ca^{2+}]$ and reached the same values as measured below 70 nM.

Active force development
In order to verify that our protocol did not alter the contractile properties of the sarcomeres, active force development with Ca^{2+} was measured under

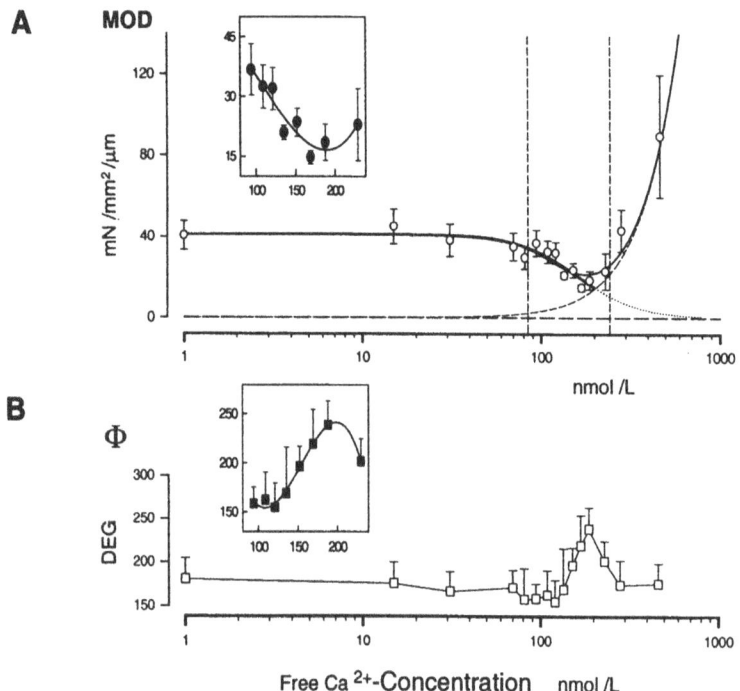

Figure 4. Sarcomere Stiffness-[Ca²⁺] relationship in skinned trabeculae: MOD and Φ were plotted against free Ca²⁺ concentration; data were expressed as mean ± S.E.M (n=17). Between 1 - 200 nM, MOD fitted through the following Hill function (function A; see thick continuous line): $MOD = MOD_{max} \cdot [Ca^{2+}] \, n_H / ([Ca^{2+}]_{50} \, n_H + [Ca^{2+}] \, n_H)$ where MOD_{max}: maximal value of MOD obtained between 1 and 200 nM (41±2 mN /mm² /μm); n_H: Hill coefficient (-2.6 ± 0.7); $[Ca^{2+}]_{50}$: [Ca²⁺] at half amplitude of the sigmoidal decay between 1 and 200 nM ($[Ca^{2+}]_{50} = 160 ± 13$ nM). It was assumed that lower part of the sigmoid (dotted part of the curve at [Ca²⁺] > 200 nM) tends to reach the stiffness of weakly attached cross-bridges (wXb-stiffness: horizontal dashed line; see text); stiffness modulus of wXb (MOD_{wXb}) was 0.06 mN /mm² /μm (see text). The "active stiffness" (function B; see ascending dashed line) was calculated from the sum of wXb-stiffness and sXb-stiffness. Development of sXb-stiffness with [Ca²⁺] was predicted from the typical active force-[Ca²⁺] relation described in Figure 5 (see text for details). Data obtained between 1 and 450 nM were fitted with the sum of function A, function B and wXb-stiffness. The lower panel shows variations of Φ in the range of [Ca²⁺]= 1-430 nM. Inserts in both upper and lower panels show on a linear horizontal scale data between 90-230 nM; third-order polynomial functions have been used to fit the data (thick continuous line) and to estimate (in a model-independent fashion) accurately variations of MOD and Φ with [Ca²⁺].

experimental conditions used in our study of stiffness. Results are reported in Figure 5, which shows that active force development was similar to data previously published from skinned cardiac trabeculae (without control of the internal sarcomere shortening, see Kentish *et al.*, (1986) and Harrison *et al.*, (1988),

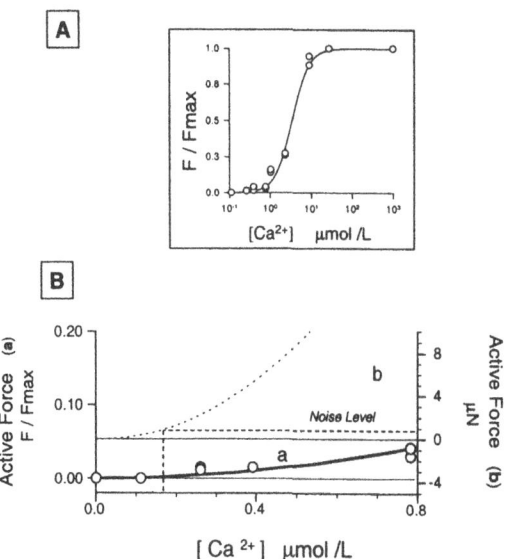

Figure 5. *The active force-$[Ca^{2+}]$ relationship. **A**: active force developed by 2 saponin skinned trabeculae under experimental conditions used for study of stiffness. Active force follows a typical Hill function when $[Ca^{2+}]$ is increased up to 1 mM: Fmax $= 56 \pm 2$ mN/mm^2, Hill coefficient $(n_H) = 2.1 \pm 0.2$, $[Ca^{2+}]_{50} = 3.4 \pm 0.3$ μM (see text for details); force was normalized to Fmax. **B**: data of Panel A at $[Ca^{2+}]$ below 800 nM represented on a linear scale. Force was expressed in μN (trace **b**) in order to facilitate the comparison with the noise level of the system (dashed line as indicated). Horizontal thin lines indicate the force zero level. The detection of force development at submicromolar $[Ca^{2+}]$ was limited by the noise in our system (0.63 μN). Given the coefficients of the Hill regression for these muscles (cross sectional area : 0.01 mm^2), we would have expected that active force equaled the noise level at $[Ca^{2+}] = 140$ nM and exceeded the noise level by a factor 4-5 at $[Ca^{2+}] = 300$ nM as it was observed experimentally (see Fig. 5B, above).*

and followed the Hill equation: $F = F_{max} \cdot [Ca^{2+}]^{n_H} / ([Ca^{2+}]_{50}^{n_H} + [Ca^{2+}]^{n_H})$; F_{max}, n_H and $[Ca^{2+}]_{50}$ were respectively 56 ± 2 mN /mm², 2.1 ± 0.2 and 3.4 ± 0.2 μM (Fig. 5A).

Theoretical analysis of the stiffness-[Ca²⁺] relationship

Stiffness - $[Ca^{2+}]$ relationship involved three steps, depending on the range of concentrations tested. Figure 4A shows that data could be accounted for if we assumed that relationship between stiffness and $[Ca^{2+}]$ resulted from the sum of 2 sigmoidal functions. One function (A) showed a sigmoidal decrease of MOD with a Kd of 160 ± 13 nM and a slope (n_A) of - 2 to -3 (-0.6 ± 0.7). The second function (B) reflected a sigmoidal increase of MOD with increasing $[Ca^{2+}]$. Function B followed directly from the active force-$[Ca^{2+}]$ relationship (Fig. 5) if one assumes that stiffness of activated cross-bridges increased in proportion to the force. It can be shown (Stuyvers *et al.*, 1997; Brenner *et al.*, 1982) that contribution of cross-bridges in weakly attached state (wXb) to stiffness at 500 Hz is 0.06 mN /mm² /mm, i.e. ~ 600 times lower than the values measured in our study (see horizontal dashed line in Fig. 4A). Stiffness of Ca^{2+}-dependent cross-bridges (in strongly attached state: sXb) followed from instantaneous release experiments. It has been demonstrated for skeletal muscle (Huxley and Simmons, 1971) and cardiac muscle (Backx and ter Keurs, 1988) that a near instantaneous release (of ~12 nm in cardiac trabeculae) of the fully activated sarcomere causes force to drop to zero. Considering that maximal force developed under our experimental conditions was approximately 60 mN /mm², stiffness of uniformly activated trabecula would have been ~5000 mN/ mm² /μm. Hence, the relationship between $[Ca^{2+}]$ and stiffness of sXb followed directly from the parameters of the Hill function fitting data of Figure 4A. Development of "active stiffness," as it is represented in Figure 4A, resulted from the sum of wXb- stiffness and sXb-stiffness. The excellent fit between experimental data and the above model suggested that stiffness-$[Ca^{2+}]$ relationship measured in our study between 1 and 450 nM resulted from the sum of the following two different Ca^{2+}-dependent mechanisms: 1) a decrease with increase of $[Ca^{2+}]$ of the stiffness generated by a system that dominates at low $[Ca^{2+}]$ (below 200 nM), i.e., at $[Ca^{2+}]$ in which intact muscle operates during diastole, 2) at $[Ca^{2+}] > 200$ nM, an increase of stiffness due to the formation of cross-bridges in the strong attached state.

Comparison between intact and skinned Trabeculae

We measured the diastolic stiffness of intact trabeculae before skinning, at 1350 ms after stimulation. We have found that, at this time, $[Ca^{2+}]$ was ~100 nM. At $[Ca^{2+}] = 100$ nM, the stiffness moduli of skinned trabeculae were nearly identical to those measured before skinning, suggesting that saponin, at steady $[Ca^{2+}]$, did not affect significantly the apparent stiffness of the sarcomeres. The

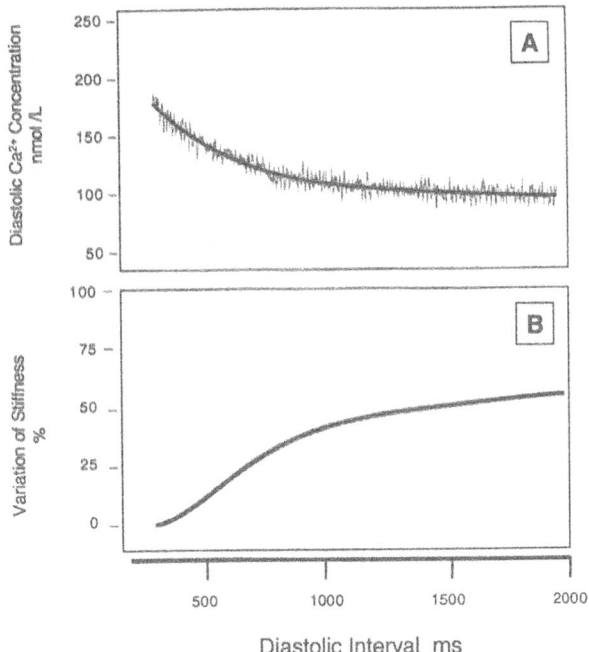

Figure 6. *Modeling of time course of stiffness during diastole from stiffness-[Ca²⁺] relation. A: variation of [Ca²⁺]ᵢ observed experimentally during diastole as represented in Figure 3. B: values of MOD calculated from values of [Ca²⁺]ᵢ of panel A using the sum of functions A and B of Figure 4.*

variation of MOD and Φ over a range of [Ca²⁺] encountered in intact muscle during diastole (230 - 90 nM) was also the same in skinned muscles as compared to intact trabeculae: MOD was inversely related to [Ca²⁺] and the variations of MOD and Φ with [Ca²⁺] were opposite. It is clear that the stiffness-[Ca²⁺] relation of the sarcomeres in skinned fibers can explain the increase of sarcomere stiffness while [Ca²⁺]ᵢ gradually declined from 200 to 90 nM during diastole of intact stimulated muscles.

Interestingly, we could reproduce faithfully the time course of stiffness observed in intact trabeculae during diastole from the stiffness-[Ca²⁺] relation determined in skinned trabeculae (see Fig. 6): values of [Ca²⁺] measured during the diastolic decline of [Ca²⁺]ᵢ in intact trabeculae (Fig. 3) were incorporated in the function based on the sum of two sigmoids and fitting the data of Figure 4A. As shown in Figure 6B, the calculated course of stiffness modulus matched well the increase seen experimentally in intact trabeculae (see Fig. 2B).

Conclusion

Our work shows that resting properties of intact cardiac sarcomeres are sensitive to $[Ca^{2+}]$ in the submicromolar range. Further analysis of these results revealed that such a Ca^{2+}-dependence originates in a visco-elastic system that is independent on formation of cross-bridges and which generates stiffness in an inverse Ca^{2+}-dependent manner. Importantly, this system operates in the range of Ca^{2+}-concentrations wherein the myocytes regulate their cytosolic Ca^{2+}-level during the diastole of cardiac muscle. Our present finding is critical because it suggests that pharmacological and pathological conditions affecting Ca^{2+}-homeostasis in the myocytes are potentially responsible for alterations of diastolic properties of the myocardium as well.

Preliminary results obtained on skinned fibers in our laboratory strongly suggest that the large endo-sarcomeric protein titin is involved in the Ca^{2+}-dependence of visco-elastic properties of cardiac sarcomeres.

References

Backx PH, Ter Keurs HEDJ. Restoring forces in rat cardiac trabeculae. *Circulation* 1988;78:II-68.

Backx PH, Ter Keurs HEDJ. Fluorescent properties of rat cardiac trabeculae microinjected with fura-2 salt. *Am J Physiol* 1993;264:H1098-H1110.

Brenner B, Schoenberg M, Chalovich JM, Greene LE, Eisenberg E. Evidences for cross-bridge attachment in relaxed muscle at low ionic strength. *Proc Natl Acad Sci USA* 1982;79:7288-7291.

Daniels MCG, Noble MIM, Ter Keurs HEDJ, Wohlfart B. Velocity of sarcomere shortening in rat cardiac muscle: relationship to force, sarcomere length, calcium, and time. *J Physiol (London)* 1984;355:367-381.

De Tombe PP, Ter Keurs HEDJ. An internal viscous element limits unloaded velocity of sarcomere shortening in rat myocardium. *J Physiol (London)* 1992;454:619-642.

Harrison SM, Lamont C, Miller DJ. Hysteresis and the length dependence of calcium sensitivity in chemically skinned rat cardiac muscle. *J Physiol (London)* 1988;401:115-143.

Huxley AF, Simmons RM. Proposed mechanism of force generation in striated muscle. *Nature* 1971;233:533-538.

Kentish JC, Ter Keurs HEDJ, Ricciardi L, Bucx JJJ, Noble MIM. Comparison between the sarcomere length-force relations of intact and skinned trabeculae from rat right ventricle. Influence of calcium concentrations on these relations. *Circ Res* 1986;58:755-768.

Stuyvers B, Miura M, Ter Keurs HEDJ. Dynamics of viscoelastic properties of rat cardiac sarcomeres during diastolic interval: involvement of Ca^{2+}. *J Physiol (London)* 1997;502:661-677.

Stuyvers B, Miura M, Jin J-P, Ter Keurs HEDJ. Ca^{2+}-dependence of diastolic properties of cardiac sarcomeres. *Prog. Biophys Mol Biol* 1998;69:425-443.

Discussion

Pollack: So that I don't disappoint you, I will ask the question about that spontaneous lengthening. Actually there were a couple of questions. One question is, have you looked at all at A-bands and what happens to A-bands during any of these processes to see if there are any dynamic length changes? The second question is whether there are any alternative hypotheses to explain the decrease in stiffness at roughly 1 mM calcium other than the one that you've given here?

TerKeurs: I have not looked at dynamic changes in the A-band width. I don't know what the highest speed at which the A-band width can change actually is. I know that in many skinned fiber studies, which are always characterized by, from the point of the view of the heart, lethal durations of contraction, then A-bands can shorten. I don't know how quick the A-band can shorten, if it does. I believe it does, but in long contractions. Secondly, did we ever look at alternatives for the increase in the modules of stiffness and the decrease in the modules of stiffness at higher external calcium concentrations? Yes we looked at all the possibilities including weak crossbridge attachment and viscous parameters derived from that. The contribution of weak crossbridges is always in the five percent range of the modulus of stiffness that we have measured. If the calcium increases, we might expect attachment of cross-bridges and an extremely steep increase of the modulus of stiffness. But we do see a decrease of the modules of stiffness while the calcium level is rising gently in the myocyte. So that would be of an opposite nature; therefore, I think this explanation has been excluded. Then, one can think about other elements in the cells, cytoskeletal elements, which might be calcium sensitive, and the only thing I can do is throw my hands up. I do not know. But that is a hypothesis to be entertained.

Pollack: Regarding my previous comment, a funny thing is Endo's finding (CSH SympXXXVII 50-510, 1972) in skeletal muscle, that at relatively low calcium concentrations, an increase of calcium results in a progressive elevation of the resting length-tension curve.

TerKeurs: That is not found in cardiac muscle.

Pollack: Ok, I guess that's the difference.

TerKeurs: If Wolfgang changes the calcium levels in the presence of BDM to paralyze the contractile apparatus, there is no change in the static stiffness properties of the muscle.

Linke: I think that is really an interesting hypothesis, the possible binding of titin and actin, and it has been clearly shown *in vitro* that it can happen. But I'm critical because we did similar, not quite these type of experiments because we didn't really add the calcium. But you know we have published a paper in 1997 in the *Biophysical Journal*. We do not see physical evidence in the intact sarcomere for actin binding to titin except for close to the Z-line. This is based on 2 sets of experiments, co-sedimentation experiments with several different types of fragments, and these were made by Mattias Gautel, based on the data by Analiesa Pastora. So he basically chose the fragments where he knew the phasing. So he didn't get any positive binding to actin with a fragment that is very close to that. Jian-Ping will know about this result and maybe talk about it later. We also did CD analysis. They have the predicted phasing. I think here you would really have to do a lot of controls. For example, you have to show you have the right folding of the protein, otherwise you might get some sort of unspecific binding. This might be a very sticky fragment you have; you may not even get it into your fiber, so you would have to show it diffuses into the trabecula, that you see a specific effect. So I think this would be important before you can come to these conclusions.

TerKeurs: Yes, but it is known for albumin that it enters the fiber. And apparently it is not sticky enough. And I am more impressed by the rather striking difference between titin I-II, versus titin I-I. They are very similar and that points more to a real effect. I think that you are quite right that one needs to do more experiments to prove it, but I think that these experiments are strong enough by themselves to entertain the hypothesis in a real way.

Linke: I think you are right. It is really a matter of getting rid of all these possible artifacts, that is what we have to think about.

TerKeurs: There is a lifetime of work here.

Sugi: I am interested in the decreasing muscle fiber stiffness at the low calcium ion concentrations. This phenomenon reminds me of so-called latency relaxation in skeletal muscle at the early stage of activation. How do you think about that?

TerKeurs: You phrased it perfectly. The way in which we have been thinking about applications of this observation is that if one would translate it into skeletal muscle work, one mounts a skeletal muscle, stretches it gently, there is therefore a slight passive force. Then one stimulates the muscle. The calcium level rises through the range where the viscoeleastic modulus decreases. That means that the passive force decreases just before active force development takes off. And if that hypothesis is correct, it may be a solution to the long standing puzzle of latency relaxation. Thank you.

Granzier: Did you look where the titin I-II fragment binds? It does not necessarily bind where the endogenous I-II binds; it may bind anywhere and compete off anything—for example, the PEVK. If you compare *in vitro* PEVK actin binding with titin I-II actin binding, PEVK binding is much stronger. I am just wondering if I-II could compete off parts of titin in the extensible region of the molecule.

TerKeurs: Well, that is work in progress. I must confess that Jian-Ping makes rarely a mistake, but he made one mistake now. He moved from Calgary to Cleveland. He will come back soon.

Labeit: Henk, I was just wondering with the titin I-II fragment, is it a histidine tagged protein or is it without any tags? I am wondering about this.

Jin: It is a non-fusion protein.

Labeit: Because if you would have histidine tags it might obviously immediately interfere with calcium. My second thing is a more common question. I think there are now so many expressed peptides from the I-band, and all in high-impact journals where things are available. You might just ask them. Rief presented I-band peptides, Carol Gregorio published several and all these peptides are available. This is a along the question that Marion Greaser asked. It wouldn't be a life-time achievement, taking 2 - 3 different peptides.

TerKeurs: I agree with that part, Siegfried. The problem is that in order to study a muscle with the resolution at the sarcomere length that we are talking about, you may go through a couple of weeks without a muscle that is good enough. But I will be a happy customer of using all the good candidate peptides to tackle this question.

Wu: My question is about the sarcomere length measurement. Since the muscle is isometric, why do sarcomeres shorten so much, to lengths even shorter than the slack lengths?

TerKeurs: Well, there are a couple of good people who can instruct you about that one, Dr. Henk Granzier, and the other one is Dr. Pollack. But to be short, in cardiac muscle, the best diffraction patterns during activation are coming from intact muscles. The reason for that is that only a fraction of muscles maintains uniformity after skinning, during contraction, enough to permit a sensible or reasonable sarcomere length measurement. The reason for that non-uniformity is not completely known, but part of it is that by skinning procedures one loses a large volume of mitochondria and that means that all the elastic constraints in the muscle, not only titin, but all the elastic constraints are released. If one uses skinned muscles and then disrupts titin, for example, by trypsin treatment, in our hands, the uniformity of the muscle as seen through the eyes of the diffraction equipment, falls apart practically completely. So titin, as the first slide of Charles showed on Thursday morning, is vital to maintenance of sarcomere integrity. You can say, all of that suggests to do away with light diffraction measurements. I don't think so because if the muscle is non-uniform you will see it in diffraction studies and you will see it in other measurements of sarcomere length. Therefore you have to deal with the problem and not with the measurement technique.

POSSIBLE CONTRIBUTION OF TITIN FILAMENTS TO THE COMPLIANT SERIES ELASTIC COMPONENT IN HORSESHOE CRAB SKELETAL MUSCLE FIBERS

Haruo Sugi, Tsuyoshi Akimoto,
Takakazu Kobayashi, Suechika Suzuki,
and Mitsuyo Shimada

*Department of Physiology, School of Medicine, Teikyo University,
Itabashi-ku, Tokyo, Japan*

Abstract: In horseshoe crab skeletal muscle fibers, the extension of SEC at the maximum isometric force P_0 is about 6% of the slack fiber length L_0 (sarcomere length, 7 μm), i.e., about 210 nm per half-sarcomere, being too large to be explained by the cross-bridge and the thin filament elasticities. Cinematographic studies of isometrically contracting myofibril bundes indicate that the highly compliant SEC mostly originates from the "elastic" thick filament misalignment in each A-band during isometric force generation. Possible contribution of the titin filaments to the "elastic" thick filament misalignment is discussed.

Introduction

The behavior of active skeletal muscle can be explained by postulating an elastic component in series with a contractile component (Hill, 1938). During the isometric force development, the contractile component shortens internally by stretching the series elastic component (SEC) until the force exerted by the

Elastic Filaments of the Cell, edited by Granzier and Pollack
Kluwer Academic/Plenum Publishers, 2000

contractile component balances with the force in the SEC. Thus, the degree of extension of the SEC under the maximum isometric force (P_0) is equal to the minimum amount of quick release required to drop the isometric force from P_0 to zero. By carefully removing the external stray compliance, Jewell and Wilkie (1958) showed that SEC in vertebrate skeletal muscle is distributed along the entire muscle length, and its degree of extension under P_0 is about 1% of the slack muscle length (L_0). This was confirmed by Huxley and Simmons (1971) on single muscle fibers, and they put forward a contraction model, in which the SEC largely resides in the cross-bridges, each having an elastic link extending from the thick filament, and the link is extended by about 10 nm, i.e., about 1% of the length of a half-sarcomere when the cross-bridge exerts its full force.

On the other hand, Suzuki and Sugi (1983) presented evidence that the SEC also resides in the thin filament in the myofilament nonoverlap region, i.e., the I-band in each sarcomere. Recently, it has been established that, in vertebrate skeletal muscle fibers, about 50% of the SEC extension at P_0 (5 nm per half-sarcomere) originates from elastic extension of the thin filaments in the I-band, while the other 50% of the SEC extension (also 5 nm per half-sarcomere) originates from the cross-bridges (Suzuki, *et al.*, 1993; Huxley *et al.*, 1994; Wakabayashi *et al.*, 1994).

In some arthropod skeletal muscles, the sarcomere length at L_0 is very long (5-10 μm) compared to that in vertebrate skeletal muscle fibers (about 2 μm) (Jahromi and Atwood, 1969). If the cross-bridge and thin filament elasticities are assumed to be the same in all types of skeletal muscle, the extension of the cross-bridges at P_0 is the same (5 nm per half-sarcomere) irrespective of the sarcomere length, while the extension of the thin filaments in the I-band is 5 x I/0.2 nm, where I/0.2 is the I-band length per half-sarcomere at L_0 relative to the corresponding value in vertebrate skeletal muscle fibers. In crayfish abdominal extensor muscle fibers, the sarcomere length at L_0 is about 9 μm while the A-band length is about 6 μm (Tameyasu and Sugi, 1979). The extension of the SEC at P_0 is estimated on the above basis to be (5 + 5 x 1.5/0.2) = 42.5 nm per half-sarcomere, i.e., about 0.9% of L_0. Contrary to this estimation, the extension of the SEC at P_0 was about 2% of L_0, i.e., about 90 nm per half-sarcomere, and this compliant SEC distributed uniformly along the fiber length, indicating that there would be an additional source of the SEC in each sarcomere (Sugi, 1979; Tameyasu and Sugi, 1979). Unfortunately, irregular striation patterns of crayfish fibers made it difficult to further study the structural origin of the SEC.

The present experiments were undertaken to investigate the structural origin of the highly compliant SEC in arthropod skeletal muscle fibers, using the telson depressor muscle fibers of a horseshoe crab *Tachypleus tridentatus*, which exhibited a reasonably uniform striation pattern along their entire length. It will be shown that the highly compliant SEC in the telson depressor muscle fibers largely originates from "elastic" misalignment of the thick filaments, caused by the elasticity of titin filaments in each sarcomere. In addition, we compared the resting tension in myofibril preparations before and after removal of myosin filaments with 0.5 M KCl, to give information about the titin filament organization in the sarcomere.

Methods

Intact fiber bundle preparation

Telson depressor muscles (Dewey, *et al.*, 1973) were isolated from the animals, and were carefully teased to obtain small fiber bundles (major diameter, 0.5-1 mm; slack length L_0, 1.6-3.4 cm). The fiber bundle preparation was mounted horizontally between the force transducer and the servo-motor. The experimental solution had the following composition (mM): NaCl, 424; KCl, 9; $CaCl_2$, 20; $MgCl_2$, 20; pH adjusted to 7.2 by $NaHCO_3$. The sarcomere length of the fibers was measured by the optical diffraction of He-Ne laser light. The preparation was made just taut at a sarcomere length of about 7 μm. For the sake of convenience, L_0 was defined as the length of the preparation with a sarcomere length of 7 μm. The muscle fibers were maximally stimulated by applying transverse alternating currents (2-4 V/cm, 30-50 Hz) through a pair of Pt plates covering two opposite walls of the experimental chamber. Quick fiber length changes (quick release, complete with 1–2 ms) were applied to the preparation at the plateau of isometric tetanus. Further details of the methods have been described elsewhere (Akimoto and Sugi, 1999).

Myofibril bundle preparation

The telson depressor muscle fibers were glycerinated by the method of Tanaka, Tanaka and Sugi (1979). Thin bundles of myofibrils (diameter, 20-40 μm; length, ≤ 0.5 cm) were dissected from the glycerinated fibers in a relaxing solution that had the following compositions (mM); KCl, 100; $MgCl_2$, 5; EGTA. 3; ATP, 4; histidine, 10 (pH 6.8). The myofibril bundle preparation was mounted horizontally in a thin layer of the relaxing solution between a glass slide and a coverslip; one end was fixed to the slide with a fast-setting glue while the other end was glued to a glass microneedle connected to a micromaniputlator. Solutions surrounding the preparation were changed by use of a micropipette for supplying solution at one side and a piece of filter paper for draining at the other side. The preparation was observed under a phase-contrast microscope (40X, objective, n.a. 0.64), and was maximally activated by iontophoretically applied Ca^{2+} through a glass micropipette (diameter, 5-15 μm) filled with 1 M $CaCl_2$. The Ca^{2+}-activated contractions were recorded with a 16 mm cine-camera at 64 frames/s (Locam, Redlake Laboratories, Inc.). Measurements of the striation pattern were made either directly on the film with the film motion analyzer, or indirectly after microdensitometer tracings of the film (Akimoto and Sugi, 1999).

The experiments were also made, in which single glycerinated muscle fibers (diameter, 40-60 μm; length, 1-1.5 cm) were mounted between the force transducer and the servo-motor with a fast-fixing glue, and the resting tension development in response to applied stretches were recorded either before and after extraction of myosin with 0.5 M KCl added to the relaxing solution.

All experiments were performed at room temperature (18-24°C).

Results

Force-extension curve of the SEC in intact fiber bundle preparations

A typical force-extension curve of the SEC of the intact fiber bundle preparation electrically stimulated at L_0 is shown in Figure 1, in which the force immediately after quick release is plotted against the amount of quick release. The minimum amount of quick release required to reduce the force from P_0 to zero was 5.8 ± 0.5% of L_0 (mean ± S.D., n = 15). The above degree of extension of the SEC at P_0 (about 6% of L_0 or 210 nm per half-sarcomere) is obviously too large for the elastic extension of the cross-bridges and the thin filaments; as the I-band length of the fibers at L_0 was about 2 μm per half-sarcomere, the extension of the SEC at P_0 can be estimated on the basis stated previously as (5 + 5 x 2/0.2) = 55 nm from per half-sarcomere, being less than 30% of the value actually measured.

High-speed cinematographic studies indicated that the length changes of the elementary fiber segments during the release were uniform along the entire length of the preparation, no appreciable fiber end compliance being observed (Akimoto and Sugi, 1999). These results indicate that the SEC is uniformly distributed in each sarcomere.

Change in striation pattern in Ca^{2+}-activated myofibrils

To explore the origin of the SEC in the sarcomere structures, a short myofibril bundle (length, about 100 μm; initial sarcomere length, 7-9 μm) with both ends fixed in position were activated with iontophoretically applied Ca^{2+}. Since the Ca^{2+}-containing pipette was placed near the middle of the myofibril bundle, the Ca^{2+}-induced contraction first occurred at the middle sarcomeres. As the result, the middle sarcomeres first shortened slowly (by 15-25%) by stretching the end sarcomeres (Fig. 2), and then stopped shortening to contract isometrically.

The change in striation pattern during the Ca^{2+}-induced contractions are summarized in Figure 3. In the middle sarcomeres, the A-band was lengthened by about 10% (P < 0.01) during the course of initial sarcomere shortening and

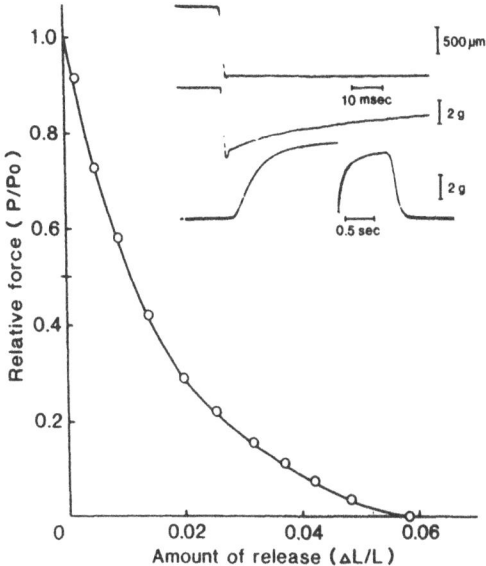

Figure 1. *Typical force-extension curve of the SEC in horseshoe crab telson depressor muscle fibers. The force immediately after a quick release (P) is plotted against the amount of quick release (ΔL). P and L are expressed relative to* P_0 *and* L_0*, respectively. Inset shows an example of the length (upper trace) and force (middle trace) records at the time of a quick release applied to an electrically stimulated preparation. The force record on a slower time base is also shown (lower trace). Arrow indicates time of onset of electrical stimulation (Akimoto and Sugi, 1999).*

then stayed constant when the sarcomeres stopped shortening. Thus, in the isometric condition attained after a period of sarcomere shortening, the A-band lengthening amounted to 5-6% of the sarcomere length. The A-band lengthening was accompanied by the corresponding shortening of the I-band, while the Z-line length showed a tendency to shorten slightly, indicating that no thin filament misalignment took place. The A-band lengthening was completely reversible: when the myofibril bundle was made to relax in relaxing solution, the A-band length as well as the sarcomere length returned to their initial value.

Comparison of myofibril resting tension before and after extraction of myosin from sarcomeres

To give information about the titin filaments, which are responsible for the development of resting tension in muscle fibers (Wang, 1985). We exam-

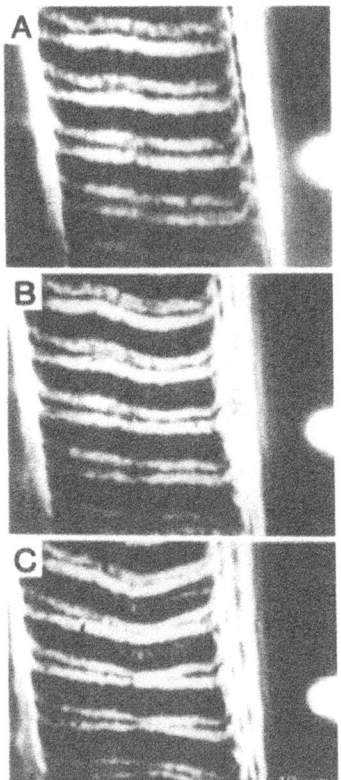

Figure 2. Selected frames from a cinefilm showing sarcomeres at the middle of a myofibril bundle (length, about 80 μm) at rest (A) and during Ca²⁺-activated contraction (B, C). Frames B and C were taken at 0.2 s and at 0.8 s after the beginning of Ca²⁺-application with a glass pipette seen at the right edge of each frame (Akimoto and Sugi, 1999).

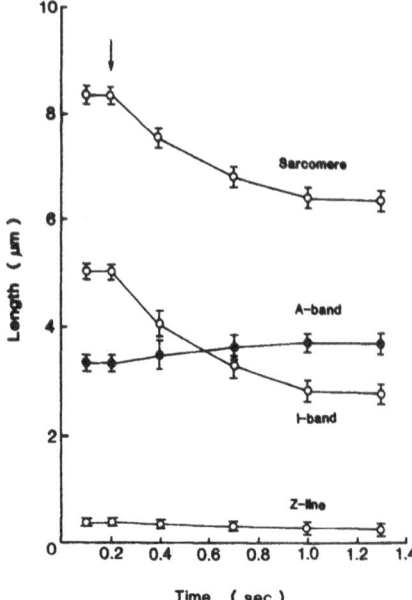

Figure 3. Length changes of the A-band, the I-band and the Z-line during Ca²⁺-induced sarcomere shortening of the myofibril bundle shown in Figure 3. Arrow indicates beginning of Ca²⁺-application. Each data point represents mean ± of measurements on 6 consecutive sarcomeres (Akimoto and Sugi, 1999).

ined the development of resting tension in response to applied stretches either before or after extraction of myosin with 0.5 M KCl. When myosin was extracted from the fiber by soaking in the relaxing solution containing 0.5 M KCl for 30-60 min, the fiber developed no mechanical response to the iontophoretic Ca^{2+} application or in response to a contracting solution (pCa, 5.0), indicating complete removal of the thick filaments. After myosin extraction, the peak resting tension developed in response to stretches (20% of L_0, at 3-5 L_0/s) was decreased to about 30% of that in the fiber before myosin extraction (Fig. 4A). The shape of the length-tension loop with clockwise rotation (Fig. 4B) remained unchanged when pCa was varied from 5 to 7.

Discussion

Elastic A-band lengthening as the main source of the compliant SEC

The present experiments have shown that, in electrically stimulated telson depressor muscle fibers, the extension of the SEC at P_0 (6% of L_0 or 210 nm per half-sarcomere) is too large to be explained by elastic extension of the cross-bridges and the thin filaments (Fig. 1), and that, in Ca^{2+}-activated myofibril bundles, the A-band is lengthened by about 10% during the isometric contraction (Figs. 2 and 3). The latter results can be taken to indicate that the highly compliant SEC mainly originates from the "elastic" A-band extension in isometrically contracting sarcomeres for the following reasons.

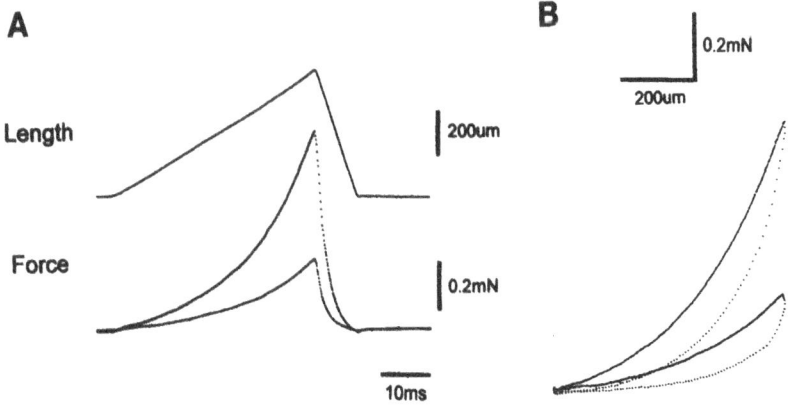

Figure 4. *Effect of extraction of myosin from single glycerinated muscle fibers on the development of resting tension. (A) Resting tension developed in response to applied stretches (20% of L_0) before (upper trace) and after (lower trace) myosin extraction. (B) Length-tension loop obtained from the records shown in A.*

As shown in Figure 3, the Ca^{2+}-activated myofibril bundle contracts iso-metrically after a period of initial shortening, and in the isometric condition the A-band length stayed constant. In 7 myofibril bundle preparations examined, the A-band lengthening amounted to 5-6% of the sarcomere length attained in the isometric conditions. This implies that the A-band in each sarcomere of electrically stimulated fibers may also be lengthened by 5-6% at P_0; if this A-band lengthening is elastic in nature, the A-band length would return to the initial value when the fibers are relaxed. The above isometric force-dependent A-band lengthening would thus show up as main part of the compliant SEC in the telson depressor fibers.

Mechanism of the elastic thick filament misalignment in the A-band

In horseshoe crab telson muscle fibers, M-line structure is absent at the center of the A-band (Dewey, *et al.*, 1973). Since M-line structure in vertebrate skeletal muscle fibers is believed to maintain alignment of the thick filaments within the A-band (Pepe, 1967), the thick filaments in the telson muscle fibers tend to move relative to one another as they are not held in position by M-line material. The thick filament misalignment, resulting in an increase of A-band length, has been thought to be effective in reducing decrease in the number of cross-bridges that can interact with the thin filaments, when the muscle fibers are markedly stretched as the telson moves from its neutral to two extreme positions (Dewey *et al.*, 1973).

Marked thick filament misalignment may occur when the fibers are made to contract isometrically; if there is an imbalance in force between the two half-sarcomeres with a single sarcomere, the stronger one stretches the weaker one, and this further increases the extent of force imbalance as the amount of myo-filament overlap increases in the stronger one. The progress of this thick fila-ment misalignment is eventually prevented when the titin filaments, connect-ing the thick filaments to the Z-line, is stretched in the weaker half-sarcomere to generate elastic force to cancel out the force imbalance between the two half-sarcomeres, as illustrated in Figure 5A. On application of a quick release (6% of L_0), the force in the fibers drop from P_0 to zero due to elastic recoil of the stretched titin filaments and slackening of the thin filaments parallel to the stretched titin filaments (Fig. 5B).

It seems possible that during the quick release, the thin filaments parallel to the stretched titin filaments do not slacken, but remain straight and slide rapidly past the thick filaments without interaction with the cross-bridges. As a matter of fact, it has been shown that, in tetanized frog skeletal muscle fibers, the thick and thin filaments can be made to slide past each other with velocities much larger than the maximum shortening velocity, in the presence of passive compressive forces (Edman, 1979).

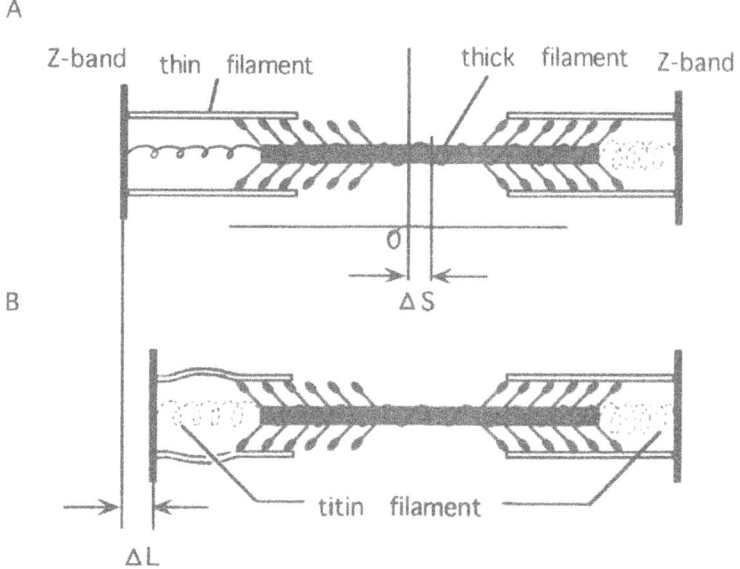

Figure 5. *Diagrams showing mechanism of thick filament misalignment in the sarcomere during isometric force generation (A), and immediately after a quick release to drop the force from P_0 to zero (B). In A, the thick-filament is displaced by ΔS from the center of the sarcomere. The resulting difference in the amount of isometric force between the left and the right half-sarcomeres is canceled out by the elastic force in stretched titin filaments in the left-half-sarcomere. In B, the stretched titin filaments are made just slack by the applied quick release (ΔL). Titin filaments generating no tension are expressed by broken lines.*

In horseshoe crab skeletal muscle fibers, the titin filaments are believed to be present not only between the Z-band and the thick filaments, but also along the entire length of the thick filaments (Wang, 1985), thereby running along the entire length of the sarcomere (Wang and Wright, 1988). The titin filaments are, however, thought to be extensible only between the Z-band and the edge of the thick filaments facing the Z-line; the titin filaments are suggested to wind around the entire thick filaments are not extensible, though they transmit tension from Z-line to Z-line through the thick filaments. If the thick filaments are removed from the sarcomere after extraction of myosin, the titin filaments would become extensible everywhere along their entire length in each sarcomere. Consequently, after the removal of thick filaments, the resting tension developed for a given amount of stretch would be reduced markedly, as the applied stretch is now taken up by the entire titin filaments in each sarcomere. In accordance with this expectation, the resting tension development in response to stretch (20% of L_0) was reduced to about one third of that in the fibers before myosin extraction (Fig. 4). If the titin filaments compliance is assumed to be uniform everywhere within each sarcomere, the above result suggests that the length of the titin filaments, winding around the thick filaments, may be about two times longer that their length between the Z-line and the edge of the thick filaments. The result that the resting tension development in the myosin-extracted fibers was not affected by changes in pCa indicates that the titin filament compliance does not change appreciably by changes in pCa of the surrounding medium.

References

Akimoto T, Sugi H. Structural origin of the series elastic component in horseshoe crab skeletal muscle fibers. *Comp Biochem Physiol* 1999;122A:139-144.

Dewey MM, Levine RJC, Colflesh DE. Structure of Limulus striated muscle: The contractile apparatus at various sarcomere length. *J Cell Biol* 1973;58:574-593.

Edman KAP. The velocity of unloaded shortening and its relation to sarcomere length and isometric force in vertebrate muscle fibres. *J Physiol* 1979;291:143-159.

Hill AV. The heat of shortening and the dynamic constants of muscle. *Proc Roy Soc* 1938;B126: 136-195.

Huxley AF, Simmons RM. Proposed mechanism of force generation in striated muscle. *Nature* 1971;233:533-538.

Huxley HE, Stewart A, Sosa H, Irving T. X-ray diffraction measurements of the extensibility of actin and myosin filaments in contracting muscle. *Biophys J* 1994;67: 2411-2421.

Jahromi SS, Atwood HL. Correlation of structure, speed of contraction, and total tension in fast and slow abdominal muscle fibers of the lobster (*Homarus americanus*). *J Exp Zool* 1969;171:25-38.

Jewell BR, Wilkie DR. An analysis of the mechanical components in frog's striated muscle. *J Physiol* 1958;143:515-540.

Pepe FA. The myosin filament I. Structural organization from antibody staining observed in electron microscopy. *J Mol Biol* 1967;27:203-225.

Sugi H. "The origin of the series elasticity in striated muscle fibers." In *Cross-bridge Mechanism in Muscle Contraction*, H Sugi, GH Pollack, eds. Tokyo: Univ. of Tokyo Press, 1979;85-102.

Suzuki S, Sugi H. Extensibility of the myofilaments in vertebrate skeletal muscle as revealed by stretching rigor muscle fibers. *J Gen Physiol* 1983;81:531-546.

Suzuki S, Oshimi Y, Sugi H. Freeze-fracture studies on the cross-bridge angle distribution at various states and the thin filaments stiffness in single skinned frog muscle fibers. *J Electron Microsc* 1993;42:107-116.

Tameyasu T, Sugi H. The origin of the series elastic component in single crayfish muscle fibres. *Experimentia* 1979;35:210-211.

Tanaka H, Tanaka M, Sugi H. The effect of sarcomere length and stretching on the rate of ATP splitting in glycerinated rabbit psoas muscle fibers. *J Biochem* 1979;86:1587-1593.

Wakabayashi K, Sugimoto Y, Tanaka H, Ueno Y, Takezawa Y, Amemiya Y. X-ray diffraction evidence for the extensibility of actin and myosin filaments during muscle contraction. *Biophys J* 1994;67:2422-2435.

Wang K. Sarcomere-associated cytoskeletal lattice in striated muscle: review and hypothesis. *Cell Muscle Motil* 1985;6:315-369.

Wang K, Wright J. Architecture of the sarcomere matrix of skeletal muscle: Immunoelectron microscopic evidence that suggests a set of parallel inextensible nebulin filaments anchored at the Z line. *J Cell Motil* 1988;107:2199-2212.

Discussion

Baatsen: In Limulus, the widening of the A-band during contraction might be associated with skewing of thick filaments with respect to each other, right?

Sugi: It is Dr. Dewey's opinion. When Jerry and I organized a meeting here in 1982, there was a debate, between Dewey's school and our school. But for example, Roger Cook said "you are right." And then I presented many evidences, and the conclusion was that in both active shortening and isometric force generation, changes in the A-band width were explained by misalignment, and not by thick filament shortening.

Baatsen: So you confirmed that it is by misalignment? Second question is why do you think misalignment occurs?

Sugi: This is due to the lack of M-line structure.

Baatsen: That is not there?

Sugi: Anyway, isometric force generation is well known to be non-stable; if one half sarcomere exerts a little more force, something should cancel this imbalance out.

Baatsen: In vertebrate muscle, the M-line connects the thick filament, right? Do you think that perhaps something connects the thick filaments, and if so, that when there is force generation, that element might also contribute to the passive tension, or to tension generation?

Sugi: Anyway, I think in vertebrate muscle the M-line may be strong enough to cancel out force imbalance. And of course, another factor to cancel out force imbalance may be stretching of cross-bridges, and so on.

Bullard: I was wondering if anybody looked at the size of the titin in these very long sarcomeres? They would be 5 microns long, in a 10-micron sarcomere.

Sugi: Oh, I have no idea. I just assumed that it may be continuous from Z-band to Z-band. But I have no concrete basis for this.

TerKeurs: I am hesitant to ask this next question because you did only two weeks of experiments. If titin is involved in the series elastic properties of the muscle, and we have heard in previous presentations that titin shows stress relaxation and, therefore, the stiffness of that element changes subject to the conditions. Would you be interested in investigating that?

Sugi: I also considered your question. One thing I know, that how slowly
            the A-band widening takes place, so I made a slower stretch. In
            that case, no appreciable stress relaxation was found. Stress
            relaxation depends greatly on the speed of stretch.

Pollack: I would like to make one comment because there is Thomas
            Neumann sitting in the back who is a bit shy, but he did some
            experiments in my lab with single isolated thick filaments from
            Limulus and also from other invertebrates. The measurements,
            which were published in the *Biophysical Journal* last year, I think,
            show that under physiological tensions that the Limulus filaments
            can extend by up to 66%, or something like that, under physiological
            forces. The change is fully reversibly. So at least with these studies
            in isolated thick filaments using the nanolever system that we have,
            we find extraordinary compliance.

Sugi: So anyway, in that case, I showed in one of my slides that restoration
            of A-band width can take place very rapidly.

SKELETAL MUSCLE-SPECIFIC CALPAIN, p94, AND CONNECTIN/TITIN: THEIR PHYSIOLOGICAL FUNCTIONS AND RELATIONSHIP TO LIMB-GIRDLE MUSCULAR DYSTROPHY TYPE 2A

Hiroyuki Sorimachi, Yasuko Ono,
and Koichi Suzuki

*Department of Molecular Biology,
Institute of Molecular and Cellular Biosciences,
The University of Tokyo, Yayoi, Bunkyo-ku, Tokyo, Japan*

Abstract: The skeletal muscle-specific calpain homologue, p94 (also called calpain 3), is essential for normal muscle function. A mutation of the p94 gene causes limb-girdle muscular dystrophy type 2A (LGMD2A), which is one type of autosomal recessive inherited disease characterized by progressive muscular degeneration. In myofibrils, p94 specifically binds to connectin/titin, and the activity of p94 is probably suppressed by this binding. Thus, we postulate that a signal transduction pathway exists, involving p94 and connectin/titin to modulate functions of skeletal muscle, and LGMD2A occurs when this signalling pathway is not properly regulated by p94. LGMD2A mutants of p94 also reveal significant information on the factors that relate structure to function in this molecule.

Introduction

Combinations of unique proteins, which are expressed in a muscle-specific manner, control normal skeletal muscle function. The protease p94, also called calpain 3, CAPN3, or nCL-1, is one such example. p94 has a number of uniquely specialized properties (Ono *et al.*, 1998a; Sorimachi *et al.*, 1997) that are fur-

Elastic Filaments of the Cell, edited by Granzier and Pollack
Kluwer Academic/Plenum Publishers, 2000

ther described below. A mutation in the p94 gene is responsible for limb-girdle muscular dystrophy type 2A (LGMD2A), which is one type of muscular dystrophy inherited as an autosomal recessive trait (Richard *et al.*, 1995).

We have shown that p94 specifically binds to another unique muscle myofibrillar protein, connectin/titin (Sorimachi *et al.*, 1995). Although a physiological significance for this specific binding has yet to be established, this is the first observation that a protease is incorporated in the myofibril. The interaction between p94 and connectin/titin implies that the molecular mechanism of LGMD2A may involve perturbation of the relationship between these molecules.

Here we review current information on p94, connectin/titin, and LGMD2A, and discuss the possible molecular mechanisms whereby they might be involved.

Results and Discussion

Calpain super family

p94 is a novel member of the calpain super family (Sorimachi *et al.*, 1989). Normally, calpain is a ubiquitously expressed intracellular cysteine protease whose proteolytic activity requires Ca^{2+} (for review, see Banik *et al.*, 1998; Carafoli and Molinari, 1998; Johnson and Guttmann, 1997; Ono *et al.*, 1998A; Sorimachi *et al.*, 1997; Suzuki and Sorimachi, 1998). Various intracellular kinases, phosphatases, phospholipases, transcription factors, and cytoskeletal proteins are processed by calpain to modulate their activities and/or structures. This strongly suggests that calpain may play a significant role in intracellular signal transduction system. In mammals, two isozymes, μ- and m-calpains, have been well studied so far, revealing the following structural features.

i) Both μ- and m-calpains consist of a distinct large subunit (ca. 80 kDa; abbreviated as μCL and mCL) and a common small subunit (ca. 28 kDa; abbreviated as 30K), forming a hetero-dimer. In humans, μCL, mCL, and 30K are encoded by independent genes. These are called, *CAPN1*, *CAPN2*, and *CAPN4*, respectively.

ii) Ca^{2+} concentrations required for the proteolytic activity of μ- and m-calpains *in vitro* are ca. 10^{-5} and 10^{-3} M, respectively. The required concentrations are lowered by autolysis, by addition of phospholipids or activator proteins, or by subunit dissociation (see also v below).

iii) μCL and mCL can be divided into four domains: pro domain (I), which is autolyzed upon activation, cysteine protease domain (II), an unknown but conserved domain (III), and calmodulin-like Ca^{2+}-binding domain (IV), which contains five EF-hand structures (see Fig. 1).

iv) 30K has two domains: N-terminal Gly-rich hydrophobic domain (V), and a calmodulin-like Ca^{2+}-binding domain (IV') that shows significant similarity to domain IV of the large subunit.

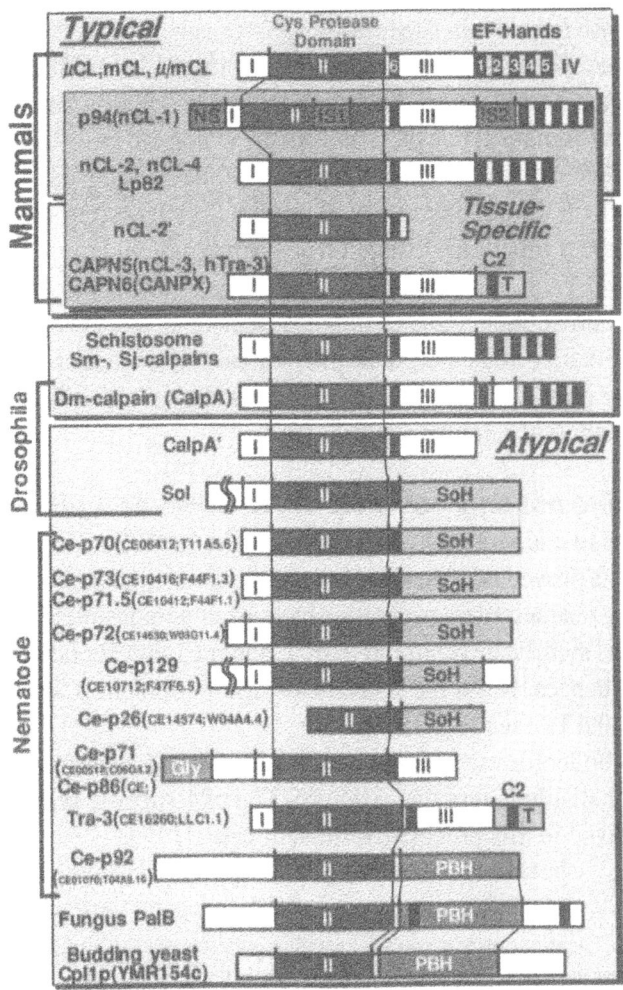

Figure 1. Schematic structure of the calpain super family.
PBH: PalB homologous domain, SoH: Sol homologous
domain, T: Tra-3 specific domain.

v) The two subunits associate through their C-terminus, *i.e.* the fifth EF-
hand structures, and dissociate in a Ca^{2+}-dependent manner. Thus, 30K functions as a regulatory subunit of calpain.

vi) 30K also serves as an intra-molecular chaperon protein of calpain that
helps final assembly of its three-dimensional structure.

vii) A specific calpain inhibitor protein, calpastatin, is also expressed ubiquitously and regulates calpain activity.

Recently, various kinds of calpain homologues have been discovered. These comprise the calpain super family (Sorimachi *et al.*, 1997). As shown in Figure 1, some species show overall similarity to μCL and mCL, which we call "typical calpain homologue," but some only have a region similar to the protease domain (domain II) of μCL and mCL, which we call "calpain-like protease." The latter often has other conserved domains, such as PBH, SoH, and domain T, which do not exist in μCL and mCL, suggesting that calpain-like proteases possess functions different from those of μ- and m-calpains. Among members of the calpain super family, Tra-3 of nematode (Barnes and Hodgkin, 1996), Sol of drosophila, PalB of fungus (Denison *et al.*, 1995), Cpl1p of budding yeast (Futai *et al.*, 1999), and p94 (calpain 3) of human (Richard *et al.*, 1995) are the only molecules whose defect is shown to be directly related to biological phenotypes, including diseases.

p94 is unique not only as a calpain, but also as a protease

p94 was first identified in 1989 at the cDNA level (Sorimachi *et al.*, 1989). Initial analysis proved troublesome for reasons described in ii below; however, the following characteristic properties have been demonstrated.

i) p94 is significantly similar to μCL and mCL but contains three specific insertion sequences: NS at the N-terminus, IS1 in domain II, and IS2 between domains III and IV (see Fig. 1).

ii) p94 undergoes very rapid and exhaustive autolysis, resulting in rapid *in vitro* (the half-life is less than 10 min). The p94-specific sequences IS1 and IS2 are involved in this process (Sorimachi *et al.*, 1993).

iii) IS2 contains a nuclear-translocation signal-like sequence, PxKKKKxKP, and p94 localizes both in the cytosol and the nucleus (Sorimachi *et al.*, 1993).

iv) Most of the effective calpain inhibitors, such as calpastatin, E-64, leupeptin, and EDTA, cannot stop the proteolytic activity of p94. Moreover, calpastatin is a good substrate of p94 (Sorimachi *et al.*, 1993).

v) Although p94 has a calmodulin-like Ca^{2+}-binding domain significantly similar to those of μCL and mCL, p94 shows apparently no Ca^{2+}-dependency.

Rapid and extensive autolysis of p94 *in vitro* makes biochemical studies at the protein level very difficult, and it is almost impossible to characterize p94 from purified protein. Thus, we have been using molecular biological methods, such as the intracellular expression system and the yeast two-hybrid system to analyze p94. As a result, we have shown that p94 specifically binds to connectin/titin (Sorimachi *et al.*, 1995). The binding site of p94 to connectin/titin is at the N-terminal part of IS2 region, and that of connectin/titin to p94 is at N2A sequences in the N_2-line region. Interestingly, p94-specific NS and IS1 regions, which are distant from IS2 region in a primary sequence, are also involved in the binding between connectin/titin N2A and p94 IS2, as described later (Ono *et al*, 1998B; Herasse *et al.*, 1999). Other data suggest that p94 also binds to the C-terminus of connectin/titin and to an unidentified protein at the Z-line, although not enough evidence for these binding sites has yet been obtained (Sorimachi *et al.*, 1995; Kinbara *et al.*, 1997).

In the same year when we reported the binding between p94 and connectin/titin, a mutation in the p94 gene was also shown to be responsible for limb-girdle muscular dystrophy type 2A (LGMD2A) (Richard *et al.*, 1995). This finding was unexpected since all muscular dystrophies, except for p94, are caused by a defect in structural proteins without apparent enzymatic activity, as described in the next section.

Limb-girdle muscular dystrophy

Muscular dystrophy is defined as a group of inherited disorders characterized by progressive muscle wasting and weakness (for reviews, see Backmann and Bushby, 1996; Bönnemann *et al.*, 1996; Cox and Kunkel, 1997; Emery, 1998; Lim and Campbell, 1998; Ozawa *et al.*, 1998; Tsuchiya and Aahata, 1997). A common feature of these diseases is the histological observation of muscle samples that typically include variations in fiber size, areas of muscle necrosis, and, ultimately, increased amounts of fat and connective tissue. As shown in Table 1, various types of muscular dystrophy have been identified genetically and/or phenotypically. Among them, limb-girdle muscular dystrophy (LGMD) is a clinically and genetically heterogeneous group of conditions. As the name indicates, muscle weakness is distributed symmetrically and selectively in the proximal limb and girdle muscles. LGMD is divided into two types according to the inheritance mode. Types 1 and 2 correspond to autosomal dominant and recessive types of LGMD, respectively. So far, three dominant and eight recessive subtypes have been identified.

Responsible gene products for LGMD2C, 2D, 2E, and 2F are γ-, α-, β-, and δ-sarcoglycans (SG), respectively, and those for LGMD1C, Duchenne-type muscular dystrophy (DMD), congenital muscular dystrophy (CMD), and Fukuyama-type congenital muscular dystrophy (FCMD) are caveolin-3, dystrophin, laminin α2 chain, and fukutin, respectively. All these products localize around the sarcolemma, and compose a large protein complex embedded into the membrane, although precise localization of fukutin has not yet been clearly elucidated (see Fig. 2). A defect in one of these components causes destruction of the whole or part of the structure, resulting in a change in the permeability of the sarcolemma. The resulting Ca^{2+} influx from outside the cell activates intracellular conventional calpains, leading to degradation of various muscle proteins, such as α-actinin, connectin/titin, etc.

Table 1. **Muscular dystrophies**

Type	D/R[*1]	Gene	Gene product	Ref.
LGMD1A	D	5q31	(lamin B1 ?)[*9]	Bartoloni *et al.*, 1998
LGMD1B	D	1q11-21	n.d.[*10]	van der Kooi *et al.*, 1997
LGMD1C	D	3p25	caveolin-3	Minetti *et al.*, 1998
LGMD2A	R	15q15.1	p94 [calpain 3][*11]	Richard *et al.*, 1995
LGMD2B, MM[*2]	R	2p13	dysferlin	Liu *et al.*, 1998; Bashir *et al.*, 1998
LGMD2C	R	13q12	γ-SG	Noguchi *et al.*, 1995
LGMD2D	R	17q12-21	α-SG [adhalin][*11]	Roberds *et al.*, 1994
LGMD2E	R	4q12	β-SG	Lim *et al.*, 1995; Bönnemann *et al.*, 1995
LGMD2F	R	5q33	δ-SG	Nigro *et al.*, 1996
LGMD2G	R	17q11-12	n.d.	Moreira *et al.*, 1997
LGMD2H	R	9q31-33	n.d.	Weiler *et al.*, 1998
DMD, BMD[*3]	R	Xp21.1	dystrophin	Hoffman *et al.*, 1987
CMD	R	6q22-23	laminin α2 [merosin]	Helbling-Leclerc *et al.*, 1995
FCMD	R	9q31	fukutin	Kobayashi *et al.*, 1998
FSMD[*4]	R	4q35-ter	n.d.	Wijmenga *et al.*, 1992
EDMD	R	Xq28	emerin	Bione *et al.*, 1994
AD-EDMD	D	1q11-23	lamin A/C	Bonne *et al.*, 1999
TMD[*5]	D	2q31	(titin ?)[*9]	Haravuori *et al.*, 1998
OPMD[*6]	D	14q11.2-13	n.d.	Brais *et al.*, 1998
DM[*7]	D	19q13.2-13.3	MtPK[*1]	Brook *et al.*, 1992
DM-2[*8]	D	3q	n.d.	Ranumn *et al.*, 1998

[*1]D: dominant, R: recessive, [*2]Miyoshi myopathy, [*3]Becker-type MD, [*4]facioscapulohumeral MD, [*5]tibial MD, [*6]oculopharyngeal MD, [*7]myotonic dystrophy, [*8]DM type 2, [*9]possible assignment, [*10]not yet determined, [*11]other familiar name; for other abbreviations, see text.

Recently, another category of muscle proteins responsible for muscular dystrophy has been identified. These are emerin and lamin A/C, responsible for X-linked recessive and autosomal dominant types of Emery-Dreifuss muscular dystrophy (EDMD and AD-EDMD), respectively (Bione *et al.*, 1994; Bonne *et al.*, 1999). Both gene products localize at the nuclear inner membrane (Bonne *et al.*, 1999; Nagano *et al.*, 1996; see Fig. 2). A defect in the emerin gene causes disappearance of the emerin protein, but dystrophin and its complex described above are normally localized, suggesting a unique mechanism for EDMD. In addition, dysferlin, whose defect is responsible for LGMD2B, has a transmembrane region at the C-terminus and several nuclear targeting signals, suggesting a localization similar to emerin (Bashir *et al.*, 1998; Liu *et al.*, 1998).

All the proteins mentioned above are structural proteins without enzyme activity, suggesting that proper cytoskeletal, extracellular, and/or nuclear struc-

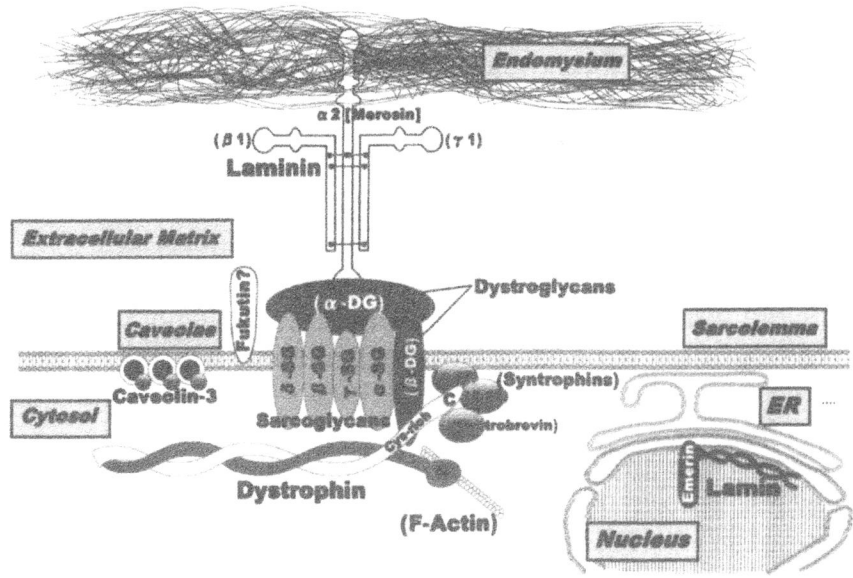

Figure 2. *Skeletal muscle structural proteins responsible for muscular dystrophies. Proteins in parentheses have not yet been identified as responsible for muscular dystrophy.*

tures of muscle cells are vital to normal function. However, p94, the product responsible for LGMD2, is a proteolytic enzyme. Localization of p94 in skeletal muscle is in the myofibrils, and nuclei. The dystrophin complex is normal in LGMD2A patients. Thus, the molecular mechanism of LGMD2A is totally different from that for other muscular dystrophies while the final symptoms are very similar. It is paradoxical that one of the calpain homologues, p94, is **inactivated** in LGMD2A while conventional calpains are **activated** in other muscular dystrophies. Some additional or alternative steps, therefore, should be presumed for LGMD2A if we are to establish a connection between inactivation of p94 and the final degradation of muscle proteins.

Nearly 100 pathogenic independent mutations in the p94 gene have been identified in LGMD2A patients so far by Richard *et al.* (1999). To examine the distinct molecular mechanism of LGMD2A, we focused on various missense point mutations of the p94 gene. We constructed 10 independent LGMD2A mutants of p94 and examined how the unique properties of p94, described previously, are involved in the molecular mechanism of LGMD2A. In theory, missense point mutations should alter one or two, but not all, of the p94 properties. Thus, the change of a property common to all point mutations should be significantly related to the LGMD2A mechanism. Therefore, we examined the autolytic activity, intra-molecular autolytic activity, binding abilities to N2A and C-terminal parts of connectin/titin, Ca^{2+}-dependency of proteolytic activity, and *in vivo* fodrin proteolytic activity (Ono *et al.*, 1998B).

To our surprise, two of the examined mutants, S744G and R769Q, retained rapid and exhaustive autolytic activity similar to the wild-type p94 (Table 2). This indicates that not all of the LGMD2A mutations necessarily disrupt protease activity.

Secondly, some mutants lost N2A binding ability, and others could not bind to the C-terminal part. However, two mutants, H334Q and R490W, showed normal connectin/titin binding ability indistinguishable from wild-type p94. This suggests that dissociation from connectin/titin is not an essential condition for LGMD2A.

Finally, after investigation of several of the biochemical properties of mutant p94, we found that *in vivo* proteolysis of fodrin, but not p94:C129S, is suppressed in all LGMD2A mutants examined. This strongly suggests that substrate proteolysis by p94 is essential for normal skeletal muscle cell function. Thus, identification of *in vivo* substrates of p94 is an urgent priority. We have found several candidates for p94 substrates, such as calpastatin and MtPK, which are now being investigated for their *in vivo* susceptibility to p94 proteolysis.

Structure-function relationship of p94

The analyses of LGMD2A mutants of p94 offered us several interesting perspectives on structure-function relationships of p94. First, not all LGMD2A

Table 2. **Properties of LGMD2A mutants of p94**

Mutation	Domain	Autolysis	N2A[1]	C-ter[2]	Ca^{2+}[3]	Cut C129S?	Cut fodrin?
Wild-type	-	++	+	+	-	yes	yes
C129S	II	-	+	+	-	no	no
S86F	I	-	-	-	-	no	no
L182Q	II	-	-	-	-	no	no
G234E	II	-	+	-	-	no	no
H334Q	II	-	+	+	-	no	no
V354G	II	-	-	-	-	no	no
P319L	IS1	-	-	+	-	no	no
R490W	III	±	+	+	+	yes	no
R572Q	III	±	+	-	+	n.d.[5]	no
S744G	IV	+++	+	-	n.d.[4]	yes	no
R769Q	IV	+++	+	-	n.d.[4]	yes	no

[1,2]: Binding ability to N2A (*1) or C-terminus (*2) of connectin/titin. [3]: Ca^{2+}-requirement for autolytic activity. [4,5]: Properties could not be determined because of too rapid autolysis (*4), or the internal autolytic fragment (*5).

mutants showed a defect in proteolytic activity when p94:C129S was used for a substrate (see Table 2). p94:C129S is considered to have a three-dimensional structure identical to that of the wild-type protein and, thus, proteolysis of p94:C129S by p94 corresponds to "inter-molecular autolysis." We had already found that rapid and exhaustive autolysis of wild-type p94 entails both "intra-molecular autolysis," meaning autolysis in its strict sense, and "inter-molecular autolysis" (Soriamchi *et al.*, 1995; Ono *et al.*, 1998B). Rapid and exhaustive autolysis requires both of these processes, since one of the LGMD2A mutations, R490W, renders p94 defective for intra-molecular, but not inter-molecular, autolysis (see Table 2). In other words, these LGMD2A mutants provide us with a tool to dissect intra- and inter-molecular autolytic processes, and this will be useful for structure-function relationship studies.

Secondly, the P319L mutation at the C-terminal part of IS1 region caused autolytic activity negative and loss of ability to bind to N_2-line connectin/titin. Since the binding site of p94 to the N_2-line connectin/titin is the N-terminal part of IS2, this property of P319L strongly suggests that IS1 region locates very close to IS2, and that they affect each other. Furthermore, our most recent studies have revealed that NS deleted p94, generated by alternative expression of exons 1 and 1' of the p94 gene, cannot bind to the N_2-line element of connectin/titin, while autolytic activity is retained (Herasse *et al.*, 1999). This suggests that both NS and IS2 are closely located in order to disclose the connectin/titin binding site of p94. Thus, we speculate that NS, IS1, and IS2 are all located close together in the p94 tertiary structure, affecting each other and thus characterizing p94 as a unique molecule.

Since P319L, as well as R490W, can proteolyze p94:C129S inter-molecularly, despite their defect in fodrin proteolysis, the recognition mechanisms of p94:C129S (*i.e.*, the IS1 region of p94) and its non-self substrate are different. In other words, p94 recognizes IS1 and other substrates probably by distinct domains. Although there is no hot point of LGMD2A mutation in the p94 gene, 20 mutations are in domain III among 53 of the missense point mutations. This number is the largest among all domains of p94, *i.e.*, 4, 18, 8, and 3 mutations in domain I (+NS), domain II, domain IV, and IS1+IS2 regions, respectively. This suggests that domain III has roles equal to or even more important than those for domain II. One possibility is that domain III is involved in the recognition mechanism of the non-self substrate of p94. This may explain why two domain III mutants of LGMD2A, R490W and R572Q, are defective for fodrin proteolytic activity despite lacking protease activity, as shown by weak autolytic activity and p94:C129S proteolyzing activity (see Table 2). It is also noteworthy that these domain III mutants have Ca^{2+}-requirement for activity, suggesting a possibility that domain III regulates the protease activity of p94 by bridging between domains II and IV.

Thirdly, mutations in domain IV also showed interesting characteristics. While both mutants, S744G and R769Q, retain rapid and exhaustive autolytic activity, they are unable to proteolyze fodrin. From several lines of evidence, we suspect that these domain IV mutations promote autolytic activity of p94 even more rapidly than that of the wild type. Thus, the reason why they cannot proteolyze fodrin might be because they are simply too unstable to access the substrate *in vivo*. Studies on the conventional calpains indicated that domain IV' of the calpain small subunit suppresses calpain protease activity. Similarly, domain IV of p94 might be involved in regulation of p94 autolytic activity. S744G and R769Q mutations make p94 unregulated, resulting in more rapid disappearance of the protein than wild type.

In conclusion, various domains of p94, have distinct functions so that p94 can function in a finely tuned manner. Thus, none of these domains are dispensable, and, moreover, they appear to function in concert to achieve proper function of the p94 molecule.

References

Banik NL, Shields DC, Ray S, Davis B, Matzelle D, Wilford G, Hogan EL. Role of calpain in spinal cord injury: effects of calpain and free radical inhibitors. *Ann N Y Acad Sci* 1998;844:131-37.

Barnes TM, Hodgkin J. The *tra-3* sex determination gene of *Caenorhabditis elegans* encodes a member of the calpain regulatory protease family. *EMBO J* 1996;15:4477-84.

Bartoloni L, Horrigan SK, Viles KD, Gilchrist JM, Stajich JM, Vance JM, Yamaoka LH, Pericak-Vance MA, Westbrook CA, Speer MC. Use of a CEPH meiotic breakpoint panel to refine the locus of limb-girdle muscular dystrophy type 1A (LGMD1A) to a 2-Mb interval on 5q31. *Genomics.* 1998;54:250-55.

Bashir R, Britton S, Strachan T, Keers S, Vafiadaki E, Lako M, Richard I, Marchand S, Bourg N, Argov Z, Sadeh M, Mahjneh I, Marconi G, Passos-Bueno MR, Moreira E de S, Zatz M, Beckmann JS, Bushby K. A gene related to *Caenorhabditis elegans* spermatogenesis factor fer-1 is mutated in limb-girdle muscular dystrophy type 2B. *Nat Genet* 1998;20:37-42.

Beckmann JS, Bushby KM. Advances in the molecular genetics of the limb-girdle type of autosomal recessive progressive muscular dystrophy. *Curr Opin Neurol* 1996;9:389-93.

Bione S, Maestrini E, Rivella S, Mancini M, Regis S, Romeo G, Toniolo D. Identification of a novel X-linked gene responsible for Emery-Dreifuss muscular dystrophy. *Nat Genet* 1994;8:323-27.

Bonne G, Di Barletta MR, Varnous S, Bécane HM, Hammouda EH, Merlini L, Muntoni F, Greenberg CR, Gary F, Urtizberea JA, Duboc D, Fardeau M, Toniolo D, Schwartz K. Mutations in the gene encoding lamin A/C cause autosomal dominant Emery-Dreifuss muscular dystrophy. *Nat Genet* 1999;21:285-88.

Bönnemann CG, McNally EM, Kunkel LM. Beyond dystrophin: current progress in the muscular dystrophies. *Curr Opin Pediatr* 1996;8:569-82.

Bönnemann CG, Modi R, Noguchi S, Mizuno Y, Yoshida M, Gussoni E, McNally EM, Duggan DJ, Angelini C, Hoffman EP. Beta-sarcoglycan (A3b) mutations cause autosomal recessive muscular dystrophy with loss of the sarcoglycan complex. *Nat Genet* 1995;11:266-73.

Brais B, Bouchard JP, Xie YG, Rochefort DL, Chrétien N, Tomé FM, Lafrenière RG, Rommens JM, Uyama E, Nohira O, Blumen S, Korczyn AD, Heutink P, Mathieu J, Duranceau A, Codère F, Fardeau M, Rouleau GA. Short GCG expansions in the PABP2 gene cause oculopharyngeal muscular dystrophy. *Nat Genet* 1998;18:164-67.

Brook JD, McCurrach ME, Harley HG, Buckler AJ, Church D, Aburatani H, Hunter K, Stanton VP, Thirion JP, Hudson T, Sohn R, Zemelman B, Snell RG, Rundle SA, Crow S, Davies J, Shelbourne P, Buxton J, Jones C, Juvnen V, Johnson K, Harper PS, Shaw DJ, Housman DE. Molecular basis of myotonic dystrophy: expansion of a trinucleotide (CTG) repeat at the 3' end of a transcript encoding a protein kinase family member. *Cell* 1992;68:799-808.

Carafoli E, Molinari M. Calpain: a protease in search of a function? *Biochem Biophys Res Commun* 1998;247:193-03.

Cox GF, Kunkel LM. Dystrophies and heart disease. *Curr Opin Cardiol* 1997;12;329-43.

Denison SH, Orejas M, Arst HN Jr. Signalling of ambient pH in *Aspergillus* involves a cysteine protease. *J Biol Chem* 1995;270:28519-22.

Emery AEH. The muscular dystrophies. *BMJ* 1998;317:991-95.

Futai E, Maeda T, Sorimachi H, Kitamoto K, Ishiura S, Suzuki K. The protease activity of a calpain-like cysteine protease in *Saccharomyces cerevisiae* is required for alkaline adaptation and sporulation. *Mol Gen Genet* 1999;260:559-68.

Haravuori H, Mäkelä-Bengs P, Udd B, Partanen J, Pulkkinen L, Somer H, Peltonen L. Assignment of the tibial muscular dystrophy locus to chromosome 2q31. *Am J Hum Genet* 1998;62:620-26.

Helbling-Leclerc A, Zhang X, Topaloglu H, Cruaud C, Tesson F, Weissenbach J, Tomé FM, Schwartz K, Fardeau M, Tryggvason K, Guicheney P. Mutations in the laminin alpha 2-chain gene (LAMA2) cause merosin-deficient congenital muscular dystrophy. *Nat Genet* 1995;11:216-18.

Herasse M, Ono Y, Fougerousse F, Kimura E, Stockholm D, Beley C, Montarras D, Pinset C, Sorimachi H, Suzuki K, Beckmann JS, Richard I. Expression and functional characteristics of calpain 3 isoforms generated through tissue-specific transcriptional and posttranscriptional events. *Mol Cell Biol* 1999;19:4047-55.

Hoffman EP, Knudson CM, Campbell KP, Kunkel LM. Subcellular fractionation of dystrophin to the triads of skeletal muscle. *Nature* 1987;330:754-758.

Johnson GV, Guttmann RP. Calpains: intact and active? *Bioessays* 1997;19:1011-18.

Kinbara K, Sorimachi H, Ishiura S, Suzuki K. Muscle-specific calpain, p94, interacts with the extreme C-terminal region of connectin, a unique region flanked by two immunoglobulin C2 motifs. *Arch Biochem Biophys* 1997;342:99-107.

Kobayashi K, Nakahori Y, Miyake M, Matsumura K, Kondo-Iida E, Nomura Y, Segawa M, Yoshioka M, Saito K, Osawa M, Hamano K, Sakakihara Y, Nonaka I, Nakagome Y, Kanazawa I, Nakamura Y, Tokunaga K, Toda T. An ancient retrotransposal insertion causes Fukuyama-type congenital muscular dystrophy. *Nature* 1998;394:388-92.

Lim LE, Campbell KP. The sarcoglycan complex in limb-girdle muscular dystrophy. *Curr Opin Neurol* 1998;11:443-52.

Lim LE, Duclos F, Broux O, Bourg N, Sunada Y, Allamand V, Meyer J, Richard I, Moomaw C, Slaughter C, Tomé FMS, Fardeau M, Jackson CE, Beckmann JS, Campbell KP. Beta-sarcoglycan: characterization and role in limb-girdle muscular dystrophy linked to 4q12. *Nat Genet* 1995;11:257-65.

Liu J, Aoki M, Illa I, Wu C, Fardeau M, Angelini C, Serrano C, Urtizberea JA, Hentati F, Hamida MB, Bohlega S, Culper EJ, Amato AA, Bossie K, Oeltjen J, Bejaoui K, McKenna-Yasek D, Hosler BA, Schurr E, Arahata K, de Jong PJ, Brown RH Jr. Dysferlin, a novel skeletal muscle gene, is mutated in Miyoshi myopathy and limb-girdle muscular dystrophy. *Nat Genet* 1998;20:31-36.

Minetti C, Sotgia F, Bruno C, Scartezzini P, Broda P, Bado M, Masetti E, Mazzocco M, Egeo A, Donati MA, Volonté D, Galbiati F, Cordone G, Bricarelli FD, Lisanti MP, Zara F. Mutations in the caveolin-3 gene cause autosomal dominant limb-girdle muscular dystrophy. *Nat Genet* 1998;18:365-68.

Moreira ES, Vainzof M, Marie SK, Sertié AL, Zatz M, Passos-Bueno MR. The seventh form of autosomal recessive limb-girdle muscular dystrophy is mapped to 17q11-12. *Am J Hum Genet* 1997;61:151-59.

Nagano A, Koga R, Ogawa M, Kurano Y, Kawada J, Okada R, Hayashi YK, Tsukahara T, Arahata K. Emerin deficiency at the nuclear membrane in patients with Emery-Dreifuss muscular dystrophy. *Nat Genet* 1996;12:254-59.

Nigro V, de Sá Moreira E, Piluso G, Vainzof M, Belsito A, Politano L, Puca AA, Passos-Bueno MR, Zatz M. Autosomal recessive limb-girdle muscular dystrophy, LGMD2F, is caused by a mutation in the delta-sarcoglycan gene. *Nat Genet* 1996;14:195-98.

Noguchi S, McNally EM, Ben Othmane K, Hagiwara Y, Mizuno Y, Yoshida M, Yamamoto H, Bönnemann CG, Gussoni E, Denton PH, Kyriakides T, Middleton L, Hentati F, Hamida MB, Nonaka I, Vance JM, Kunkel LM, Ozawa E. Mutations in the dystrophin-associated protein gamma-sarcoglycan in chromosome 13 muscular dystrophy. *Science* 1995;270:819-22.

Ono Y, Sorimachi H, Suzuki K. Structure and physiology of calpain, an enigmatic protease. *Biochem Biophys Res Commun* 1998a;245:289-94.

Ono Y, Shimada H, Sorimachi H, Richard I, Saido T C, Beckmann J S, Ishiura S, Suzuki K. Functional defects of a muscle-specific calpain, p94, caused by mutations associated with limb-girdle muscular dystrophy type 2A (LGMD2A). *J Biol Chem* 1998b;273:17073-78.

Ozawa E, Noguchi S, Mizuno Y, Hagiwara Y, Yoshida M. From dystrophinopathy to sarcoglycanopathy: evolution of a concept of muscular dystrophy. *Muscle Nerve* 1998;21:421-38.

Ranum LP, Rasmussen PF, Benzow KA, Koob MD, Day JW. Genetic mapping of a second myotonic dystrophy locus. *Nat Genet* 1998;19:196-98.

Richard I, Broux O, Allamand V, Fougerousse F, Chiannilkulchai N, Bourg N, Brenguier L, Devaud C, Pasturaud P, Roudaut C, Hillaire D, Passos-Bueno M-R, Zatz M, Tischfield J A, Fardeau M, Jackson CE, Cohen D, Beckmann JS. Mutations in the proteolytic enzyme calpain 3 cause limb-girdle muscular dystrophy type 2A. *Cell* 1995;81:27-40.

Richard I, Roudaut C, Saenz A, Pogue R, Grimbergen J E M A, Beley C, Cobo A-M, de Diego C, Eymard B, Gallano P, Ginjaar H B, Lasa A, Pollitt C, Topaloglu H, de Visser M, van der Kooi A, Bushby K, Bakker E, Lopez de Munain A, Fardeau M, and Beckmann J S. Calpainopathy - A survey of mutations and polymorphisms. *Am J Hum Genet* 1999;64:1524-40.

Roberds SL, Leturcq F, Allamand V, Piccolo F, Jeanpierre M, Anderson RD, Lim LE, Lee JC, Tomé FMS, Romero NB, Fardeau M, Beckmann JS, Kaplan J-C, Campbell KP. Missense mutations in the adhalin gene linked to autosomal recessive muscular dystrophy. *Cell* 1994;78:625-33.

Sorimachi H, Imajoh-Ohmi S, Emori Y, Kawasaki H, Ohno S, Minami Y, Suzuki K. Molecular cloning of a novel mammalian calcium-dependent protease distinct from both m- and μ-types. Specific expression of the mRNA in skeletal muscle. *J Biol Chem* 1989;264:20106-11.

Sorimachi H, Toyama-Sorimachi N, Saido TC, Kawasaki H, Sugita H, Miyasaka M, Arahata K, Ishiura S, Suzuki K. Muscle-specific calpain, p94, is degraded by autolysis immediately after translation, resulting in disappearance from muscle. *J Biol Chem* 1993;268:10593-605.

Sorimachi H, Kinbara K, Kimura S, Takahashi M, Ishiura S, Sasagawa N, Sorimachi N, Shimada H, Tagawa K, Maruyama K, Suzuki K. Muscle-specific calpain, p94, responsible for limb-girdle muscular dystrophy type 2A, associates with connectin through IS2, a p94-specific sequence. *J Biol Chem* 1995;270:31158-62.

Sorimachi H, Ishiura S, Suzuki K. Structure and physiological function of calpains. *Biochem J* 1997;328:721-32.

Suzuki K, Sorimachi H. A novel aspect of calpain activation. *FEBS Lett* 1998;433:1-4.

Tsuchiya Y, Arahata K. Emery-Dreifuss syndrome. *Curr Opin Neurol* 1997;10:421-25.

van der Kooi AJ, van Meegen M, Ledderhof TM, McNally EM, de Visser M, Bolhuis PA. Genetic localization of a newly recognized autosomal dominant limb-girdle muscular dystrophy with cardiac involvement (LGMD1B) to chromosome 1q11-21. *Am J Hum Genet* 1997;60:891-95.

Weiler T, Greenberg CR, Zelinski T, Nylen E, Coghlan G, Crumley MJ, Fujiwara TM, Morgan K, Wrogemann K. A gene for autosomal recessive limb-girdle muscular dystrophy in Manitoba Hutterites maps to chromosome region 9q31-q33: evidence for another limb-girdle muscular dystrophy locus. *Am J Hum Genet* 1998;63:140-47.

Wijmenga C, Hewitt JE, Sandkuijl LA, Clark LN, Wright TJ, Dauwerse HG, Gruter AM, Hofker MH, Moerer P, Williamson R, *et al.* Chromosome 4q DNA rearrangements associated with facioscapulohumeral muscular dystrophy. *Nat Genet* 1992;2:26-30.

Discussion

Jin: I noticed in your histology slides that you compared 40 weeks normal wild-type mice with 106 weeks transgenics. Did you compare normal mice and transgenics of the same age as well?

Sorimachi: That is a very good point. We actually examined the same age control mice and identified significantly less, lobulated fibers.

Jin: So the next question is what causes this dominant effect, because you still have the endogenous wild-type gene there?

Sorimachi: That is a very complicated point. First, I showed in my slides that wild type p94 is very unstable, and its half life is only about 10 minutes. However, preliminary data suggest that the *in vivo* half-life is much longer than that *in vitro*. So we imagine that p94 is normally bound to connectin/titin and connectin/titin stabilizes p94. But p94 C129S protein has a similar 3-D structure but no protease activity, so it is very stable, but nothing else is different from wild type p94. So this can take over the wild-type p94 age dependently. So maybe, accumulation of p94 C129S protein occurs age dependently and then shows a dominant negative effect of this protein.

Keller: I am interested in the conventional calpain cleavage of titin. Do you have a function for that?

Sorimachi: Well, one likely function is for turnover of connectin/titin. Calpain may cut connectin/titin in one or two locations and after that, the connectin/titin may be destabilized and then other proteases will further degrade connectin/titin. Maybe the turnover regulation of connectin/titin is one of the roles of conventional calpain proteases.

Keller: Right. So in the platelet system we have evidence that the platelet titin is also sensitive to a μ-calpain. It happens after activation and we don't quite know why yet, but there is some suggestion that it may have a physiological effect other than simply turnover. So in terms of the turnover process then, there must be some additional regulation of the p94 that would allow that to occur, or do you think that it just occurs at some low endogenous level?

Sorimachi: That is a good question. I always can use my imagination, but one possible regulation of p94 activity is from connectin/titin. What I mean is that the binding site of connectin/titin to p94 is very close to the PEVK region. Although it is not significant, PEVK region has very weak similarity to calpastatin, which is the conventional calpain inhibitor. But as I mentioned, calpastatin itself has no inhibitory effect for p94, but possibly the connectin/titin PEVK region can act as an inhibitor for p94. It is one possibility. Another possibility is that MEF2 transcription factor, which plays a very important role for muscle specific gene expression, may be involved in the p94 transcription activation. So at the transcription level p94 activity may also be regulated.

Greaser: Do I understand right that you think that there is a p94 on every titin/connectin. Is that right?

Sorimachi: I think so. But as I mentioned, the wild-type p94 is very unstable so it is very difficult to quantify the amount of p94.

Greaser: So there is no way to take whole muscle and stop the proteolysis before you do an SDS gel?

Sorimachi: Well, one drastic method is to add mercury. It is a strong inhibitor for cysteine proteases. But the high toxicity of mercury makes these experiments too dangerous.

Greaser: It is surprising that you do not get much staining of the protein.

Sorimachi: Well I think it is because of the process of fixation. *In vivo* p94 is stable, but once it is frozen or fixed, it becomes unstable.

Greaser: If the half-life is of the order of 10 minutes, I would think that you would be able to make SDS gels and find p94. You agree? But you don't, so something is wrong.

TerKeurs: If p94 is involved in muscular dystrophy because it allows more rapid breakdown of the filament complex in the sarcomere, one would assume that there is always turnover of sarcomeres with breakdown and rebuilding of the structure. So, you would expect that accelerated breakdown in the abnormal p94 situation would cause increased protein turnover in patients with the limb girdle dystrophy. Has that been described?

Sorimachi: This is a very good point. The muscular dystrophy, especially limb girdle muscular dystrophy, develops only at high age. Maybe the first occurrence, can be as late as age 50 or 60 years, or something like that. So maybe proteolysis is promoted, but the difference is very small compared to normal human. Thus, it is very difficult to examine.

Labeit: Just coming back to this question, patients also have an increased creatine kinase activity which argues for increased turnover. My question is, would it be possible by breeding to bring your mutant p94 and the *mdx* mouse together to test if dystrophy and p94 are part of the same sarcomere destruction cascade?

Sorimachi: Yes, it is possible, and a very interesting experiment. Actually we are now constructing a knockout mouse expressing only p94:C129S protein. So we are going to make that knockout and then cross with the *mdx* mouse.

Andrew: The muscular dystrophy is it a dominant or recessive muscular dystrophy? And secondly, in MEF2 is there any evidence that it is cleaved in order to go to the nucleus, or is that just a hypothesis?

Sorimachi: The first question, is the limb muscular dystrophy recessive? Well I did intentionally hide that. The second question, significance of the proteolytic cleavage of MEF2 is under investigation in our laboratory. At present, it is just a hypothesis.

Andrew: Do I understand that you are not going to tell me whether it is dominant or recessive?

Sorimachi: I left it out because it takes too much time. But OK. LGMD2A is a recessive disorder, and this is a bit inconsistent with the fact that the transgenic mouse showed a phenotype. What are the reasons? First of all, the transgenic mice show mild symptoms, as in humans. That is one possibility. The second possibility is that the C129S protein is a very unique protein. As I mentioned, the p94:C129S protein has a completely identical 3-D structure compared to the wild type, so the protein turnover regulating system cannot recognize C129S protein as abnormal protein. This is the difference between the other MD2A mutants. Over 100 mutations are identified, but no C129S mutant was found. That is, C129S mutation might inherit as dominant.

GENERAL DISCUSSION
(Led by Henk ter Keurs)

TerKeurs: As part of the general discussion, I would like to indicate that we all enjoyed the congenial and pleasant character of this meeting. I thank Jerry (Pollack) and Henk Granzier again for organizing such a beautiful meeting. I must say I came here with the explicit intention to learn about titin, and I think I did. It started off with a marvelous historical overview by Charles. I would like to start this general discussion, more or less, at that point, because of the initial statement that Charles made criticizing the conventional textbook picture of a sarcomere where myosin is suspended in mid-air between two actin filaments, and nobody knows where it will go. That has always reminded me of a man wearing large trousers and no suspenders clearly walking around in an unstable state. So the question of stability of the sarcomere immediately comes comes to mind, especially after hearing Siegfried's very interesting introduction and promise for a multitude of isoforms that we are going to learn about in the near future. Add to that Henk Granzier's discussion this morning about the construction of different isoforms of titin in one sarcomere. I would like to ask the audience what their opinion is about anticipated stability of the sarcomere. If four stiff titins envelop myosin at one side plus 2 compliant titins at the opposite side, is stability of the muscle enhanced, or is the gent wearing his pants in a skewed fashion?

Granzier: I believe that the evidence suggests that at the myocardial tissue level there is co-expression of isoforms. We now have preliminary results that suggest co-expression at the level of the myocyte and even at the level of the sarcomere. However, the A-band location may still be stable as on average the sarcomere may contain the same number of stiff and compliant molecules at one side of the A-band than at the other side. One point of clarification is that, in my viewpoint, titin may not be able to keep the A-band centered during prolonged contractions, but that instead, titin resets the A-band position during the relaxation phase of a contraction-relaxation cycle. So titin is a spring that resets the A-band position, and having different isoforms within a sarcomere in parallel, in my opinion, will still make it possible for titin to perform this function.

Sanger: What is the half life of titin in adult muscle, and how long does it
             take to synthesize a single titin molecule? When you are looking
             at live cells, where we can follow the same myofibrils, it is very
             dramatic when a cardiomyocyte decides to go into cell division,
             just all the myofibrils are disassembled within a matter of 20
             minutes. If you get an interface cell, where you are looking at
             some myofibrils, you just see one myofibril disassemble while
             others are being formed. We have no idea why that particular
             myofibril disassembled, because in that case mitosis is not taking
             place. I think, like for example, thick filaments are very stable at
             low ionic strength. If you are looking at them and yet when you
             look at the live cell, the muscle is capable of taking molecules that
             shouldn't be very soluble at physiological ionic strength, and
             moving them about. So I would like to say that we shouldn't think
             that titin filaments are somehow making sarcomeres in cement.
             They have a tremendous amount of stability and dynamics. That is
             why I asked Carol when she did her micro-injection experiments
             how fast it induced disassembly. That indicates to me that the titin
             really is very dynamic, and you might be capturing it while it is
             loose. So I was thinking of Haruo's work done on this P94, that it
             almost might be a chaperone, that as soon as this titin molecule is
             synthesized, if it has a very sensitive calpain site, this P94 could sit
             on it and ride it in to the sarcomere, to protect it in some way. But
             anyway, it would be good if we could get some ideas on what the
             half life of titin is in adult muscle and how long it actually takes to
             synthesize a molecule.

Greaser: Calculation indicates that synthesis of a titin molecule takes on the
             order of about 5 or 6 hours. Now that may be before the molecular
             weight doubled. In terms of the turnover, there is no data out there
             that I am aware of. The turnover studies done earlier on myosin
             and actin in heart muscle suggest a half life of typically 3-5 days,
             but there is no data out there at all on the titin.

TerKeurs: The half-life studies or incorporation studies I am aware of are
             overall protein incorporation in heart muscle being 10% per day.

Pollack: I would like to return to the problem of the suspenders that you
             talked about. What you neglected in your discussion is the modern
             invention of the belt to go along with the suspenders. The issue of
             stability is really a critical one, but the issue may transcend what
             happens with one thick filament and titins because if the titins or

the thick filaments are interconnected to one another, then what matters really is not the individual thick filament or the individual titin, but the whole sarcomere.

Granzier: I agree with that.

TerKeurs: In the beginning of this meeting, two days ago, I sensed a discussion about how many titin molecules are actually surrounding thick filament. Has the debate been resolved during the 2 days, and is it 6 per half thick filament?

Keller: I'll just comment on that also, and I mentioned this to John. It's probably totally coincidental, but the ratio of 12 titins to a thick filament in muscle is approximately the same as the ratio of titin to fewer myosins in the small bipolar filaments that we find in nonmuscle systems. Maybe that's coincidental, or maybe it is a result of some sort of evolutionary relationship.

Linke: I'd like to follow up on what Jerry said. I think we are still believing in titin too much as in a straight or singular molecule. There might be much more to this than we anticipate. I just talked with John about this. Consider the end filament, which may be a composite filament consisting of several titin molecules. Also, as Carol has pointed out, there might be interactions between titin and other proteins, including the thin filament proteins, and there might be other interactions we still don't know about. We probably have to rethink the model of titin as a singular protein.

TerKeurs: I think that everybody in the room, who has observed the beautiful freeze fracture electromicrographs of muscle will agree that this system is not a simple filament system. It looks more like a meticulously built three dimensional micro computer with a mixture of function and structure. For example, simultaneous enzyme function and structural function providing interaction between different enzymes and their substrates.

Vigoreaux: I just want to state the obvious and make a brief comment. When we talk about things like turnover and ratios of titin to thick filament, let us please mention which muscle type we are referring to and which organism because these things are likely to be very different among different muscle types and different organisms.

TerKeurs: I would like a summary for the 2 major muscle types, cardiac and skeletal muscle, about the following questions. Suppose that one stretches one of his muscles over its dynamic range with which you and I use it. Sarcomere length range in the heart goes from 1.7 to 2.2 micron; beyond that is rare. So which element in titin is extended and which element is not? And of course, the words extended and not extended are not black and white, but please summarize.

Granzier: Let's start with a slack sarcomere where there is no external force. Titin's extensible region is now in a contracted state. If the sarcomere shortens to below the slack length then the extensible region straightens. The same happens if the sarcomere is stretched above slack. The extension of the extensible region is not uniform. Initially, the Ig segments extend predominantly. Ig domains, however, remain folded; the individual domains have some rotational freedom with respect to each other so that they can be part of a collapsed structure (slack sarcomeres) as well as a straightened filament. Once the tandem Ig segments start to approach their contour lengths, the PEVK segment extension becomes dominant. I would not exclude that the PEVK has structure or that electrostatic interactions occur along its polypeptide chain, but ultimately the PEVK will reach a length that is only consistent with it being fully unfolded. In the heart, the unique N2B sequence then starts to elongate. I think that is all that will happen in the heart under physiological conditions which, depending on the species, contains sarcomeres that may extend to perhaps 2.3 - 2.4 microns. In skeletal muscle where this large unique N2B sequence is not present, you only deal with those first two mechanisms: tandem Ig segment extension and PEVK extension. The tandem Ig and the PEVK are much longer than in the heart, and they can accommodate a wide SL range without requiring Ig domain unfolding.

Linke: I think that was a beautiful summary. I would like to add maybe two points. One is a comment, one is a question to Charles. The first is, when we look at single myofibrils isolated from cardiac and skeletal muscle, we always find that the slack length is larger in skeletal muscles. In the single myofibrils there is no surrounding tissue so it must be a property of the sarcomere. Why is this? Why is the slack sarcomere length longer in a skeletal myofibril compared to cardiac? I have a proposal which is debatable, but it has to do with the wormlike chain behavior of the immunoglobulin domain regions. This is relatively easily explained and I mentioned that today already once. If you have a contour length to persistence length ratio, which is relatively low, about 10 to 1, it may be–at least that's what the elasticity theory predicts–that this entropic chain does not collapse entirely. It might be a semi-flexible chain. Therefore, the sarcomere does not shorten down to 1.7 microns or 1.6, as it should if there was complete collapse of titin. Titin may really stay somewhat extended, and perhaps the difference between muscle types then is the amount of Ig domains in the molecule. It's just a proposal; I don't have any experimental evidence for that. The second is a question to Charles. What do we do with the region between the Z-line center and the T12 antibody? Your micrographs and also our data show that there is some extensibility, but we usually ignore it. What do you say about this? Is there any extension in the physiological sarcomere length range?

Trombitás: At physiological lengths, there is very little extension. But if we exceed the physiological length, for example, about 5 micron extension for the whole sarcomere, then you can see that T12 moves away from the Z-line towards the A-band. We just had a very interesting experiment from the human soleus muscle. As we overstretched the muscle, we could see almost regular extension of the T12. And as I showed in my original paper in '93, if the sarcomere length was around 6 micron in frog muscle, which is very long in the frog muscle, then the T12 elongated, although not regularly.

Granzier: Can I respond to your first issue? Entropic elasticity theory does not predict that the chain fully collapses. The mean-square end-to-end distance is 2AL (A: persistence length; L: contour length). So indeed you would not expect that the end-to-end length goes down to zero. And if you look at the slack lengths in mammalian heart, it is not 1.8 microns; that is where you would expect it to be if you only take the A-band and inextensible region near the Z-line into account, it is a little bit more than that. And in skeletal muscle where the contour length of titin's extensible region is longer than in cardiac muscle, the slack will thus be slightly longer as well

Labeit: Basically I would like to know how Charles feels.

TerKeurs: It would be good if we ask Charles in a moment to give his summary from connectin filaments to co-expression to titin isoforms.

As an interlude, concluding or stopping this discussion and going after that to Charles' presentation. I thought about the nice puzzle that John Trinick has given us. That is, if we have a hexagonal arrangement in the A-band region and we have a foursome Z-band region, how do you make the simplest connection between the A-band structures and the Z-band structures? Maybe there is one simple solution. A square is composed of two triangles which share two corners of the square. That means that you can take a square apart and have two triangles, and you have a hexagonal element. The only thing you need to do is to fuse two corners of those six elements and link up the elements of the hexagon in the A-band to the elements in the square. That means the only thing you need to communicate while you are building the Z-band is that you have to build twice as many titin filaments as in the closest adjacent corner, and for the next one it's half as much and for the next one it should be twice as much again. That means that it works out to alternatingly 4 filaments and 2 filaments on the corners. This requirement gives the proper number and the proper building blocks and only one rule so that even a simple masoner can do the job.

Trinick: If your proposal was correct, I suppose that if the titin does join the thin filaments near the edge of the Z-line, one can certainly see a distinct thickening of the thin filaments in that region. So if your proposal was correct, then thin filament positions that were diametrically opposed, you'd have two of the pairs in a square that would be significantly thicker because they would have twice as many titin molecules. I'm not sure that there is any evidence of that.

TerKeurs: Not necessarily because at the other side of the Z-band, you do the same thing but with a 90 degree twist.

Trinick: I suppose if you cut sections that were sufficiently thin but they didn't include the others. Well you couldn't do that cause they overlap.

FROM CONNECTING FILAMENTS TO CO-EXPRESSION OF TITIN ISOFORMS

Károly Trombitás[1], Alexandra Freiburg[2],
Marion Greaser[3], Siegfried Labeit[2],
and Henk Granzier[1]

[1]*Department of Veterinary and Comparative Anatomy, Pharmacology, and Physiology, Washington State University, Pullman, WA*
[2]*European Molecular Biology Laboratory, Meyerhofstrasse 1, Heidelberg, Germany*
[3]*Muscle Biology Laboratory, University of Wisconsin, Madison, WI*

Abstract:　The molecular basis of elasticity in insect flight muscle has been analyzed using both the mechanism of extensibility of titin filaments (Trombitás *et al., J. Cell Biol.* 1998;140:853-859), and the sequence of projectin (Daley *et al., J. Mol. Biol.* 1998;279:201-210). Since a PEVK-like domain is not found in the projectin sequence, it is suggested that the sarcomere elongation causes the slightly "contracted" projectin extensible region to straighten without requiring Ig/Fn domain unfolding. Thus, the extensible region of the projectin may be viewed as a single entropic spring. The serially linked entropic spring model developed for skeletal muscle titin was applied to titin in the heart. The discovery of unique N2B sequence extension in physiological sarcomere length range (Helmes *et al., Circ. Res.* 1999;84:1339-1352) suggests that cardiac titin can be characterized as a serially linked three-spring system. Two different cardiac titin isoform (N2BA and N2B) co-exist in the heart. These isoforms can be differentiated by immunoelectron microscopy using antibody against sequences C-terminal of the unique N2B sequence, which is present in both isoforms. Immunolabeling experiments show that the two different isoform are co-expressed within the same sarcomere.

Elastic Filaments of the Cell, edited by Granzier and Pollack
Kluwer Academic/Plenum Publishers, 2000

As this conference has shown convincingly, titin's flexible nature is central to understanding the molecular mechanism of extensibility. Titin's flexibility is apparent from early micrographs of isolated and purified titin molecules that revealed that the titin molecule is not straight but instead forms nodules of contracted protein (e.g., Maruyama, 1984; Trinick, 1984; Wang, 1984). This 'contracted' state of the molecule results from thermally induced bending motions that shorten the end-to-end length to much less than its contour length. Direct evidence for titin's entropic spring property was obtained either in intact myofibrils by using a freeze-break method in which titin was first stretched and then mechanically broken (Trombitás *et al.*, 1990, 1993, 1997), or *in vivo* by stretching myocytes beyond overlap and subsequently releasing. Results reveal that either the broken titin filaments retracted towards the Z-line in the freeze break case, or the overstretched titin filaments contracted crumpling the thin filaments near the Z-lines (Fig. 1). In the intact muscle, a similar retraction occurs when muscle is released to its slack length, a phenomenon that occurs in both cardiac and skeletal muscles (Trombitás *et al.*, 1995; 1998a). For example, in human soleus muscle, the end-to-end length of the extensible region in slack sarcomeres is ~100 nm while the contour length of tandem Ig segments is ~500 nm (human soleus titin). Thus, due to titin's flexible nature, thermally induced bending motions tend to shorten the molecule both *in vitro* an *in vivo*.

When slack sarcomeres are stretched or shortened, titin's conformational entropy is lowered and titin straightens (Helmes *et al.*, 1996). However, titin's extensible region does not behave uniformly. When the so-called PEVK sequence, rich in proline, glutamate, valine and lysine residues, was discovered within the central I-band region of titin, this stimulated the suggestion that due to its unusual amino acid composition (high content of proline and charged amino acids) the PEVK sequences may not form stable structures. Therefore, this domain may provide an important source of extensibility of the titin filament (Labeit and Kolmerer, 1995a). Based on immunofluorescent studies using a titin antibody (N2A), which labeled an epitope at the N-terminal of the unique sequence PEVK segment (Linke *et al.*, 1996), or immunoelectron microscopic studies using another titin antibody (SM1), which labeled an epitope N- terminal of the PEVK segment (Gautel and Goulding, 1996), a model was proposed in which the tandem Ig segments and the PEVK segment extend sequentially.

Experiments using sequence-specific titin antibodies that mark the boundaries of the tandem Ig and PEVK segments, made it possible to follow the extension of these segments in the sarcomere. These experiments were performed on soleus muscle fibers, to take advantage of the known sequence of titin in this muscle type (Labeit and Kolmerer, 1995a). In slack sarcomeres the

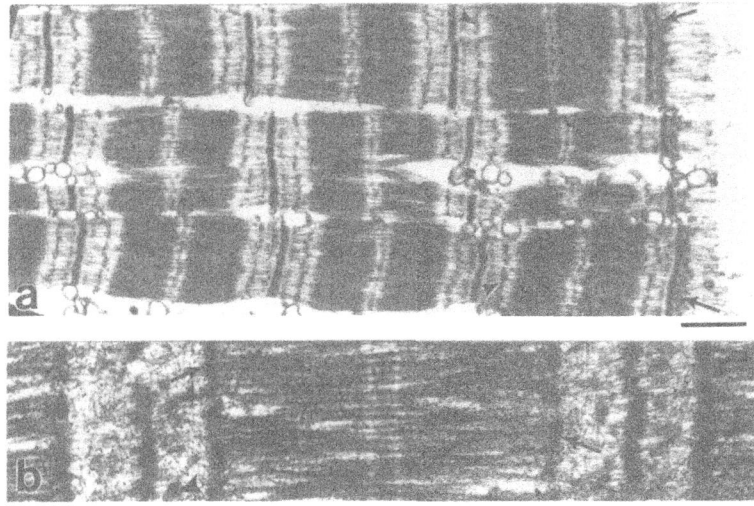

Figure 1. *Titin filaments behave as an entropic spring.* **a:** *Stretched bundles of few fibers from rabbit psoas muscle were quickly frozen in liquid nitrogen and broken in a direction perpendicular to the fiber axis. The broken segments were thawed very slowly in formaldehyde fixative at 0 C°. Then the segments were labeled with RT13 titin antibodies (Wang et al., 1991). The antibody labeled a titin epitope in the intact I band near the A-I junction (arrowheads). This epitope retracted to the T12 epitope position in sarcomeres broken at the A-I junction illustrating the titin filament "contraction" (arrows) and revealing the entropic spring nature of titin in the intact muscle. Scale bare: 1 μm.* **b:** *Myocyte from bovine atrium, stretched beyond overlap then released. The overstretched titin filaments shortened the sarcomeres during release, but the thin filaments could not slide back to the A-band crumpling the thin filaments in the I-band. The myocyte was labeled with titin antibodies: MIR (directed to an immunogenic epitope close to the edge of the A-band, arrows) and anti-I18-19 (arrowheads). Scale bar: 1 μm.*

tandem Ig and PEVK segments were in a contracted state, and as the sarcomere was stretched, the contracted tandem Ig segments extended first, then the PEVK extension became dominant (Fig. 2; Trombitás *et al.*, 1998a; Linke *et al.*, 1998a,b).

Early molecular models of titin's elasticity proposed that stretching titin results in unfolding of immunoglobulin and fibronectin domains, extending each domain from ~4 nm to ~30 nm as it unravels (Soteriou *et al.*, 1993; Erickson, 1994). The unfolding model was challenged by Politou *et al.* (1995; 1996),

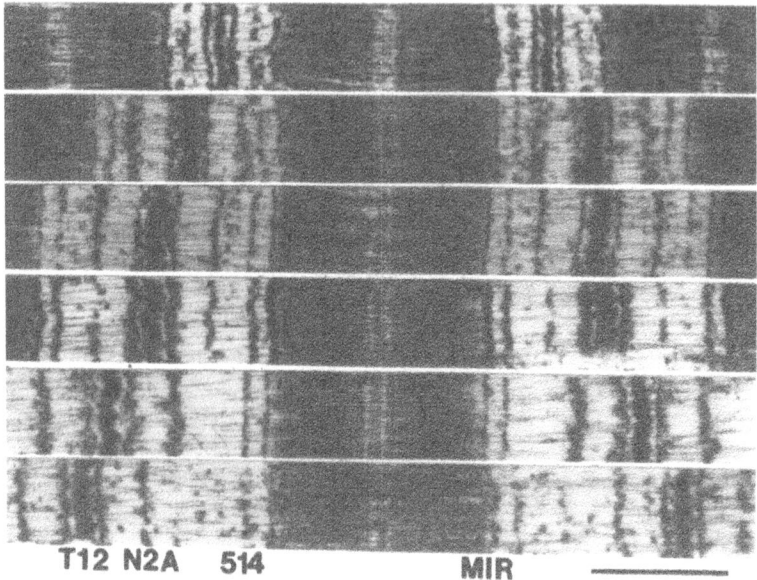

Figure 2. *Human soleus sarcomeres labeled with sequence-specific titin antibodies (T12, N2A, 514 and MIR) that mark the tandem Ig and PEVK segments. With increasing sarcomere length, the PEVK segment extension (N2A-514) becomes dominant. Scale bar: 1 μm.*

who suggested that the domains' stability is so high that unfolding may not occur under physiological condition. Our work on human soleus muscle is consistent with this. Modeling tandem Ig and PEVK segments as entropic springs with different bending rigidities (Trombitás *et al.*, 1998a) suggested that at physiological sarcomere lengths the Ig-like domains of tandem Ig segments remain folded. Thus, the extension of the tandem Ig segments that occurs upon stretching slack sarcomeres results from alignment of the Ig domains, while the domains maintain their tertiary structure.

Sequential extension of tandem Ig and PEVK segments results from their different bending rigidities. The tandem Ig segment contains relatively large sub-domains and thus has a high bending rigidity and low conformational entropy. As a result, the tandem Ig segments extend under low force. The PEVK segment is likely to be largely unfolded and therefore has a low bending rigidity and high conformational entropy— the PEVK segment extends under high force. Thus, the extensible region of skeletal muscle titin may be viewed as serially linked entropic springs with different bending rigidities (Fig. 3).

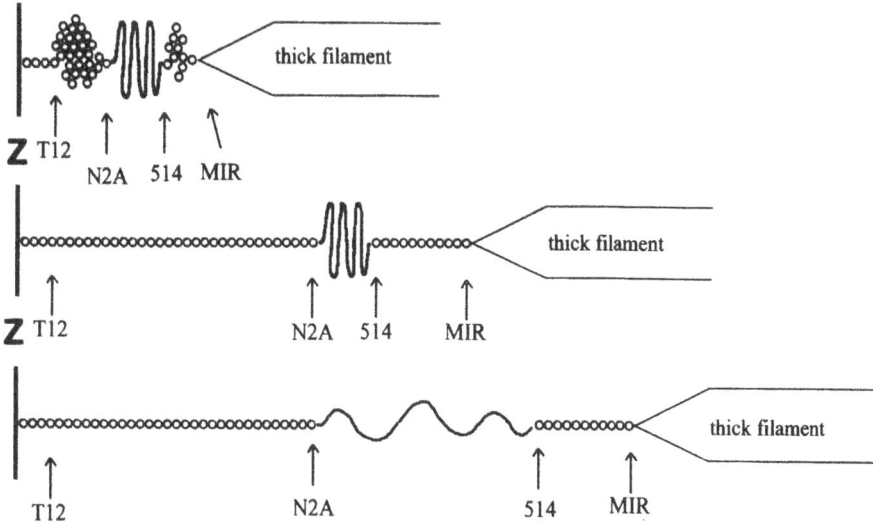

Figure 3. *Schematic diagram of sequential extension of vertebrate skeletal muscle titin (adapted from Horowits, 1999).*

The molecular basis of elasticity in insect flight muscle

Can insights obtained from work on skeletal muscle titin be applied to the connecting filaments? It is well known that the passive tension of insect flight muscles (IFM) increases steeply with sarcomere stretch and that it reaches a plateau at only ~10% stretch (Pringle, 1967; White, 1983; Granzier and Wang, 1993). The sarcomere can be extended further than this limit, but the passive tension – sarcomere length relation – is not reversible and structural damage to the sarcomere may occur. For example, in Lethocerus muscle fibers, thick filaments translocate (White, 1983). In the honey bee, sarcomeres can be stretched to beyond overlap and connecting filaments can reach a length of ~1 μm (Fig. 4) An upper limit of ~1 μm was also found for the connecting filaments in Lethocerus, based on stretching fibers in rigor (Withe, 1983).

The connecting filament protein – named projectin, mini-titin, invertebrate connectin – has been purified from different species (e.g., bee: Saide *et al.*, 1989; Nave and Weber, 1990; Lethocerus: Lakey *et al.*, 1990; Locusta migrata: Nave and Weber, 1990; *Drosophila*: Saide *et al.*, 1989; Nave and Weber, 1990; Crayfish claw muscle: Hu *et al.*, 1990). For simplicity, in the following, we refer to the connecting filament protein as projectin. The molecular weight of projectin is uncertain, and varies with species. In SDS page, the bee and Lethocerus projectin mobility is slightly less than nebulin (sequencing studies have shown a minimal molecular weight of 773 kDa for human soleus muscle nebulin, possibly co-expressed with larger splice isoforms; Labeit and

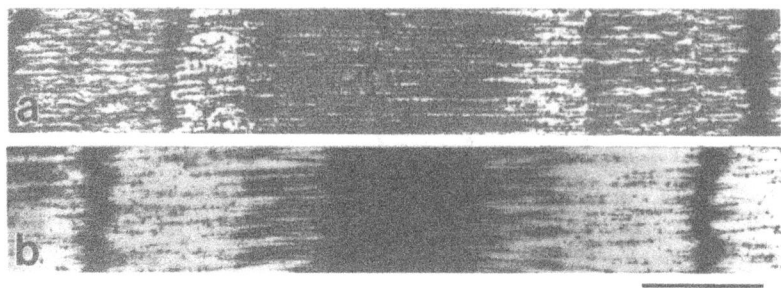

Figure 4. *Overstretched bee flight muscle. Compare the fully extended*
connecting filament lengths in the two sarcomeres stretched either in relaxed
(a), or in rigor state (b). They have a similar length of about 1 μm.
a: *the muscle was stretched in relaxed condition. When the connecting filaments*
reached their extensible limit, the thick filament started to elongate creating an
elongation zone gap. The dense regions at the end of the thin filaments represent
the residual overlap zones of thin and elongated thick filaments. Thus the fully
elongated connecting filaments are little shorter, than the thin filaments.
b: *The same muscle was stretched in rigor, then activated. Because of the rigor*
bonds, all the thin filaments were broken at the Z line level. The external force
in rigor condition elongated the connecting filaments until they reached their
extensible limit. In this case the thick filaments were not elongated, since when
the force exceeded a critical limit, new sarcomeres were gradually broken and
kept the force almost constant. Note that the activation pushed all the broken
thin filaments towards the center of the sarcomere. When the complete double
overlap zone developed, the thin filament sliding movement stopped, indicating
that the bridges can not exert more force (Trombitás and Tigyi-Sebes, 1977,
1984, 1985). Applying this phenomenon to the partial double overlap, the active
force should decrease with increasing width of double overlap zones. Namely,
bridges in the double overlap zones can not generate force. This bridge property
may be important in the heart muscle, where double overlap zone develops in
part of the contraction cycle. Scale bar: 1 μm.

Kolmerer, 1995b). From mobility on SDS gel, the estimated molecular weight
of projectin ranges from 600 to 1200 kDa (Bullard and Leonard, 1996). The
length of the purified (proteolitically cleaved) molecule was measured to be
260 nm (Nave and Weber, 1990), the contour length of intact molecule should
be slightly longer. All the antibodies against projectin uniformly labeled the I-
band and the lateral part of the A-band of insect flight muscle, so seemingly the
projectin molecules span about 300 nm, starting from the edge of the Z-line,
running through the I-band to the end region of the thick filaments (e.g., Saide,
1981; Saide *et al.*, 1989, 1990; Nave and Weber, 1990; Hu *et al.*, 1990; Lakey
et al., 1990; Vigoreaux *et al.*, 1991). At the distal end of the A-band, the projectin

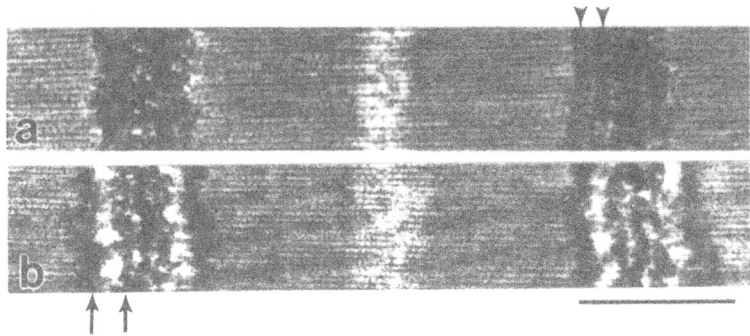

Figure 5. *Lethocerus flight muscle labeled with projectin antibody.* **a:** *In sarcomere near rest length there are two distinct antibody binding regions in every half sarcomere, one which labels the I-band, the other labels the distal end of the A band (arrowheads).* **b:** *Sarcomere extended about 10% above rest length. The first epitope remained near the Z line, as above, but the second epitope moved with the edge of the A band (arrows) showing for the first time that the contracted projectin extended. Scale bar: 1 μm.*

molecules firmly attach to the thick filaments (Reedy, 1971; Trombitás and Tigyi-Sebes, 1977, 1984, 1985; White, 1983). Figure 5 shows an example of anti-projectin labeled Lethocerus flight muscle.

Drosophila projectin was recently sequenced by Daley *et al.* (1998). It consists of immunoglobulin-like and fibronectin-like type domains that repeat in patterns similar to that of twitchin. A kinase domain is present as well, but a PEVK-like domain was not found. Based on the known number of Ig/Fn domain (39 Fn and 31 Ig domains) and assuming a domain spacing in a fully extended molecule of 5 nm results in a predicted contour length of ~350 nm (the NH2 terminal end of the molecule has not yet been fully sequenced).

It seems reasonable to assume that projectin fully penetrates the ~100 nm wide Z-disc, as is the case for titin (Gregorio *et al.*, 1998), and that it extends ~ 50 nm into the A-band (Nave and Weber, 1990). This leaves ~200 nm for projectin's extensible I-band segment. In slack sarcomeres, the I-band is ~70-100 nm wide (Fig. 6) and, thus, projectin's extensible region is likely to be in a slightly 'contracted' state. Considering that the sarcomere stretch amplitude that gives rise to reversible force response is limited to ~10%, or ~125 nm per half sarcomere, it seems likely that projectin's extension can be accommodated by straightening of the extensible region without requiring Ig/Fn domain unfolding. Whether straightening can explain the high passive tension developed by IFM remains to be established. A 10% sarcomere stretch results in a large increase in the fractional extension of projectin's extensible region, and it is possible that the increase in entropic force is sufficient to explain the mea-

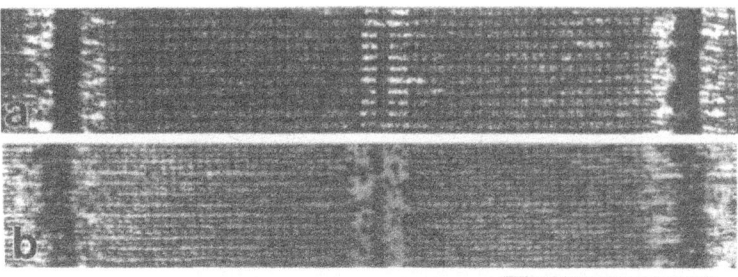

Figure 6. *The length of the I-bands and the H-zone in bee flight muscle at rest length. The muscle was fixed in the intact thorax. The one filament layer sections contain: a) thick and thin filament layer, b) thin filament layer only. Scale bar: 1 μm.*

sured force increase, especially if the base-line fractional extension in slack sarcomeres is relatively high. Thus, the extensible region of the projectin may be viewed as a single entropic spring.

Currently, it can not be excluded that some of the unique sequences within the extensible region of projectin are extensible and involved in setting the passive tension level, analogous to what has been found recently for cardiac titin (Helmes *et al.,* 1999; Trombitás *et al.,* 1999). Finally, it is also possible that during non physiological sarcomere extension some of the Ig/Fn domains of projectin unfold. In the Lethocerus and bee IFM the connecting filaments can be extended to a length of ~1 μm (see Fig. 4; White, 1983; Trombitás and Tigyi-Sebes, 1977, 1984). It is difficult to envision how such a long length can only be attained without unfolding of Fn/Ig domains. Based on results with skeletal and cardiac titin, which indicate that unfolding is unlikely to occur under physiological conditions, it may be considered unlikely that unfolding takes place under physiological conditions as physiological stretch amplitudes of IFM are small. To more firmly elucidate the molecular mechanism of force development by the connecting filaments, immunoelectron microscopic studies with sequence-specific antibodies will be required.

Although separation of insect flight muscle proteins on SDS gels detect a titin-sized protein with molecular mass of ~3000 kDa, (Lakey *et al.,* 1993; Bullard and Leonard, 1996; Machado *et al.,* 1998; Andrew *et al.,* in this volume), a full-size titin has never been isolated from insect flight muscle. Therefore, further studies are required to clarify if insects indeed express a titin-like protein that acts as an elastic filament system, comparable to the situation in vertebrates.

Co-expression of titin isoforms in the heart

Can the serially linked entropic springs model developed for skeletal muscle titin be applied to cardiac titin? The N2B titin isoform of heart muscle has a short proximal tandem Ig segment and a very short PEVK segment (Fig. 7). These segments reach their contour length at a sarcomere length of ~2.3 μm (Granzier *et al.*, 1997). However, N2B titin also contains a large unique sequence within its N2B element (Labeit and Kolmerer, 1995a) and it was recently shown that this sequence elongates in sarcomeres longer than ~2.2 μm, (Fig. 8; Helmes *et al.*, 1999; Linke *et al.*, 1999; Trombitás *et al.*, 1999). At a sarcomere length of 2.8 μm, the unique N2B sequence elongation reaches 150 nm (partial extension), while the PEVK segment is fully elongated (60 nm) in the N2B titin isoform (Fig. 9). Thus, cardiac titin can be characterized as a serially linked three-spring system (Fig. 10).

Study of the elasticity of cardiac titin filament system is complicated by the expression of a range of different titin isoforms. Recently, two major cardiac titin isoforms, referred to as N2B and the N2BA titins, have been discovered (Freiburg *et al.*, unpublished results). These isiforms, generated by differential splicing, contain a different number of Ig domains within their extensible region and PEVK segments that differ in length (Labeit and Kolmerer, 1995a; Centner *et al.*, in this volume). Evidence for co-expression of titin isoforms was obtained in western blot experiments on myocardium of large mammals and in immunofluorescence studies on single cardiac myocytes isolated from large mammals (Cazorla *et al.*, 2000). Since these isoforms have different I-band segments, myocytes that contain predominantly the N2B isoform are expected to have different mechanical properties from the ones that express predominantly the N2BA isoform. Therefore, we speculate that the co-expression of different titin isoforms may modulate the passive stiffness of cardiac myocytes.

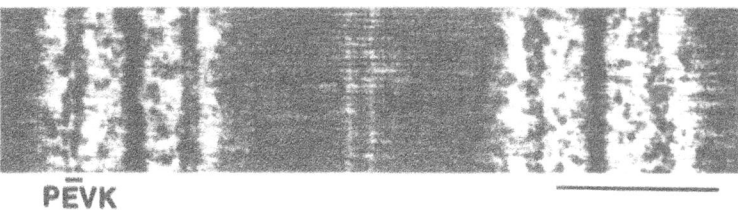

PEVK

Figure 7. Rabbit heart muscle labeled with titin antibodies that mark the PEVK segment (anti-I18-19, anti-I76-77). Since the PEVK segment length did not exceed 60 nm even in overstretched sarcomere (SL 2.6), the titin is expected to be the N2B isoform. Scale bar: 1 μm.

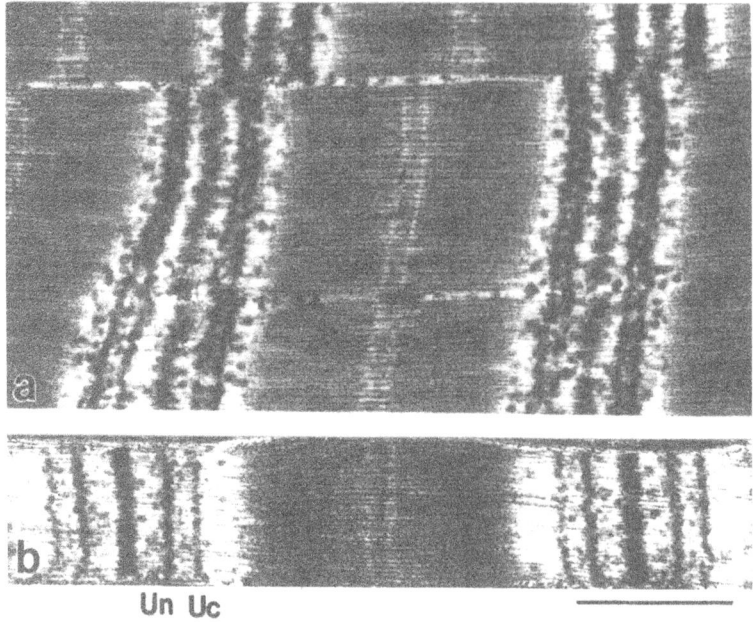

Un Uc

Figure 8. *Rabbit heart muscle labeled with titin antibodies that mark the N- and C-terminal end (named Un and Uc subsequently) of unique N2B sequence (anti-I16-17, anti-I18-19).* ***a:*** *Based on the limited distance between the Uc epitope and the A-I junction, the myocyte is expected to contain N2B titin isoform. Note the gradual unique sequence extension, as a function of sarcomere length (the unique sequence extension reaches 150 nm at SL of 2.8 μm).* ***b:*** *At very long sarcomere (SL 3> μm) the extension of unique sequence increases slightly (170 nm at 3.1 SL), but the distance between the C-terminal end of unique sequence and the A-I junction changes dramatically. Scale bar: 1 μm.*

The two different cardiac isoforms can be differentiated by immunoelectron microscopy using antibodies against sequences C-terminal of the unique N2B sequence. This unique sequence is present in both cardiac isoforms (Fig. 11, Freiburg *et al.*, unpublished), but due to the much longer tandem Ig and PEVK segments of the N2BA isoform, the C-terminal end of the unique sequence will be closer to the Z-line in the N2BA titins. When isoforms are co-expressed within the same sarcomere. two epitopes are expected, one near the Z-line (derived from N2BA titin) and one further from the Z-line (derived from N2B titin). Indeed, in our recent experiments on cardiac muscle such double epitope was found (Fig. 12). Thus, not only is cardiac titin a three spring system, different isoforms that contain length variants of the spring are co-expressed within the same sarcomere. Titin is an extraordinary spring indeed!

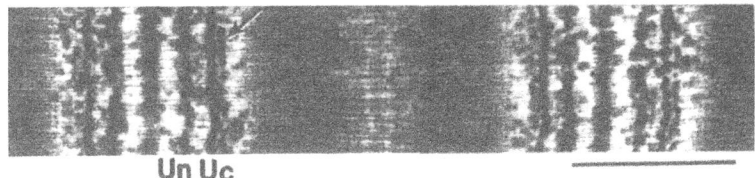

Un Uc

Figure 9. *Rabbit heart muscle labeled with titin antibodies that mark the N and C-terminal end of unique N2B sequence and PEVK segment (anti-I16-17: Un, anti-I18-19: Uc, anti-I76-77: arrow). In this overstretched sarcomere (Sl=2.7 μm), the fully extended short PEVK segment shows the N2B nature of titin. The PEVK and unique sequence extension (60 and 150 nm subsequently) directly demonstrate the serially linked three-spring nature of the heart titin. Scale bar: 1 μm.*

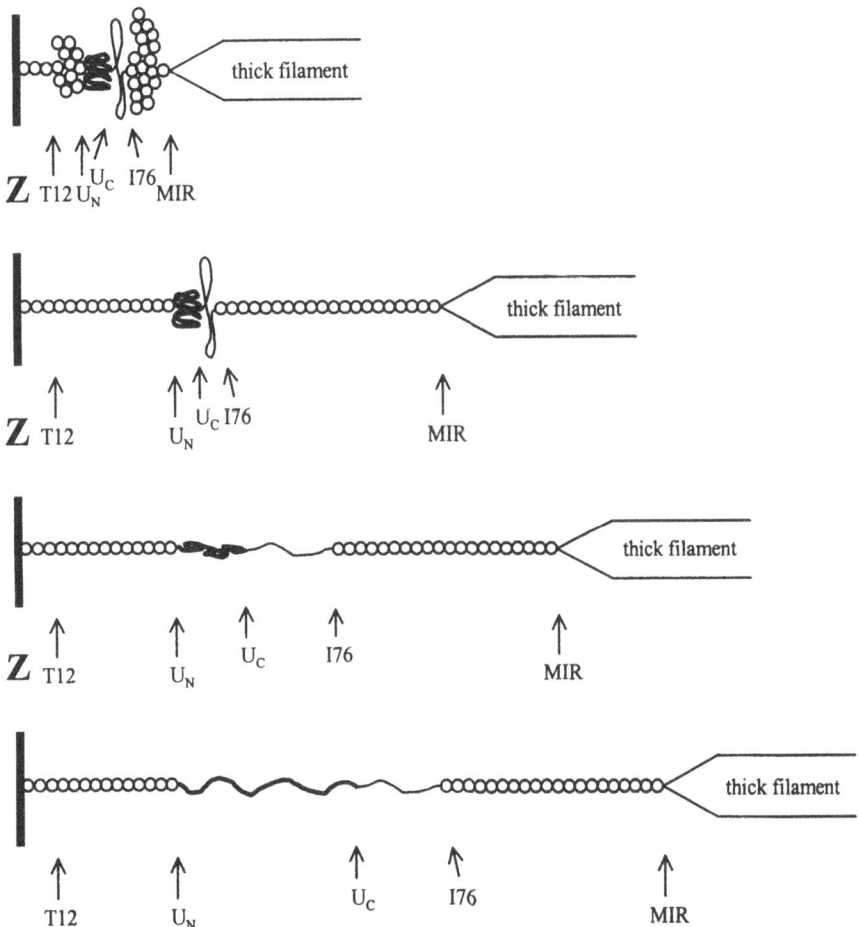

Figure 10. *Schematic diagram of the extension of heart N2B titin as a three-spring system.*

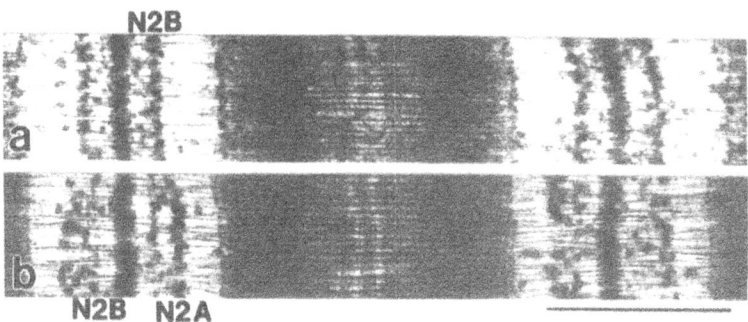

Figure 11. *Heart muscles from rat (a) and pig (b) co-labeled with N2B and N2A unique sequences associated titin antibodies. The rat heart contains predominantly N2B titin isoform, the pig heart also contains N2BA isoform. Scale bar: 1 µm.*

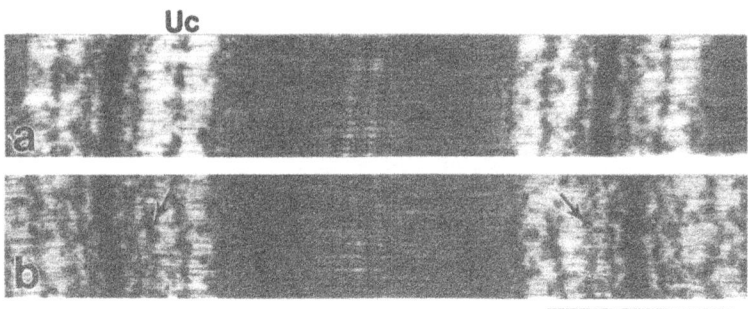

Figure 12. *Rabbit heart labeled with T12, anti-I18-19 (C terminal of N2B unique sequence), and MIR titin antibodies. a: The expected three epitopes are clearly observable. b: An extra epitope appeared close to the T12 epitope. This epitope is likely to correspond to the C-terminal end of the N2B unique sequence in the N2BA isoform. Scale bar: 1 µm.*

Acknowledgments

We thank Danielle Higgins and Mark Macnabb for technical support.   This work was supported by grants from the National Institutes of Health, National Heart, Lung, and Blood Institute (HL61497 and HL62881 to H.G., HL47053 to M.G.), the Human Frontier Science Program (to S.L.), the Forschungsfond für Klinische Medizin Mannheim (to S.L.), the Deutsche Forschungsgemeineschaft (La668/5-1 to S.L. and SFB 320 to S.L.).

References

Bullard B, Leonard K. Modular proteins of insect muscle. *Adv Biophys* 1996;33:211-221.

Cazorla O, Freiburg A, Helmes M, Centner T, McNabb M, Wu Y, Trombiteas K, Labeit S, Granzier H. Differential expression of cardiac titin isoforms and modulation of cellular stiffness. *Circ Res* 2000;86:59-67.

Daley J, Southgate R, Ayme-Southgate A. Structure of the *Drosophila* projectin protein: isoforms and implication for projectin filament assembly. *J Mol Biol* 1998;279:201-210.

Erickson HP. Reversible unfolding of fibronectin type III and immunoglobulin domains provides the structural basis for stretch and elasticity of titin and fibronectin. *Proc Natl Acad Sci USA* 1994;91:10114-10118.

Gautel M, Goulding D. A molecular map of titin/connectin elasticity reveals two different mechanisms acting in series. *FEBS Lett* 1996;385:11-14.

Granzier HL, Wang K. Passive tension and stiffness of vertebrate skeletal and insect flight muscles: the contribution of weak cross-bridges and elastic filaments. *Biophys J* 1993;65:2141-2159.

Granzier H, Kellermayer M, Helmes M, Trombitás K. Titin elasticity and mechanism of passive force development in rat cardiac myocytes probed by thin-filament extraction. *Biophys J* 1997;73(4):2043-2053.

Gregorio CC, Trombitás K, Centner T, Kolmerer B, Stier G, Kunke K, Suzuki K, Obermayr F, Herrmann B. The NH2 terminus of titin spans the Z-disk: its interaction with a novel 19-kD ligand (t-cap) is required for sarcomeric integrity. *J Cell Biol* 1998;143:1013-1027.

Helmes M, Trombitás K, Centner T, Kellermayer M, Labeit S, Linke WA, Granzier H. Mechanically driven contour-length adjustment in rat cardiac titin's unique N2B sequence titin is an adjustable spring. *Circ Res* 1999;84:1339-1352.

Helmes M, Trombitás K, Granzier H. Titin develops restorin force in rad cardiac myocytes. *Circ Res* 1996;79(3):619-626.

Horowits R. The physiological role of titin in striated muscle. *Rev Physiol Biochem Pharm* 1999;138:57-96.

Hu DH, Matsuno A, Terakado K, Matsuura T, Kimura S, Maruyama K. Projectin is an invertebrate connectin (titin): isolation from crayfish claw muscle and localization in crayfish claw muscle and insect flight muscle. *J Muscle Res Cell Motil* 1990;11:497-511.

Labeit S, Kolmerer B. Titins: giant proteins in charge of muscle ultrastructure and elasticity. *Science* 1995a;270:293-296.

Labeit S, Kolmerer B. The complete primary structure of human nebulin and its correlation to muscle structure. *J Mol Biol* 1995b;248:308-315.

Lackey A, Ferguson C, Labeit S, Reedy M, Larkins A, Butcher G, Leonard K, Bullard B. Identification of high molecular weight proteins in insect flight and leg muscle. *EMBO J* 1990;9:3459-3467.

Lackey A, Labeit S, Gantel M, Fergusson J, Barlow DP, Leonard K, Bullard B. Kettin, a large molecular protein in the Z-disc of insect muscles. *EMBO J* 1993;12:2863-2871.

Linke W, Ivemeyer M, Olivieri M, Kolmerer B, Rüegg J, Labeit S. Towards a molecular understanding of the elasticity of titin. *J Molec Biol* 1996;261:62-71.

Linke WA, Stockmeier MR, Ivemayer M, Hosser H, Mundel P. Characterizing titin's I-band Ig domain region as an entropic spring. *J Cell Sci* 1988a; 111:1567-1574.

Linke WA, Ivemayer M, Mundel P, Stockmaier MR, Kolemer B. Nature of PEVK-titin elasticity in skeletal muscle. Proc Natl Acad Sci USA 1988b;95:8052-8057.

Linke W, Rudy D, Centner T, Gautel M, Witt C, Labeit S, Gregorio C. I-band titin in cardiac muscle is a three-element molecular spring and is critical for maintaining thin filament structure. *J Cell Biol* 1999;156:631-644.

Machado C, Sunkel CE, Andrew DJ. Human autoantibodies reveal titin as a chromosomal protein. *J Cell Biol* 1998;141:321-333.

Maruyama K, Kimura S, Yoshidomi H, Sawada H, Kikuch Masako. Molecular size and shape of B-connectin, an elastic protein of striated muscle. *Biochem* 1984;95:1423-1433.

Nave R, Weber K. A myofibrillar protein of insect muscle related to vertebrate titin connects Z band and A band: purification and molecular characterization of invertebrate mini-titin. *J Cell Sci* 1990;85:535-544.

Politou AS, Gautel M, Improta S, Vangelista L, Panstore A. The elastic I-band region of titin is assembled in a "modular" fashion by weakly interacting Ig-like domains. *J Mol Biol* 1996;255:604-616.

Politou AS, Thomas DJ, Pastore A. The folding and stability of titin immunoglobulin-like modules, with implications for the mechanism of elasticity. *Biophys J* 1995;69:2601-2610.

Pringle JW. The Contractile mechanism of insect fibrillar muscle. *Prog Biophys Mol Biol* 1967;17:1-60.

Reedy MK. "Electron Microscope Observations Concerning the Behavior of the Cross-Bridge in Striated Muscle." In *Contractility of Muscle Cells and Related Processes*, RJ Podolsky, ed. Englewood Cliffs, NJ: Prentice-Hall, 1971.

Saide JD. Identification of a connecting filament protein in insect fibrillar flight muscle. *J Molec Biol* 1981;153:661-679.

Saide JD, Chin-Bow S, Hogan-Sheldon J, Busquets-Turner L, Vigoreaux SO, Valgeirsdottir K, Pardue ML. Characterization of components of Z-bands in the fibrillar flight muscle of *Drosophila* melanogaster. *J Cell Biol* 1989;109:2157-2167.

Saide JD, Chin-Bow S, Hogan-Sheldon J, Busquets-Turner L. Z-band proteins in the flight muscle and leg muscle of the honeybee. *J Muscle Res Cell Motil* 1990;11:125-136.

Soteriou A, Clarke A, Martin S, Trinick J. Titin folding energy and elasticity. *Proc R Soc Lond B Biol Sci* 1993;254:83-86.

Trinick J, Knight P, Whiting A. Purification and properties of native titin. *J Mol Biol* 1984;180:331-356.

Trombitás K, Tigyi-Sebes A. Fine structure and mechanical properties of insect muscle. In *Insect Flight Muscle*, RT Tregear, ed. Amsterdam: Elsevier/North Holland Biomedical Press, 1977.

Trombitás K, Tigyi-Sebes A. Cross-bridge interaction with oppositely polarized actin filaments in double-overlap zones of insect flight muscle. *Nature* 1984;309:168-170.

Trombitás K, Tigyi-Sebes A. How actin filament polarity affects crossbridge force in doubly-overlapped insect muscle. *J Muscle Res Cell Motil* 1985;126:2285-2288.

Trombitás K, Pollack GH, Wright J, Wang K. Elastic behavior and arrangement of titin filaments in the I-band. *Proc XIIth Intl Congress for EM* 1990;478-479.

Trombitás K, Pollack GH, Wright F, Wang K. Elastic properties of titin filaments demonstrated using a freeze-fracture technique. *Cell Motil Cytoskel* 1993a;24:274-283

Trombitás K, Jin JP, Granzier H. The mechanically active domain of titin in cardiac muscle. *Circ Res* 1995;77:856-861.

Trombitás K, Greaser ML, Pollack GH. Interaction between titin and thin filaments in intact cardiac muscle. J Muscle Res Cell Motil 1997; 18: 345-351.

Trombitás K, Greaser M, Labeit S, Jin JP, Kellermayer M, Helmes M, Granzier H. Titin extensibility in situ: entropic elasticity of permanently folded and permanently unfolded molecular segments. *J Cell Biol* 1998a;140:853-859.

Trombitás K, Greaser M, French G, Granzier H. PEVK extension of human soleus muscle titin revealed by immunolabeling with the anti-titin antibody 9D10. *J Struct Biol* 1998b;122:188-196.

Trombitás K, Freiburg A, Centner T, Labeit S, Granzier H. Molecular dissection of N2B cardiac titin's extensibility. *Biophys J* 1999;77:3186-3196.

Vigoreaux JO, Saide J, Pardue ML. Structurally different *Drosophila* striated muscles utilize distinct variants of Z-band associated proteins. *J Muscle Res Cell Motil* 1991;12:340-354.

Wang K. "Cytoskeletal matrix in striated muscle: The role of titin, nebulin and intermediate filaments." In *Contractile Mechanisms in Muscle*, GH Pollack, H Sugi, eds. New York, NY: Plenum Press, 1984.

Wang K, McCarter R, Wright J, Beverly J, Ramirez-Mitchell R. Regulation of skeletal muscle stiffness and elasticity by titin isoforms: a test of the segmental extension model of resting tension. *Proc Natl Acad Sci USA* 1991;88:7101-7105.

White DCS. The elasticity of relaxed insect fibrillar flight muscle. *J Physiol* 1983;343:31-57.

INDEX

Symbols

A

B

C

W

Z

The manufacturer's authorised representative in the EU is Springer
Nature Customer Service Centre GmbH, Europaplatz 3, 69115 Heidelberg,
Germany. If you have any concerns regarding our products, please
contact ProductSafety@springernature.com

Printed and bound by CPI Group (UK) Ltd, Croydon, CR0 4YY
23/04/2026
02095607-0018